To Ellen
with spe

DR. G. J. BRAEM
6.2.'99

Paphiopedilum

Paphiopedilum

Eine Monographie aller Frauenschuh-Orchideen
der asiatischen Tropen und Subtropen

A monograph of all tropical and subtropical
Asiatic slipper-orchids

Dr. Guido J. Braem

BRÜCKE-VERLAG KURT SCHMERSOW · D-3200 HILDESHEIM

CIP-Titelaufnahme der Deutschen Bibliothek

Braem, Guido J.:
Paphiopedilum: e. Monographie aller Frauenschuh-Orchideen d. asiat. Tropen u. Subtropen / Guido J. Braem. –
Hildesheim: Brücke-Verl. Schmersow, 1988

ISBN 3-87105-021-0

Gesamtherstellung: Hagemann-Druck, 3200 Hildesheim, Arnekenstraße 22-25 · Printed in W.-Germany

ISBN 3-87105-021-0

Veröffentlicht/Published 1988

Inhaltsverzeichnis / Index

Alphabetische Liste der besprochenen Arten und Naturhybriden

Alphabetic list of the species and natural hybrids discussed

Paphiopedilum acmodontum
Paphiopedilum appletonianum
Paphiopedilum argus
Paphiopedilum armeniacum
Paphiopedilum barbatum
Paphiopedilum bellatulum
Paphiopedilum bullenianum
Paphiopedilum callosum
Paphiopedilum chamberlainianum
Paphiopedilum charlesworthii
Paphiopedilum ciliolare
Paphiopedilum concolor
Paphiopedilum dayanum
Paphiopedilum delenatii
Paphiopedilum druryi
Paphiopedilum elliottianum
Paphiopedilum emersonii
Paphiopedilum fairrieanum
Paphiopedilum glanduliferum
Paphiopedilum glaucophyllum
Paphiopedilum × godefroyae
Paphiopedilum haynaldianum
Paphiopedilum hirsutissimum
Paphiopedilum hookerae
Paphiopedilum insigne
Paphiopedilum javanicum
Paphiopedilum kolopakingii
Paphiopedilum lawrenceanum
Paphiopedilum lowii
Paphiopedilum malipoense
Paphiopedilum mastersianum
Paphiopedilum micranthum
Paphiopedilum niveum
Paphiopedilum papuanum
Paphiopedilum parishii
Paphiopedilum philippinense
Paphiopedilum purpuratum
Paphiopedilum randsii
Paphiopedilum robinsonii
Paphiopedilum rothschildianum
Paphiopedilum sanderianum
Paphiopedilum sangii
Paphiopedilum spicerianum
Paphiopedilum stonei
Paphiopedilum sukhakulii
Paphiopedilum supardii
Paphiopedilum superbiens
Paphiopedilum tonsum
Paphiopedilum urbanianum
Paphiopedilum venustum
Paphiopedilum victoria-regina
Paphiopedilum villosum
Paphiopedilum violascens
Paphiopedilum × wardii

Vorwort

Die Anfänge dieses Buches gehen auf 1983 zurück. Sie waren keineswegs in dem Ehrgeiz begründet, eine *Paphiopedilum*-Monographie zu schreiben, sondern basierten auf der leidigen Erkenntnis, daß es kein modernes und umfassendes Werk über diese Gattung gab.

Sehr schnell aber wurde mir klar, daß es unendlich schwierig sein würde, alle Literatur zu sammeln und sowohl allen Richtungen der Wissenschaft als auch den Bedürfnissen der Orchideenliebhaber und der kommerziellen Händler gerecht zu werden. Ich hoffe, daß ein vernünftiger Kompromiß gefunden werden konnte.

Leider haben die Vorbereitungen sich über mehrere Jahre hingezogen. Aus den Plänen einer ersten deutschsprachigen Arbeit entwickelte sich die Idee, auch dieses Werk, wie meine bisher erschienenen Teile der *Cattleya*-Monographie, zweisprachig zu veröffentlichen. Außerdem waren die Orchideensucher nicht erfolglos, so daß das Manuskript mehrfach ergänzt werden mußte.

Ich habe soweit wie möglich die Neubeschreibungen, die während der Arbeit veröffentlicht wurden, eingearbeitet. Irgendwann muß man aber einen Strich ziehen und ein Buch abschließen. Somit konnten aber die letzten drei Neubeschreibungen von *Paphiopedilum*-Arten nicht berücksichtigt[1)]werden.

Dieses Werk hätte niemals zustande kommen können ohne die uneigennützige Teilnahme von einer Vielzahl von Personen, die entweder Korrektur gelesen und/oder Bildmaterial zugesteuert haben und/oder dem Autor in vielen Gesprächen und Diskussionen konstruktive Kritik und Anregungen entgegengebracht haben. Für die letztendlich verfaßten Meinungen und Ansichten zeichnet aber nur der Autor verantwortlich.

In erster Linie danke ich meiner Frau Gudrun für viele Anregungen und Ermutigungen sowie für die Durchsicht der deutschen Texte.

Ganz besonderer Dank gebührt Herrn H. SANG (Essen) für die Durchsicht zweier Fassungen des deutschen Textes sowie für die überwiegende Zahl der Illustrationen; Herrn N. POPOW (Wolfsburg-Fallersleben) für viele anregende Diskussionen und Pflanzenmaterial sowie Herrn Dr. H. AZADEHDEL (Long Eaton, U.K.) für seine Bemühungen, bisher unbekannte Arten zur wissenschaftlichen Aufarbeitung zur Verfügung zu stellen.

[1)] **Schlechteriana 1** (1): (1987) und **Schlechteriana 1** (2): (1988).

Preface

The idea to write a book on slipper orchids dates back to 1983. It did not arise from the ambition to write a monograph on *Paphiopedilum* but out of the fact that no modern and comprehensive scientific treatment of the genus was available.

It wasn't long before I became aware that it would be immensely troublesome to collect all the literature and that it would nearly be impossible to compile a book satisfying the needs in respect to all aspects of science as well as those of the amateur and the commercial orchidist. I hope to have found a workable compromise.

Unfortunately, the preparation of this book has taken several years. The idea of writing the first German treatise of the genus developed in the task to publish the book bilingual just like the two already published parts of my *Cattleya* monograph. Furthermore, the orchid hunters were not without success and their discoveries made it necessary to update the manuscript several times.

I have, as far as possible, incorporated all new descriptions. However, at a certain stage one must set a full stop after each and every task and pass on the manuscript to the printers. For that reason, the last three descriptions of *Paphiopedilum*-species[1)] could not be included.

This work could have never been accomplished without the unselfish contributions of many people. Some have read various parts of the text, others have contributed illustrative material, a few have done both. I had many discussions with several colleges and friends yielding quite some constructive criticism and fresh ideas. However, only the author signs responsible for the meanings and opinions as published here.

Primarily, my gratitude goes to my wife Gudrun for the many years of help and encouragement as well as for the proof reading of the German texts.

Special gratitude is due Mr. H. SANG (Essen) for the reading of two drafts of the German text and for the majority of the illustrations; Mr. N. POPOW (Wolfsburg-Fallersleben) for many exciting discussions and his generosity in supplying plant material; Dr. H. AZADEHDEL (Long Eaton, U.K.) for his efforts in discovering new species and making them available to science.

I also thank Dr. Pamela BURNS-BALOGH, Mr. J. COMBER, Mr. W. ERRULAT, Dr. M. LIEBICH,

[1)] **Schlechteriana 1** (1): (1987) und **Schlechteriana 1** (2): (1988).

Mein aufrichtiger Dank gebührt auch Frau Dr. P. BURNS-BALOGH sowie Herrn J. COMBER, Herrn W. ERRULAT, Herrn Dr. M. LIEBICH, Herrn U. LÖB, Herrn R. JENNY, Herrn Dr. R. MARWITZ, Herrn O. Möller, Herrn C. OAKLEY, Herrn U. PATERNY, Herrn H. PINKEPANK, Herrn J. PINSKE, Herrn O. RIEGLER, Herrn A. SCHOCK, Herrn Dr. K.-H. SENGHAS, Herrn W. STUMME.

Auch danke ich dem Brücke-Verlag für die gewohnt gute Zusammenarbeit.

Lahnau, Januar 1988
Dr. Guido J. Braem

Mr. U. LÖB, Mr. R. JENNY, Dr. R. MARWITZ, Mr. O. MÖLLER, Mr. C. OAKLEY, Mr. U. PATERNY, Mr. H. PINKEPANK, Mr. J. PINSKE, Mr. O. RIEGLER, Mr. A. SCHOCK, Dr. K.-H. SENGHAS and Mr. W. STUMME.

Last, certainly not least, I thank the editor, Brücke-Verlag for a fruitful cooperation.

Lahnau, January 1988
Dr. Guido J. Braem

Einführung und geschichtlicher Überblick

Die Frauenschuhorchideen gehören zu den am längsten bekannten Pflanzen der Orchideenfamilie. In seiner **Species Plantarum** von 1753 begründete LINNÉ unter anderem eine Gattung *Cypripedium*[1] für die auch in Mitteleuropa bekannte Art *Cypripedium calceolus.* Bis Ende des 19. Jahrhunderts wuchs die Gattung *Cypripedium* L. auf etwa 100 Arten an, wovon etwa ein Viertel aus gemäßigten Klimazonen und drei Viertel aus den Tropengebieten Asiens und Amerikas stammten.

1838[2] stellte RAFINESQUE in **Flora Telluriana** die Gattung *Stimegas* mit Typus *Stimegas venustum* (Wallich ex Sims) Rafinesque und die Gattung *Cordula* mit Typus *Cordula insignis* (Wallich ex Lindley) Rafinesque auf. Die Arbeiten RAFINESQUEs wurden bis vor kurzem aus den verschiedensten Gründen kaum beachtet. Auch die LINDLEYsche Gattung *Uropedium,* aufgestellt 1848 in **Orch. Linden.** mit Typus *Uropedium lindenii* Lindley fand kaum Anerkennung, ebensowenig die von H. G. REICHENBACH begründete Gattung *Selenipedium,* welche dieser 1854 für zwei Frauenschuharten aus dem tropischen Teil Amerikas aufstellte.

Auch heute noch gibt es einige hartnäckige Gärtner, die alle Frauenschuharten unter der Gattung *Cypripedium* belassen wollen.

PFITZER stellte in seiner 1886 erschienenen Arbeit **Morphologische Studien über die Orchideenblüte,** zusätzlich zu der REICHENBACHschen Gattung *Selenipedium,* die Gattung *Paphiopedilum*[3] auf. Er bezeichnete damit die Arten aus den tropischen Gebieten Asiens.

1896 überarbeitete R. A. ROLFE im **Orchid Review** alle Frauenschuharten und erhob die PFITZERsche Sektion *Phragmipedium* in den Rang einer autonomen Gattung.

[1] *Cypris* = Zypern, der Aphrodite geweihte Insel und *pedilon* = Pantoffel.

[2] Obwohl in der **Flora Telluriana** das Jahr 1836 als Erscheinungsdatum für alle Teile des Werkes angegeben ist, ist nachgewiesen, daß die tatsächlichen Erscheinungsdaten der einzelnen Teile wie folgt sind:
Teil 1 bis 3: 1837; Teil 4: 1838.

[3] *Paphos* = eine Insel mit Venustempel und *pedilum* = Pantoffel.

Introduction and historical Review

The "Slipper Orchids" belong to the oldest known groups of the family. In his **Species Plantarum,** LINNÉ (1753) created, among others, *Cypripedium*[1] to accommodate a plant from Central Europe which he named *Cypripedium calceolus.*

By the end of the 19th. century, the genus *Cypripedium* L. included some 100 species. Of these, approximately 25 % were known from the temperate regions. The rest were found in the Asian and American tropics and sub-tropics.

In his **Flora Telluriana,** published in 1838[2] RAFINESQUE described *Stimegas* with the type species *Stimegas venustum* (Wallich ex Sims) Rafinesque and the genus *Cordula* with the type species *Cordula insignis* (Wallich ex Lindley) Rafinesque. Until recently, the work of RAFINESQUE was ignored by most authors for a number of reasons. The same can be said for the genus *Uropedium,* established by LINDLEY in **Orch. Linden.** in 1848, as well as H. G. REICHENBACHs genus *Selenipedium* which he published in 1854 to accommodate two slipper orchids from the American tropics.

Even today, there are a number of stubborn growers who insist keeping all slipper orchids in the genus *Cypripedium.*

It was E. PFITZER, a German botanist working in Heidelberg, who reviewed the slipper orchids in his **Morphologische Studien über die Orchideenblüte** (published 1886). In the course of this work he separated all the tropical species from *Cypripedium* and combined them in the new autonomous genus *Paphiopedilum.*[3]

In the **Orchid Review,** R. A. ROLFE (1896) put those species of slipper orchids which originated from the American tropics and subtropics into the autonomus genera *Phragmipedium* and *Selenipedium.* As a result of this review, the genus *Paphiopedilum* is now reserved for species from the Asian continent.

[1] *Cypris* = Cyprus, an island devoted to the goddess Aphrodite and *pedilon* = a slipper.

[2] Although on all parts of Rafinesque's **Flora Telluriana** the date of publication is indicated as being 1836, it has been conclusively proven that parts 1, 2 and 3 were published in 1837 and that part 4 was published in 1838.

[3] *Paphos* = island with Venus-temple and *pedilum* = slipper.

Die nunmehr vier nebeneinander existierenden Gattungen lassen sich wie folgt aufschlüsseln:

1. Blätter mehrfach gefaltet, wechselständig. Rhizom deutlich ausgeprägt. Sprosse mit deutlich sichtbaren Internodien 2

1a. Blätter einfach gefaltet, 2zeilig angeordnet. Rhizom vorhanden, aber meist weniger deutlich ausgeprägt. Laubtragende Sprosse ohne auffällige Internodien 3

2. Laubtragende Sprosse meist verzweigt. Pflanze bis zu fünf Meter hoch. Blüten klein und zahlreich, Fruchtknoten 3fächrig, Pflanzen aus den amerikanischen Tropen . . *SELENIPEDIUM*

2a. Laubtragende Sprosse immer unverzweigt. Pflanzen maximal ein Meter hoch. Blüten mittelgroß und wenig an der Zahl. Fruchtknoten 1fächrig. Pflanzen aus den gemäßigten Klimazonen der nördlichen Erdhalbkugel und den subtropischen Gebieten Nord- und Mittelamerikas sowie Asiens . . *CYPRIPEDIUM*

3. Sepalen sich dachziegelartig deckend. Fruchtknoten 1fächrig. Pflanzen aus Asien. Nördliche Verbreitungsgrenze vom Himalaya bis Südchina, südliche Arealgrenze von Südindien über Java nach Neuguinea und den Salomon-Inseln . . . . *PAPHIOPEDILUM*

3a. Sepalen sich klappenförmig deckend. Fruchtknoten 3fächrig. Pflanzen aus Mittel- und Südamerika . . . . *PHRAGMIPEDIUM*

The existing four genera accommodating slipper orchids can be differentiated by the following key.

1. Leaves plicate, alternate. Rhizome distinct. Stems with distinct internodes 2

1a. Leaves conduplicate, distichous. Rhizome present, but generally less distinct. Leaf bearing stems without distinct internodes . . 3

2. Leaf bearing stems generally branched. Plants up to 5 meters high. Flowers multiple and small. Ovary 3-celled. Plants from the American tropics . . . . *SELENIPEDIUM*

2a. Leaf bearing stems never branched. Plants up to 1 meter high. Flowers few and of intermediate size. Ovary 1-celled. Plants from the temperate regions of the northern hemisphere and of the subtropical areas of the American and Asian continents . . . *CYPRIPEDIUM*

3. Sepals imbricate. Ovary 1-celled. Plants from Asia. Northern limits from the Himalaya to Southern China, southern limits from the south of India to the Salomon Archipelago . . . . *PAPHIOPEDILUM*

3a. Sepals valvate. Ovary 3-celled. Plants from Central and South America . . . . *PHRAGMIPEDIUM*

Die Gattung *Paphiopedilum* Pfitzer

Da RAFINESQUE bereits 1838 in seiner **Flora Telluriana** zwei bis dahin als *Cypripedium*-Arten bekannte Taxa zu eigenständigen Gattungen erhob, gäbe es zwei Gattungsnamen, die, der Prioritätsregel zufolge, dem Namen *Paphiopedilum* vorgezogen werden müßten. In diesem Fall hat die Internationale Botanische Kommission es aber vorgezogen[1], den durch PFITZER 1886 geprägten Gattungsnamen *Paphiopedilum* zu „konservieren"[2].

Die gültige Nomenklatur der Gattung ist somit wie folgt:

PAPHIOPEDILUM Pfitzer, *nomen conservandum*

Morph. Stud. Orchideenbl.: 11 (1886)
Typus:
Paphiopedilum insigne (Wallich ex Lindley) Pfitzer

Nomenklatorisches Synonym (Synonymer Gattungsname, basierend auf gleichem Typus):

Cordula Rafinesque
Fl. Tell. Teil **4:** 46 (Nr. 934) 1838

Taxonomisches Synonym (Synonymer Gattungsname, der auf einem anderen Typus basiert):

Stimegas Rafinesque
Fl. Tell. Teil **4:** 45 (Nr. 933) 1838
Typus: *Stimegas venustum* (Wallich ex Sims) Rafinesque

Seit der Begründung der Gattung *Paphiopedilum* durch PFITZER im Jahr 1886 sind nur wenige umfassende Bearbeitungen dieser Orchideengruppe veröffentlicht.

1896 überarbeitete R. A. ROLFE alle Frauenschuharten, 1897 erfolgte eine Besprechung der Gattung durch HALLIER anläßlich seiner Beschreibung von *Paphiopedilum amabile.* 1903 dann eine Bearbeitung durch PFITZER, der die Gattung in drei mehr oder weniger natürliche Gruppen aufteilte und 47 Arten sowie eine Vielzahl von Varietäten behandelte.

Obwohl diese Pflanzengruppe seit den ersten Importen nach England Anfang des vorigen Jahrhunderts allgemein großes Interesse fand, wurde der Gattung, nach PFITZER und ROLFE, sehr lange, auch von den Wissenschaftlern, nur geringe Aufmerksamkeit zuteil. Erst seit den 60er Jahren dieses Jahrhunderts wurde wieder mehr über *Paphiopedilum*-Arten geschrieben.

[1] **Taxon 8** (7): 242 (1959).

[2] Siehe **Internationaler Code der Botanischen Nomenklatur** (Ausgabe 1983) Artikel 14 und 15.

The genus *Paphiopedilum* Pfitzer

At the time PFITZER (1886) published his concept of the genus *Paphiopedilum*, two species belonging to the same group of orchids had already been separated from the genus *Cypripedium* and set up as two monospecific genera under the name of *Cordula* and *Stimegas* (RAFINESQUE 1838).

This implied, that according to the international rules of botanical nomenclature, the first published name *(Stimegas)* would have to be given priority.

In this case however, the International Botanical Committee[1] has chosen to protect the name PFITZER chose for the genus against the earlier generic names given by RAFINESQUE.[2]

The valid nomenclature of the genus is therefore as follows:

PAPHIOPEDILUM Pfitzer, *nomen conservandum*

Morph. Stud. Orchideenbl.: 11 (1886).
Type species:
Paphiopedilum insigne (Wallich ex Lindley) Pfitzer

Nomenclatoral synonym (name of equal rank based on the same type):

Cordula Rafinesque
Fl. Tell. Part **4:** 46 (No. 934), 1838

Taxonomic synonym (name of equal rank based on a different type):

Stimegas Rafinesque
Fl. Tell. Part **4:** 45 (No. 933), 1838
Type species: *Stimegas venustum* (Wallich ex Sims) Rafinesque

Since the description of the genus *Paphiopedilum* by PFITZER, very few comprehensive revisions have been published.

In 1896, R. A. ROLFE discussed all slipper orchids and, in 1897, a discussion of the genus by HALLIER was published accompanying his description of *Paphiopedilum amabile.* In 1903, PFITZER revised the genus and divided it into three, more or less natural, groups with a total of 47 species and numerous varieties.

This group of plants was given a lot of general interest since the first imports to England. Nevertheless, for a considerable period, the genus attracted little interest from amateurs and scientists.

It took about sixty years for the interest in *Paphiopedilum* to grow again, until, in the nineteen sixties publications about the genus became more abundant.

[1] **Taxon 8** (7): 242 (1959).

[2] See **International Code of Botanical Nomenclature** (Edition 1983) Articles 14 and 15.

So ist es vor allem das Verdienst von Dr. med. J. FOWLIE, daß viel über *Paphiopedilum* in der von ihm geleiteten Zeitschrift **Orchid Digest** erschien.

Durch ihn wurden auch andere motiviert, so der Industrielle BIRK und der Zoologe ASHER.

Schon im letzten Jahrhundert wurde der Unart gefrönt, Varietäten als eigenständige Arten zu beschreiben. Vor allem der jüngere REICHENBACH war Spezialist auf diesem Gebiet. So entstanden *Paphiopedilum laevigatum* und *Paphiopedilum roebbelinii,* in Wirklichkeit Varietäten von *Paphiopedilum philippinense,* und so entstanden auch *Paphiopedilum petri* und *Paphiopedilum burbidgei,* in Wirklichkeit Formen von *Paphiopedilum dayanum,* um nur zwei Beispiele zu nennen.

In diesem Buch soll nun der Versuch gemacht werden, anhand der neuesten, wissenschaftlich gesicherten Erkenntnisse, bestehende Probleme der Taxonomie innerhalb der Gattung *Paphiopedilum* zu bearbeiten und bestehende Fehler zu berichtigen. Sicherlich verbleibt trotz allem noch genügend Spielraum für persönliche Ansichten und Interpretationen. In solchen Fällen habe ich mich für eine konservative Lösung entschieden.

Leider wurden zu oft Unterschiede in der Blattmarmorierung, vegetativer Größe, Blütenfärbung und Blütenbehaarung sowie geringfügige Unterschiede in der Morphologie der Staminodiumplatte als Begründung für die Abgrenzung von Arten herangezogen. Inzwischen wissen wir aber, daß innerhalb der Populationen einzelner Arten recht große Unterschiede auftreten können. So ist unter anderem die Blattmarmorierung bei *Paphiopedilum purpuratum, Paphiopedilum sukhakulii* und *Paphiopedilum venustum* (siehe Tafel Seite 139, E, F, G) äußerst variabel. Ebenso die Staminodien bei *Paphiopedilum niveum, Paphiopedilum hookerae* und bei der *Paphiopedilum appletonianum*-Verwandtschaft. Diese Beispiele ließen sich noch beliebig fortsetzen.

Einige Arten mit internationaler Verbreitung wurden mehrmals als eigenständiges Taxon beschrieben. Ein Beispiel hierfür sind *Paphiopedilum dianthum* und *Paphiopedilum parishii.*

Ordnung in ein solches Wirrwarr zu bringen, war und ist nicht einfach. Oft ist es schwierig herauszufinden, WAS WO und durch WEN veröffentlicht wurde. Häufig haben die Autoren sich nicht die Mühe gemacht, bestimmte Typusexemplare zu untersuchen. So wurde (und wird) *"Paphiopedilum schmidtianum"* zu der *Paphiopedilum appletonianum*-Verwandtschaft gerechnet, obwohl ein einziger Blick auf den in Kopenhagen deponierten

Dr. J. FOWLIE, especially, wrote a lot about the genus *Paphiopedilum* in the journal **Orchid Digest** of which he is the editor.

Through his articles, others, for example the zoologist, ASHER, and the industrialist, BIRK, were motivated to study the genus and write about it.

As early as the last century varieties were repeatedly described as autonomous species. H. G. REICHENBACH was a specialist in this field. He described, for example, *Paphiopedilum laevigatum* and *Paphiopedilum roebbelinii* which are, in fact, mere synonyms or, at the most, forms of *Paphiopedilum philippinense,* and *Paphiopedilum petri* and *Paphiopedilum burbidgei* which are conspecific with *Paphiopedilum dayanum.*

In this book the genus is discussed in view of the latest, scientifically-founded data. In this way, I have attempted to solve some of the existing problems of the taxonomy of this group of orchids.

Of course, even when using modern data, enough room remains for personal interpretation and points of view. In such cases, I habe chosen to search for a conservative solution.

Too often, differences in the venation of the leaves, the size of the plants, the colour of the flowers or the intensity with which the flowers or single parts thereof are covered with hairs, as well as some minimal differences in the morphology of the staminodal plate are used to support the descriptions of new species.

We now know, however, that considerable variation can occur within the populations of one single species. Thus, for example, is the venation of the vegetative leaves of *Paphiopedilum purpuratum, Paphiopedilum sukhakulii,* and *Paphiopedilum venustum* (see Plate page 139, E, F, G) very variable. The same can be said for the staminodal plate within populations of *Paphiopedilum niveum, Paphiopedilum hookerae* and within the *Paphiopedilum appletonianum*-complex.

Some species found in several different countries have been described more than once as separate species, for example *Paphiopedilum parishii* and *Paphiopedilum dianthum.*

To clear up a taxonomic mess like that was not very easy. Very often there is considerable difficulty in finding WHAT has been published WHERE and by WHOM. Very often the fact that authors do not care to examine type species is the cause for misinterpretations. *"Paphiopedilum schmidtianum",* for example, was and often still is considered to be a species of the *Paphiopedilum appletonianum*-

Typus klarstellt, daß *Cypripedium schmidtianum* – welches übrigens (korrekterweise) niemals zu *Paphiopedilum* umkombiniert wurde – eindeutig als Synonym zu *Paphiopedilum callosum* anzusehen ist.

Die einzige bisher erschienene und ernstzunehmende moderne Revision der Gattung ist die von KARASAWA und SAITO (1982).[1]

Sicherlich spielen bei KARASAWA und SAITO, und auch bei mir, Blütenstruktur, Blütenfärbung, vegetative Größe, Blattmarmorierung und geographische Verbreitung eine Rolle. Aber diese Merkmale werden immer als Teil mehrerer Merkmalskombinationen interpretiert, und Taxa basieren nie auf einem einzelnen kleinen – subjektiven – Merkmalsunterschied.

Was meine Auffassung der *Paphiopedilum*-Taxonomie anbelangt, so folge ich großenteils der taxonomischen Untergliederung von KARASAWA und SAITO, jedoch mit einigen kleineren, mir berechtigt erscheinenden Korrekturen.

[1] Während der Drucklegung dieser Monographie erschien außerdem eine Bearbeitung der Gattung durch P. J. CRIBB.

complex although one short look at the type species (deposited at Copenhagen) proves that *Cypripedium schmidtianum* (which, by the way, was never transferred to *Paphiopedilum*) is, without the slightest doubt, to be considered conspecific with *Paphiopedilum callosum.*

The revision of KARASAWA and SAITO (1982) is the only one which can claim serious scientific value[1].

Of course, flower structure, flower colour, vegetative size, leaf venation and geographic distribution do play a role in the work of KARASAWA and SAITO as well as in mine. But we always consider these characteristics embedded in a complex and they are always interpreted in that sense, and taxa are never based on one single – subjective – character variation.

As far as my view of the taxonomy of the genus *Paphiopedilum* is concerned, I follow the taxonomic subdivision of KARASAWA and SAITO to a considerable extent, merely making some smaller amendments which I believe to be indicated.

[1] Whilst this monograph was being printed, a treatise of the genus by P. J. CRIBB was published.

Taxonomische Untergliederung der Gattung *Paphiopedilum* mit Schlüssel zu den Untergattungen, Sektionen und Untersektionen.

PAPHIOPEDILUM Pfitzer, *nomen conservandum*
Morphologische Studien über d. Orchideenblüthe: 11, 1886
Typus: *Paphiopedilum insigne* (Wall. ex Lindley) Pfitzer

1. Laubblätter marmoriert, Infloreszenz meist einblütig, 2n = 26. Laterale Petalen deutlich größer als die Fahne, horizontal abstehend. Oberer Rand des Lippenhauptlappens deutlich nach innen gefaltet

i) Pollinien granulös, Lippe rundlich

UNTERGATTUNG PARVISEPALUM
Karasawa & Saito
Bull. of the Hiroshima Botanical Garden Nr. 5: 31, 1982
Typus:
Paphiopedilum delenatii Guillaumin
Zu der Untergattung gehörige Arten:
Paphiopedilum armeniacum Chen & Liu
Paphiopedilum delenatii Guillaumin
Paphiopedilum emersonii Koopowitz & Cribb
Paphiopedilum malipoense Chen & Tsi
Paphiopedilum micranthum Tang & Wang

ii) Pollinien viskös, Lippe länglich

UNTERGATTUNG BRACHYPETALUM
Hallier
Ann. Jard. Bot. Buitenzorg 14: 18–52 (34), 1897
Typus:
Paphiopedilum concolor (Bateman) Pfitzer
Zu der Untergattung gehörige Arten:
Paphiopedilum bellatulum (Rchb. f.) Stein
Paphiopedilum concolor (Bateman) Pfitzer
Paphiopedilum niveum (Rchb. f.) Stein
Zu der Untergattung gehörige Naturhybride:
Paphiopedilum × *godefroyae* (Godefroy) Stein (pro sp.)

Taxonomic subdivision of the genus *Paphiopedilum* with keys to the subgenera, sections and subsections.

PAPHIOPEDILUM Pfitzer, *nomen conservandum*
Morphologische Studien über d. Orchideenblüthe: 11, 1886
Type species: *Paphiopedilum insigne* (Wall. ex Lindley) Pfitzer

1. Leaves tesselate; inflorescence generally uniflorous; 2n = 26; lateral petals much larger than the upper sepal, spreading horizontally; upper margin on the main lobe of the labellum distinctly incurved.

i) Pollinia granulous, lip distinctly calceolate

SUBGENUS PARVISEPALUM
Karasawa & Saito
Bull. of the Hiroshima Botanical Garden No. 5: 31, 1982
Type species:
Paphiopedilum delenatii Guillaumin
Species belonging to this subgenus:
Paphiopedilum armeniacum Chen & Liu
Paphiopedilum delenatii Guillaumin
Paphiopedilum emersonii Koopowitz & Cribb
Paphiopedilum malipoense Chen & Tsi
Paphiopedilum micranthum Tang & Wang

ii) Pollinia viscous; lip ovoid

SUBGENUS BRACHYPETALUM
Hallier
Ann. Jard. Bot. Buitenzorg 14: 18–52 (34), 1897
Type species:
Paphiopedilum concolor (Bateman) Pfitzer
Species belonging to this subgenus:
Paphiopedilum bellatulum (Rchb. f.) Stein
Paphiopedilum concolor (Bateman) Pfitzer
Paphiopedilum niveum (Rchb. f.) Stein
Natural hybrid belonging to this subgenus:
Paphiopedilum × *godefroyae* (Godefroy) Stein (pro sp.)

2. Laubblätter einfarbig grün, dick, fleischig; Infloreszenz mehrblütig, 2n = 26

UNTERGATTUNG POLYANTHA
(Pfitzer) Brieger
in SCHLECHTER: **Die Orchideen,**
3. Auflage: 175, 1971

Synonyma:
Sektion Polyantha Pfitzer
in ENGLER, A.: **Bot. Jahrb. 19:** 41, 1894

Untergattung Anotopedilum Pfitzer
in ENGLER, A.: **Das Pflanzenr. IV** (50): 58, 1903

Typus:
Paphiopedilum lowii (Lindley) Pfitz.

i) Fahne und Synsepalum fast identisch in Größe, Form und Farbgebung. Oberer Rand des Lippenhauptlappens deutlich nach innen gefaltet, keine lateralen Aurikeln vorhanden (außer bei *Paphiopedilum randsii*).

SEKTION MASTIGOPETALUM Hallier
Ann. Jard. Bot. Buitenzorg 14: 18–52 (43), 1897

Synonym:
Sektion Gonatopetalum Pfitzer
in ENGLER, A.: **Das Pflanzenr. IV** (50): 62, 1903

Typus:
Paphiopedilum stonei (Hooker) Pfitzer

Zu der Sektion gehörige Arten:

Paphiopedilum elliottianum (O'Brien) Stein
Paphiopedilum glanduliferum (Blume) Stein
Paphiopedilum kolopakingii Fowlie
Paphiopedilum philippinense (Rchb. f.) Stein
Paphiopedilum randsii Fowlie
Paphiopedilum rothschildianum (Rchb. f.) Stein
Paphiopedilum sanderianum (Rchb. f.) Stein
Paphiopedilum stonei (Hooker) Stein
Paphiopedilum supardii Braem & Löb

ii) Fahne und Synsepalum mit deutlichem Unterschied in Größe, Form und/oder Farbgebung. Oberer Rand des Lippenhauptlappens nicht nach innen gefaltet, sondern gerade oder leicht nach außen gerichtet. Aurikeln deutlich vorhanden.

2. Leaves uni-coloured green, not tesselate, thick, fleshy; Inflorescence pluriflorous, 2n = 26

SUBGENUS POLYANTHA
(Pfitzer) Brieger
in SCHLECHTER: **Die Orchideen,**
3. ed.: 175, 1971

Synonyma:
Section Polyantha Pfitzer
in ENGLER, A.: **Bot. Jahrb. 19:** 41, 1894

Subgenus Anotopedilum Pfitzer
in ENGLER, A.: **Das Pflanzenr. IV** (50): 58, 1903

Type species:
Paphiopedilum lowii (Lindley) Pfitzer

i) Upper and lower sepals almost the same in shape, size and colour patterns; upper margin of the main lobe of the labellum strongly incurved, no lateral auricles (except in *Paphiopedilum randsii*).

SECTION MASTIGOPETALUM Hallier
Ann. Jard. Bot. Buitenzorg 14: 18–52 (43), 1897

Synonym:
Section Gonatopetalum Pfitzer
in ENGLER, A.: **Das Pflanzenr. IV** (50): 62, 1903

Type species:
Paphiopedilum stonei (Hooker) Pfitzer

Species belonging to this section:

Paphiopedilum elliottianum (O'Brien) Stein
Paphiopedilum glanduliferum (Blume) Stein
Paphiopedilum kolopakingii Fowlie
Paphiopedilum philippinense (Rchb. f.) Stein
Paphiopedilum randsii Fowlie
Paphiopedilum rothschildianum (Rchb. f.) Stein
Paphiopedilum sanderianum (Rchb. f.) Stein
Paphiopedilum stonei (Hooker) Stein
Paphiopedilum supardii Braem & Löb

ii) Upper and lower sepals clearly differentiated in shape, size and/or colour patterns; upper margin of the main lobe of the labellum not incurved but straight or directed slightly outwards; lateral auricles well developed.

I) Seitliche Petalen zur Spitze hin breiter werdend, schmal löffelförmig, leicht nach unten gerichtet, nur leicht gedreht.

SEKTION POLYANTHA

Typus:
Paphiopedilum lowii (Lindley) Stein

Zu der Sektion gehörige Arten:

Paphiopedilum haynaldianum (Rchb. f.) Stein
Paphiopedilum lowii (Lindley) Stein

II) Seitliche Petalen zur Spitze hin schmaler werdend, stark gedreht und deutlich nach unten gerichtet.

SEKTION MYSTROPETALUM Pfitzer in ENGLER, A.: **Das Pflanzenr. IV** (50): 65, 1903

Typus:
Paphiopedilum parishii (Rchb. f.) Stein

Zu der Sektion gehörige Art:

Paphiopedilum parishii (Rchb. f.) Stein

3. Laubblätter einfarbig grün, Infloreszenz einblütig; Lippe mit deutlich ausgebildeten Aurikeln am Oberrand

UNTERGATTUNG PAPHIOPEDILUM

Typus:
Paphiopedilum insigne (Wallich ex Lindley) Pfitzer

i) Seitliche Petalen zur Spitze hin breiter werdend

SEKTION PAPHIOPEDILUM

Typus:
Paphiopedilum insigne (Wallich ex Lindley) Pfitzer

Zu der Sektion gehörige Arten:

Paphiopedilum charlesworthii (Rolfe) Pfitz.
Paphiopedilum insigne (Wallich ex Lindley) Pfitzer
Paphiopedilum villosum (Lindley) Stein

ii) Seitliche Petalen an der Spitze nicht breiter als in der Mitte

I) Seitliche Petalen linealisch-lanzettförmig, herabhängend, Spitze deutlich nach hinten gebogen

SEKTION CERATOPETALUM Hallier **Ann. Jard. Bot. Buitenzorg 14:** 18–52 (39), 1897

Typus:
Paphiopedilum fairrieanum (Lindley) Stein

I) Lateral petals gradually widening towards the apex, narrowly spoon-shaped, spreading slightly downward, only slightly twisted.

SECTION POLYANTHA

Type species:
Paphiopedilum lowii (Lindley) Stein

Species belonging to this section:

Paphiopedilum haynaldianum (Rchb. f.) Stein
Paphiopedilum lowii (Lindley) Stein

II) Lateral petals gradually narrowed towards the apex, strongly twisted and drooping.

SECTION MYSTROPETALUM Pfitzer in ENGLER, A.: **Das Pflanzenr. IV** (50): 65, 1903

Type species:
Paphiopedilum parishii (Rchb. f.) Stein

Species belonging to this section:

Paphiopedilum parishii (Rchb. f.) Stein

3. Leaves uni-coloured green, not tesselate; inflorescence uni-flowered; upper margins of the labellum with well developed lateral ears.

SUBGENUS PAPHIOPEDILUM

Type species:
Paphiopedilum insigne (Wallich ex Lindley) Pfitzer

i) Lateral petals gradually widening towards the apex

SECTION PAPHIOPEDILUM

Type species:
Paphiopedilum insigne (Wallich ex Lindley) Pfitzer

Species belonging to this subsection:

Paphiopedilum charlesworthii (Rolfe) Pfitzer
Paphiopedilum insigne (Wallich ex Lindley) Pfitzer
Paphiopedilum villosum (Lindley) Stein

ii) Apical portion of the lateral petals not wider than the middle portion

I) Lateral petals linear-lanceolate, drooping, strongly distorted backwards at the apex

SECTION CERATOPETALUM Hallier **Ann. Jard. Bot. Buitenzorg 14:** 18–52 (39), 1897

Type species:
Paphiopedilum fairrieanum (Lindley) Stein

Zu der Sektion gehörige Art:

Paphiopedilum fairrieanum (Lindley) Stein

II) Seitliche Petalen lanzett- bis spatelförmig, Spitze nicht nach hinten gebogen

a) Seitliche Petalen abstehend, mehr oder weniger deutlich gedreht, Fahne und seitliche Petalen ohne ausgeprägte, dicke Mittellinie

SEKTION STICTOPETALUM Hallier
Ann. Jard. Bot. Buitenzorg 14: 18–52 (40), 1897

Typus:
Paphiopedilum hirsutissimum (Lindley ex Hooker) Stein

Zu der Sektion gehörige Art:

Paphiopedilum hirsutissimum (Lindley ex Hooker) Stein

b) Seitliche Petalen nach innen (vorne) gekrümmt, Fahne und seitliche Petalen mit ausgeprägter, dicker Mittellinie

SEKTION THIOPETALUM Hallier
Ann. Jard. Bot. Buitenzorg 14: 18–52 (39), 1897

Typus:
Paphiopedilum druryi (Beddome) Stein

Zu der Sektion gehörige Arten:

Paphiopedilum druryi (Beddome) Stein
Paphiopedilum spicerianum (Rchb. f.) Stein

4. Laubblätter marmoriert, Infloreszenz einblütig, ausgeprägte Aurikeln am Lippenoberrand, Synsepalum kleiner und weniger gestreift als die Fahne

UNTERGATTUNG SIGMATOPETALUM Hallier
Ann. Jard. Bot. Buitenzorg 14: 18–52 (36), 1897

Typus:
Paphiopedilum venustum (Wallich ex Lindley) Pfitzer

i) Untere Laubblattseite über die ganze Oberfläche dunkel purpur getönt, obere Seite blaß graugrün und tief purpurgrün marmoriert

SEKTION SIGMATOPETALUM

Typus:
Paphiopedilum venustum (Wallich ex Lindley) Pfitzer

Species belonging to this section:

Paphiopedilum fairrieanum (Lindley) Stein

II) Lateral petals lanceolate to spathulate, apical portion not distorted backwards

a) Lateral petals spreading directly, more or less twisted; upper sepal and lateral petals without a distinct thick line in the middle.

SECTION STICTOPETALUM Hallier
Ann. Jard. Bot. Buitenzorg 14: 18–52 (40), 1897

Type species:
Paphiopedilum hirsutissimum (Lindley ex Hooker) Stein

Species belonging to this section:

Paphiopedilum hirsutissimum (Lindley ex Hooker) Stein

b) Lateral petals more or less strongly incurved; upper sepal and lateral petals with a distinct thick line at the middle.

SECTION THIOPETALUM Hallier
Ann. Jard. Bot. Buitenzorg 14: 18–52 (39), 1897

Type species:
Paphiopedilum druryi (Beddome) Stein

Species belonging to this section:

Paphiopedilum druryi (Beddome) Stein
Paphiopedilum spicerianum (Rchb. f.) Stein

4. Leaves tesselate; inflorescences uni-flowered; lateral ears of upper margin of the lip well developed; synsepalum much smaller and less striped than the upper sepal

SUBGENUS SIGMATOPETALUM Hallier
Ann. Jard. Bot. Buitenzorg 14: 18–52 (36), 1897

Type species:
Paphiopedilum venustum (Wallich ex Lindley) Pfitzer

i) Under surface of leaves tinged with dark purple throughout, upper surface tesselate with pale greyish green parts and dark purplish green parts.

SECTION SIGMATOPETALUM

Type species:
Paphiopedilum venustum (Wallich ex Lindley) Pfitzer

Zu der Sektion gehörige Art:

Paphiopedilum venustum (Wallich ex Lindley) Pfitzer

ii) Untere Laubblattseite grün, nur an der Basis manchmal purpurn getönt. Obere Seite graugrün, weißlich und grün marmoriert

I) Seitliche Petalen sichelförmig, an der Spitze mehr oder weniger verbreitert. Ränder im Mittelteil leicht gewellt. Fahne grünlich, mit stark nach innen gedrehten Rändern in der oberen Hälfte und undeutlicher Längsstreifung

SEKTION SPATHOPETALUM Pfitzer in ENGLER, A.: **Das Pflanzenr. IV** (50): 78, 1903

Typus:
Paphiopedilum appletonianum (Gower) Rolfe

a) Staminodium im Vergleich zu der Blütengröße sehr groß, fast kreisrund; basale Ränder der seitlichen Petalen lang behaart, basaler und mittlerer Teil der seitlichen Petalen klein gefleckt

UNTERSEKTION MACRONODIUM Karasawa & Saito
Bull. of the Hiroshima Botanical Garden Nr. 5: 1–69, 1982

Typus:
Paphiopedilum hookerae (Rchb. f.) Stein

Zu der Untersektion gehörige Art:

Paphiopedilum hookerae (Rchb. f.) Stein

b) Staminodium kleiner, länglich rauten- bis herzförmig; Ränder der seitlichen Petalen ohne lange Haare, aber mit mehr oder weniger deutlichen großen Flecken.

UNTERSEKTION SPATHOPETALUM

Typus:
Paphiopedilum appletonianum (Gower) Rolfe

Zu der Untersektion gehörige Arten:

Paphiopedilum appletonianum (Gower) Rolfe
Paphiopedilum bullenianum (Rchb. f.) Pfitzer
Paphiopedilum robinsonii (Ridley) Ridley

II) Seitliche Petalen nicht sichelförmig, Ränder nicht gewellt, Fahne weiß oder purpurn,

Species belonging to this section:

Paphiopedilum venustum (Wallich ex Lindley) Pfitzer

ii) Under surface of leaves pale green throughout, or slightly tinged with deep purple at the base, upper surface greyish green, tesselate with white and green parts

I) Lateral petals knife-shaped, more or less widened in the apical portion, margins slightly undulate in the middle; upper sepal greenish, margin of the upper half strongly incurved, longitudinal stripes indistinct

SECTION SPATHOPETALUM Pfitzer in ENGLER, A.: **Das Pflanzenr. IV** (50): 78, 1903

Type species:
Paphiopedilum appletonianum (Gower) Rolfe

a) Staminodal plate well developed in comparison to the overall size of the flower, large, rounded; lateral petals usually with long, black hairs on basal margin; basal and middle part of the lateral petals with fine spots.

SUBSECTION MACRONODIUM Karasawa & Saito
Bull. of the Hiroshima Botanical Garden No. 5: 1–69, 1982

Type species:
Paphiopedilum hookerae (Rchb. f.) Stein

Species belonging to this subsection:

Paphiopedilum hookerae (Rchb. f.) Stein

b) Staminodal plate smaller, oblong rhomboid to cordate; lateral petals with larger, brownish spots on the basal margin; basal margin with short, minute cilia.

SUBSECTION SPATHOPETALUM

Type species:
Paphiopedilum appletonianum (Gower) Rolfe

Species belonging to this subsection:

Paphiopedilum appletonianum (Gower) Rolfe
Paphiopedilum bullenianum (Rchb. f.) Pfitzer
Paphiopedilum robinsonii (Ridley) Ridley

II) Lateral petals shaped otherwise, margins not undulate throughout; upper sepal white

Ränder im oberen Teil nicht nach innen gedreht, fast gerade und leicht nach hinten gebogen, Längsstreifen deutlich sichtbar

a) Seitliche Petalen ohne große, deutliche Flecken oder Warzen, fast gleich breit von der Basis bis zur Spitze, obere (apikale) Hälften dunkler gefärbt als die unteren (basalen)

SEKTION BLEPHAROPETALUM Pfitzer in ENGLER, A.: **Das Pflanzenr. IV** (50): 82, 1903

Typus:

Paphiopedilum mastersianum (Rchb. f.) Stein

Zu der Sektion gehörige Arten:

Paphiopedilum mastersianum (Rchb. f.) Stein

Paphiopedilum papuanum (Ridley ex Rendle) L. O. Williams

Paphiopedilum sangii Braem

Paphiopedilum violascens Schlechter

b) Seitliche Petalen mit großen, mehr oder weniger verdickten, dunkelpurpurnen Flecken, Petalen über die ganze Fläche fast einheitlich gefärbt

1) Seitliche Petalen kaum bewimpert, schmal spatelförmig, zur Spitze hin allmählich verbreitert; Blüten blaßbraun gefärbt

SEKTION PUNCTATUM Karasawa & Saito

Bull. of the Hiroshima Botanical Garden Nr. 5: 1–69, 1982

Typus:

Paphiopedilum tonsum (Rchb. f.) Stein

Zu der Sektion gehörige Art:

Paphiopedilum tonsum (Rchb. f.) Stein

2) Seitliche Petalen mehr oder weniger einheitlich bewimpert, schmal länglich bis linealisch; Blüten nicht blaßbraun gefärbt

A) Seitliche Petalen schmal-länglich, fast einheitlich breit von der Basis bis zur Spitze, nicht gedreht, horizontal abstehend; Staminodium mehr oder weniger halbmondförmig

SEKTION PLANIPETALUM Karasawa & Saito

Bull. of the Hiroshima Botanical Garden Nr. 5: 1–69, 1982

Typus:

Paphiopedilum sukhakulii Schoser & Senghas

or purplish, margins of upper half not incurved almost straight or slightly recurved, longitudinal stripes distinct

a) Lateral petals without large, more or less swollen, distinct spots, almost same width from base to apex, upper halves usually deeper in colour than lower halves

SECTION BLEPHAROPETALUM Pfitzer in ENGLER, A.: **Das Pflanzenr. IV** (50): 82, 1903

Type species:

Paphiopedilum mastersianum (Rchb. f.) Stein

Species belonging to this section:

Paphiopedilum mastersianum (Rchb. f.) Stein

Paphiopedilum papuanum (Ridley ex Rendle) L. O. Williams

Paphiopedilum sangii Braem

Paphiopedilum violascens Schlechter

b) Lateral petals with large, more or less swollen deep purple spots, almost same colour throughout

1) Lateral petals hardly ciliate, almost entire throughout, narrow spathulate, gradually widened towards the apex; flowers pale brown in colour

SECTION PUNCTATUM Karasawa & Saito

Bull. of the Hiroshima Botanical Garden No. 5: 1–69, 1982

Type species:

Paphiopedilum tonsum (Rchb. f.) Stein

Species belonging to this section:

Paphiopedilum tonsum (Rchb. f.) Stein

2) Lateral petals regularly or irregularly ciliate, narrow oblong or linear oblong; flowers not pale brown

A) Lateral petals narrow oblong, almost same width from base to apex, not twisted, horizontally spreading; staminodal plate more or less lunate

SECTION PLANIPETALUM Karasawa & Saito

Bull. of the Hiroshima Botanical Garden No. 5: 1–69, 1982

Type species:

Paphiopedilum sukhakulii Schoser & Senghas

Zu der Sektion gehörige Arten:

Paphiopedilum purpuratum (Lindley) Stein
Paphiopedilum sukhakulii Schoser & Senghas

Zu der Sektion gehörige Naturhybride:

Paphiopedilum × wardii Summerhayes (pro sp.)

B) Seitliche Petalen linealisch, zur Spitze hin allmählich schmaler werdend, leicht nach unten gerichtet; Staminodium halbrund oder länglich

SEKTION BARBATA Kränzlin
Orchidaceanum Genera et Species: 53–73, 1901

Typus:
Paphiopedilum barbatum (Lindley) Pfitzer

1) Fahne ausgeprägt, groß, mehr oder weniger gerundet, oberer Teil nicht nach innen gedreht, sondern leicht nach hinten gebogen. Ränder der seitlichen Petalen mit großen, lang behaarten Warzen

UNTERSEKTION BARBATA

Typus:
Paphiopedilum barbatum (Lindley) Pfitzer

Zu der Untersektion gehörige Arten:

Paphiopedilum argus (Rchb. f.) Stein
Paphiopedilum barbatum (Lindley) Stein
Paphiopedilum callosum (Rchb. f.) Stein
Paphiopedilum lawrenceanum (Rchb. f.) Stein

2) Fahne eiförmig, deutlich langgespitzt und deutlich gekielt, Ränder der seitlichen Petalen kaum gefleckt

a) Ränder der seitlichen Petalen mit einheitlicher, schwarzer Behaarung über die ganze Länge; Staminodium halbrund, dicht und kurz behaart

UNTERSEKTION LORIPETALUM Karasawa & Saito
Bull. of the Hiroshima Botanical Garden Nr. 5: 1–69, 1982

Species belonging to this section:

Paphiopedilum purpuratum (Lindley) Stein
Paphiopedilum sukhakulii Schoser & Senghas

Natural hybrid belonging to this section:

Paphiopedilum × wardii Summerhayes (pro sp.)

B) Lateral petals linear oblong, gradually narrowed toward the apex, slightly declivitous; staminodal plate semicircular or oblong

SECTION BARBATA Kränzlin
Orchidacearum Genera et Species: 53–73, 1901

Type species:
Paphiopedilum barbatum (Lindley) Pfitzer

1) Upper sepal well developed, large, more or less rounded; apical portion not incurved, but gently recurved; margins of the lateral petals with large, slightly swollen, deep purple blotches with long, blackish cilia

SUBSECTION BARBATA

Type species:
Paphiopedilum barbatum (Lindley) Pfitzer

Species belonging to this subsection:

Paphiopedilum argus (Rchb. f.) Stein
Paphiopedilum barbatum (Lindley) Stein
Paphiopedilum callosum (Rchb. f.) Stein
Paphiopedilum lawrenceanum (Rchb. f.) Stein

2) Upper petal ovate, sharply acute and strongly carinate at the apex; blotches hardly present on the margins of the lateral petals

a) Margins of the lateral petals with black, regularly arranged long cilia throughout; staminodal plate semiorbicular, surface densely covered with short cilia

SUBSECTION LORIPETALUM Karasawa & Saito
Bull. of the Hiroshima Botanical Garden No. 5: 1–69, 1982

Typus:
Paphiopedilum ciliolare (Rchb. f.) Stein

Zu der Untersektion gehörige Arten:

Paphiopedilum ciliolare (Rchb. f.) Stein
Paphiopedilum dayanum (Rchb. f.) Stein
Paphiopedilum superbiens (Rchb. f.) Stein

b) Ränder der seitlichen Petalen nur an den Warzen oder nur schwach behaart; Staminodiumoberfläche fast glatt

UNTERSEKTION CHLORONEURA
Hallier
Ann. Jard. Bot. Buitenzorg 14: 18–52 (36), 1897

Typus:
Paphiopedilum javanicum (Reinwardt ex Blume) Stein

Zu der Untersektion gehörige Arten:

Paphiopedilum acmodontum Schoser ex Wood
Paphiopedilum javanicum (Reinwardt ex Blume) Stein
Paphiopedilum urbanianum Fowlie

5. Laubblätter einheitlich graugrün oder leicht marmoriert; Infloreszenz mehrblütig. Blüten sich nacheinander öffnend; seitliche Petalen linealisch, deutlich mehrfach gedreht, horizontal abstehend, unregelmäßig gefleckt, mit einheitlicher Randbehaarung

UNTERGATTUNG COCHLOPETALUM
(Hallier) Karasawa & Saito
Bull. of the Hiroshima Botanical Garden Nr. 5: 1–69, 1982

Typus:
Paphiopedilum victoria-regina (Sander) Wood

Zu der Untergattung gehörige Arten:

Paphiopedilum chamberlainianum (Sander) Stein
Paphiopedilum glaucophyllum (J. J. Smith) J. J. Smith
Paphiopedilum victoria-reginae (Sander) Wood

Type species:
Paphiopedilum ciliolare (Rchb. f.) Stein

Species belonging to this subsection:

Paphiopedilum ciliolare (Rchb. f.) Stein
Paphiopedilum dayanum (Rchb. f.) Stein
Paphiopedilum superbiens (Rchb. f.) Stein

b) Margins of the lateral petals ciliate only from the swollen blotches, or faintly hairy; surface of the staminodal plate almost smooth

SUBSECTION CHLORONEURA
Hallier
Ann. Jard. Bot. Buitenzorg 14: 18–52 (36), 1897

Type species:
Paphiopedilum javanicum (Reinwardt ex Blume) Stein

Species belonging to this subsection:

Paphiopedilum acmodontum Schoser ex Wood
Paphiopedilum javanicum (Reinwardt ex Blume) Pfitzer
Paphiopedilum urbanianum Fowlie

5. Leaves greyish green throughout or faintly tesselate; inflorescence pluri-flowered; flowers spreading out separately; lateral petals linear, strongly twisted, horizontally spreading, irregularly and roughly blotched and regularly ciliate at the margins

SUBGENUS COCHLOPETALUM
(Hallier) Karasawa & Saito
Bull. of the Hiroshima Botanical Garden No. 5: 1–69, 1982

Type species:
Paphiopedilum victoria-regina (Sander) Wood

Species belonging to this subgenus:

Paphiopedilum chamberlainianum (Sander) Stein
Paphiopedilum glaucophyllum (J. J. Smith) J. J. Smith
Paphiopedilum victoria-reginae (Sander) Wood

Untergattung PARVISEPALUM
Karasawa & Saito

Die Untergattung **PARVISEPALUM** wurde durch KARASAWA und SAITO anläßlich ihrer Revision der Gattung *Paphiopedilum* aufgestellt,[1] um den damals drei bekannten Arten *Paphiopedilum armeniacum, Paphiopedilum delenatii* und *Paphiopedilum micranthum* gerecht zu werden. In der Tat weichen diese Orchideenarten und die inzwischen entdeckten Arten *Paphiopedilum emersonii* und *Paphiopedilum malipoense* so weit von den übrigen Arten der Gattung ab, daß die Aufstellung einer selbständigen Untergattung geboten schien.

Die Pflanzen dieser Untergattung stammen allesamt aus dem südlichen Teil Chinas oder dem nördlichen Teil Indochinas. Obwohl eine enge Verwandtschaft zu den Arten der Untergattung **BRACHYPETALUM** durch die marmorierten Blätter, die ein- bis zweiblütigen Infloreszenzen und eine Chromosomenzahl von 2n = 26 angedeutet ist, unterscheiden sich die Arten der Untergattung **PARVISEPALUM** durch die granulösen Pollinien, die verlängerten und deutlich gestielten Stigmata, die Ausbildung von Stolonen, ein im Vergleich zu den seitlichen Petalen deutlich kleineres dorsales Sepalum (Fahne), und eine ausgeprägt rundliche, dünne und leicht gefaltete Lippe. Es ist vor allem das letzterwähnte Merkmal, das sofort die Frage nach der Verwandtschaft zu der Gattung *Cypripedium* aufkommen läßt, und in der Tat wurde bereits mehrmals gemutmaßt, daß diese *Paphiopedilum*-Gruppe ein Bindeglied darstelle.

Schlüssel zu den Arten der Untergattung PARVISEPALUM

1. Blätter nur sehr undeutlich marmoriert, einfarbig grün erscheinend . . . . *Paphiopedilum emersonii*

1a. Blätter deutlich marmoriert 2

2. Sepalen und Petalen weißlich, mehr oder weniger lila überhaucht, mit tief purpurnen Adern. Schuh im Vergleich zu den übrigen Blütenteilen wesentlich größer . . . . *Paphiopedilum micranthum*

2a. Sepalen und Petalen anders gefärbt, Schuh nicht wesentlich größer im Vergleich zu den übrigen Blütenteilen 3

[1] **Bull. of the Hiroshima Botanical Garden** Nr. 5: 1–69 (1982).

Subgenus PARVISEPALUM
Karasawa & Saito

When revising the genus *Paphiopedilum*[1] KARASAWA and SAITO created the subgenus **PARVISEPALUM** to accomodate *Paphiopedilum armeniacum, Paphiopedilum delenatii* and *Paphiopedilum micranthum*, the three species of the group known at that time. Since then, two further species obviously belonging to the same group, *Paphiopedilum emersonii* and *Paphiopedilum malipoense,* have been described. These species do indeed differ from all other taxa of the genus in such a way as to warrant them to be considered part of a separate autonomous subgenus.

All plants of this subgenus originate from the southern parts of China or the northern parts of Indochina. Although the presence of tesselate leaves, the uni-florous (at the outmost 2-flowered) inflorescence and the mitotic chromosome count of 2n = 26 indicate a close relationship to the species of the subgenus **BRACHYPETALUM**. The species of the subgenus **PARVISEPALUM** differ from their relatives by their granulous pollinia, their much elongated and long-stalked stigmata, the occurrence of stolons, an upper sepal much smaller than the lateral petals, and a strongly calceolate, thin and faintly plicate labellum. The latter characteristic especially brings up the question about the relationship to the genus *Cypripedium*. In fact, on several occasions, this group of slipper orchids have been postulated to be a link between the genera *Cypripedium* and *Paphiopedilum*.

Key to the species of the subgenus PARVISEPALUM

1. Leaves only slightly tesselate, appearing unicoloured green . *Paphiopedilum emersonii*

1a. Leaves distinctly tesselate 2

2. Sepals and petals whitish, with a more or less intense pink hue and deep purple veins. Lip extremely large in comparison to the other flower parts . . *Paphiopedilum micranthum*

2a. Sepals and petals coloured otherwise. Lip not oversized in comparison to the other flower parts 3

[1] **Bull. of the Hiroshima Botanical Garden** No. 5: 1–69 (1982)

3. Blüte grünlich, Petalen und Sepalen purpurn geadert, Laubblätter breit-elliptisch . . . . *Paphiopedilum malipoense*

3a. Sepalen und Petalen einfarbig, ohne deutliche Aderung 4

4. Sepalen, Petalen und Lippe einheitlich aprikosen- oder schwefelgelb, basaler Teil der Innenseite mit kleinen purpurnen Flecken überzogen . . . . *Paphiopedilum armeniacum*

4a. Sepalen, Petalen und Lippe einheitlich leicht rosa-weiß, Innenseite der Lippe nicht gefleckt . . . . *Paphiopedilum delenatii*

3. Flower greenish; petals and sepals veined purple; leaves broadly elliptic . . . . *Paphiopedilum malipoense*

3a. Sepals and petals unicoloured, without distinct veins 4

4. Sepals, petals and lip uniformly coloured apricot to sulphur-yellow; basal portion of the inner surface of the lip dotted with minute, purple spots . *Paphiopedilum armeniacum*

4a. Sepals, petals and lip uniformly pale pinkish white; inner surface of the lip not dotted . . . . *Paphiopedilum delenatii*

Paphiopedilum armeniacum Chen & Liu

Acta Botanica Yunnanica 4 (2): 163–167 (1982)

Über die eigentliche Entdeckungsgeschichte dieser Paphiopedilumart ist wenig bekannt. Die Pflanze wurde wahrscheinlich von dem Chinesen A. L. ZHANG im Jahre 1979 erstmals gesammelt und durch CHEN und LIU, wie oben erwähnt, beschrieben.

Die Entdeckung des gelbblühenden *Paphiopedilum armeniacum*[1] in der an Vietnam grenzenden chinesischen Provinz Yunnan hat nicht nur die Diskussion über die Verbreitung des sicherlich eng verwandten *Paphiopedilum delenatii* wieder aufblühen lassen, sondern auch Spekulationen über die Farbvariationen innerhalb dieser letztgenannten Art in Gang gebracht.

In der Tat wurde in vielen Orchideenliebhaber- und -fachkreisen seit der Entdeckung von *Paphiopedilum armeniacum* davon ausgegangen, daß diese Pflanze, obwohl als eigenständige Art beschrieben, „nur" eine gelbblühende Variante von *Paphiopedilum delenatii* Guill. darstelle. Noch im März 1983 schrieb P. J. CRIBB,[2] nunmehr Kurator des Orchideenherbars in Kew (England): „... wäre nicht die unterschiedliche Blütenfarbe, so würde ich nicht zögern, sie *(Paphiopedilum armeniacum* und *Paphiopedilum delenatii)* als eine Art zu betrachten."

Nun sind sich alle Orchideenexperten einig darüber, daß Blütenfarbe allein kaum ein Artabgrenzungskriterium sein kann (siehe meine einleitenden Betrachtungen). In der Zwischenzeit sind Pflanzen von *Paphiopedilum armeniacum* über Hong Kong und Japan in den Westen gelangt, so daß mehr lebendes Material untersucht werden konnte. Dabei sind einige mehr oder weniger gravierende Unterschiede zu *Paphiopedilum delenatii* erarbeitet worden. Der wohl wichtigste Unterschied liegt in Form und Größe des Staminodiums, welches bei *Paphiopedilum delenatii* ziemlich rautenförmig ist und in der Vorderansicht der Blüte vollständig sichtbar wird. Das Staminodium von *Paphiopedilum armeniacum* dagegen ist etwas größer und hat die Form eines verlängerten Dreiecks. Es ist

[1] *armeniacum* = aprikosenfarben.

[2] Further observations on the genus *Paphiopedilum* (Orchidaceae) in China. **The Plantsman 4** (4): 254 (1983).

Paphiopedilum armeniacum Chen & Liu

Little is known about the discovery of this species. The first plants were most probably collected by the Chinese botanist A. L. ZHANG in 1979 and described by CHEN & LIU as cited above.

The discovery of the yellow-flowered *Paphiopedilum armeniacum*[1] in the Chinese province Yunnan adjacent to Vietnam has nourished the discussion about the true distribution of the related *Paphiopedilum delenatii* and the colour variations within this latter species.

Quite a number of orchid amateurs as well as scientists have, ever since *Paphiopedilum armeniacum* was discovered, considered this plant to be the yellow coloured variant of *Paphiopedilum delenatii,* independent of its description as an autonomous species. In March 1983,[2] P. J. CRIBB, Curator of the Orchid Herbarium at Kew, wrote: "Comparison of *Paphiopedilum armeniacum* and

[1] *armeniacum* = apricot coloured.

[2] Further observations on the genus *Paphiopedilum* (Orchidaceae) in China. **The Plantsman 4** (4): 254 (1983).

Paphiopedilum armeniacum Chen & Liu

etwas nach unten gerichtet, so daß der untere Teil (die Spitze des Dreiecks) vom Labellum verdeckt wird.

Die Blüten von *Paphiopedilum delenatii* sind unwesentlich kleiner als die von *Paphiopedilum armeniacum.* Ein anderer Unterschied (wohl der auffallendste) ist die Blütenfarbe und die innere Beflekkung des Schuhs von *Paphiopedilum armeniacum.*

Bemerkenswert ist, daß die Originalbeschreibung von *Paphiopedilum armeniacum* von einer aprikosenfarbenen Blüte spricht, obwohl alle Pflanzen, welche bisher aufgetaucht sind, schwefelgelbe Blüten aufwiesen. Wie dem auch sei, weder bei *Paphiopedilum delenatii* (wobei fast alle in Kultur befindlichen Pflanzen von einigen wenigen, vielleicht sogar von einem einzigen Klon abstammen) noch bei dem als *Paphiopedilum armeniacum* beschriebenen Taxon ist irgend etwas Eindeutiges über Blütenvariationen bekannt, so daß man die Blütenfärbung grundsätzlich nicht überbewerten sollte.

Paphiopedilum delenatii shows them to be very closely allied. Indeed, but for their different flower colour, I would have no hesitation in considering them conspecific."

In the mean time, it is agreed by all orchid experts that the colour of the flower alone can not be used to differentiate between two taxa at the species level (see also my introduction to this book). In the mean time, quite a number of plants of *Paphiopedilum armeniacum* have reached the West, mainly via Hong Kong. It was therefore possible to examine enough living material, and these examinations have shown that there are a few important differences between *Paphiopedilum armeniacum* and *Paphiopedilum delenatii.*

The most important difference is in the form and size of the staminodal plate. With *Paphiopedilum delenatii* this is rather rhomboid and completely visible when looking upon the flower frontally. The staminodal plate of *Paphiopedilum armeniacum* however is somewhat larger and its form is that of a prolonged triangle. It is somewhat directed downwards so that the bottom part (the tip of the triangle) is covered by the labellum. The flowers of *Paphiopedilum delenatii* are slightly smaller than those of *Paphiopedilum armeniacum.* Another difference, certainly the most striking one, is the flower colour and the spotted inside of the lip of *Paphiopedilum armeniacum.*

It is quite remarkable that the original description of the species notes the flowers to be apricot coloured whereas all plants examined up till now have shown surphureous blooms. Anyway, confirmed information about the colour variation of the flowers of *Paphiopedilum delenatii* (whereby it is quite possible that all plants of this species known to be in culture originate from one single collection) and of those of *Paphiopedilum armeniacum* is not available. Therefore the importance of the flower colour in this complex should not be stressed.

Letztendlich wird die Frage nach dem eigentlichen Unterschied zwischen *Paphiopedilum armeniacum* und *Paphiopedilum delenatii* erst dann geklärt werden können, wenn wir eine größere (statistische) Zahl von Klonen der beiden Arten untersucht haben. Bis dahin werde ich *Paphiopedilum armeniacum* als eigenständige Art, als die es auch beschrieben wurde, behandeln.

Paphiopedilum armeniacum hat ein verlängertes, 2 bis 3 mm dickes Rhizom, welches 5 bis 7 Blätter trägt, die 6 bis 12 cm lang, 1,8 bis 2,3 cm breit und relativ spitz zulaufend sind. Ihre Oberseite ist weißlich grün mit auffallend tiefgrüner Marmorierung; unterseitig sind sie mit dunklen, unregelmäßigen, purpurroten Flecken überzogen und deutlich gekielt.

Die ca. 20 bis 25 cm hohe, meist aufrecht stehende, grüne Infloreszenz ist über die gesamte Länge purpurn gepunktet und bräunlich rot behaart.

Die Braktee ist bis ca. 1,8 cm lang, hellgrün und wie der Infloreszenzschaft intensiv purpurrot gefleckt.

Der Fruchtknoten ist 3 bis 3,5 cm lang, hell fahlgrün, auf den Kanten braun gefleckt, insgesamt zottig weiß behaart.

Die Blüte ist aprikosenfarbig bis schwefelgelb, bei vollständiger Ausbreitung 6,5 bis 8 cm im Durchmesser.

Die Fahne ist eiförmig bis länglich eiförmig, 2,2 bis 3,8 cm lang und 1,4 bis 2,2 cm breit. Sie ist mehr oder weniger spitz, innen kahl. Die Ränder sind im oberen Teil nach innen gerollt, so daß die Fahne ausgesprochen spitz endend erscheint.

Das Synsepalum ist ähnlich, etwas kleiner, doppelt gekielt, mit stumpfem, einspitzigem Ende, und, genau wie die Fahne, innen kahl. Die Petalen sind breit eiförmig bis fast kreisrund, 2,8 bis 5,3 cm lang, 2,5 bis 3 cm breit, mit mehr oder weniger abgerundeten Enden, von 10 bis 20 Adern durchzogen, innen an der Basis weißlich behaart, außen glatt, die Ränder fein bewimpert, aprikosenfarben bis gelb, mit rot gepunkteter Basis.

Die Lippe ist elliptisch bis eiförmig, deutlich sackartig ausgebildet, 4 bis 5 cm lang und 3,2 bis 4 cm breit. Sie hat eine kurze, fingernagelförmige Basis und nach innen geschlagene Ränder. Sie ist innen weißlich behaart und auf der inneren Rückseite purpurrot gepunktet. Die länglich dreieckige Staminodiumplatte ist 1 bis 2 cm lang und 1 bis 1,5 cm breit, manchmal mit etwas nach hinten gebogener unterer Spitze; gelb mit rötlicher Mittelzeichnung.

It will only be possible to convincingly answer the question about the real difference between *Paphiopedilum delenatii* and *Paphiopedilum armeniacum* after scientific examination of a large number of plants originating from different clones of both species.

Until such data is available, I will consider *Paphiopedilum armeniacum* to be a good autonomous species, as described.

Paphiopedilum armeniacum has a prolonged rhizome which is 2 to 3 mm thick and which bears 5 to 7 leaves. The leaves are 6 to 12 cm long, 1.8 to 2.3 cm wide, and, more or less, acute. Their upper surface is greenish-white and distinctly tesselate with dark green. The under surface is prominently keeled and covered with dark, irregular, purple-red spots.

The inflorescence is erect, green, up to 25 cm high, spotted purple throughout and covered with reddish brown hairs.

The bract is up to about 1.8 cm long, green and, just as the scape, spotted purple-red.

The ovary is 3.0 to 3.5 cm long, green, spotted-brown on the ridges, completely covered by white hairs.

The (usually single) flower is apricot- to sulphureous, 6.5 to 8 cm in diameter when spread.

The dorsal sepal is ovate to elliptic, 2.2 to 3.8 cm in length and 1.4 to 2.2 cm wide; more or less acute; and glabrous on the inside. The upper part of the margins is curved inwards giving the impression of acuteness.

The synsepalum resembles the dorsal petal but is somewhat smaller, with two keels on the back and a single rounded apex; the inner surface glabrous.

The petals are broadly ovate to suborbicular, 2.8 to 5.3 cm long and 2.5 to 3 cm wide; the apex more or less rounded; with 10 to 20 veins; the basal part of the inner surface covered with white hairs, the outer surface glabrous; the margins finely-ciliate; apricot to sulphureous with reddish spots near the base.

The lip is elliptic to ovate, distinctly calceolate, 4 to 5 cm long and 3.2 to 4 cm wide, with a short unguiculate base and infolded margins. The inner surface is covered by short white hairs and spotted reddish on the rear. The elongated, triangular staminodal plate is 1 to 2 cm long and 1 to 1.5 cm wide, the apex sometimes curved somewhat backwards; yellow with a red pattern in the middle.

Diese Paphiopedilumart wurde als Humusepiphyt[1] auf Felsen in der chinesischen Provinz Yunnan gefunden. Das Klima in dieser Region entspricht dem Nordvietnams.

Paphiopedilum armeniacum sollte im warmen Gewächshaus unter konstant hoher Luftfeuchtigkeit gehalten werden. Die Pflanzen sind mäßig feucht zu halten mit einer etwas kühleren und trockeneren Ruheperiode zwischen August und Oktober.

Die Blütezeit ist von Januar bis April.

This species of *Paphiopedilum* was found growing as a humus epiphyte[1] on rocks in the Chinese province of Yunnan. In this region, the climate is the same as in the northern part of Vietnam indicating that *Paphiopedilum armeniacum* should be cultivated in the warm house with constant high humidity. The plants should be kept medium moist and given a short rest between August and October with lower temperatures and restricted waterings.

Paphiopedilum armeniacum generally flowers between January and April.

[1] Die meisten Autoren betrachten alle Arten der **Cypripedioideae** im allgemeinen als terrestrisch wachsende Pflanzen. Nur die Pflanzen, die unterirdische Speicherorgane aufweisen (welchen Ursprungs auch immer) können als „terrestrisch" (im Sinne der Wuchsform) betrachtet werden. Innerhalb der **Cypripedioideae** gilt dies nur für die Gattung ***Cypripedium.*** Alle Arten der Gattungen ***Paphiopedilum, Phragmipedium*** und ***Selenipedium*** hingegen wachsen grundsätzlich als **Epiphyten,** manchmal in Humus auf dem Waldboden oder auf Felsen.

[1] Most authors indicate the species of the **Cypripedioideae** in general to be terrestrial. However, only those plants having subterrestrial storage organs (of whatever origin) can be considered to be growing "terrestrially". Within the **Cypripedioideae** this is only the case with species of the genus ***Cypripedium.*** All species of the genera ***Paphiopedilum, Phragmipedium*** and ***Selenipedium*** grow as **epiphytes,** sometimes in vegetable mould (humus) near, or on the ground or on rocks.

Paphiopedilum delenatii Guillaumin

Bulletin Soc. Bot. France 71: 554/558 (1924)

Synonym: *Cypripedium delenatii* (Guill.) C. H. Curtis **Gardeners' Chronicle,** 3. Ser., **89:** 208 (1931)

Paphiopedilum delenatii Guill.

Paphiopedilum delenatii Guill. wurde um 1913 von einem französischen Offizier in Tonkin (Vietnam) gefunden. Die Pflanzen wurden an Herrn DELENAT, dem damaligen Leiter des Botanischen Garten in Saint Germain-en Laye, in der Nähe von Saint Cloud, gesandt, der sie kultivierte und 1924 in Paris ausstellte. Diese Pflanzen waren dann auch die Grundlage für die Beschreibung durch GUILLAUMIN. Im Jahre 1922 wurden Pflanzen dieser Art von dem Orchideensammler POILANE in Annam gefunden. Obwohl viele Autoren davon ausgehen, daß *Paphiopedilum delenatii* lediglich in Vietnam beheimatet ist, ist es doch viel wahrscheinlicher, daß die Art innerhalb Indochinas ein weit größeres Gebiet besiedelt. Diese Hypothese wird auch durch das Auffinden der verwandten *Paphiopedilum armeniacum* und *Paphiopedilum emersonii* in China bekräftigt. Leider waren bis jetzt Informationen über die Verbreitung der Orchideen in diesem, für westliche Wissenschaftler fast unzugänglichen Gebiet nur spärlich in der Literatur vertreten. Ein bedauernswerter Zustand, der sich leider nur sehr zaghaft bessert.

Paphiopedilum delenatii ist eine als Humusepiphyt wachsende, krautige Pflanze, deren kurzer, aufrecht stehender, etwa 3 cm langer Sproß bis zu sieben gedrängt stehende Blätter trägt.

Paphiopedilum delenatii Guill. was discovered by a French army officer in Tonkin (Vietnam) about the year 1913. The plants were sent to Mr. DELENAT who was, at that time, the director of the botanical garden at Saint Germain-en Laye near Saint Cloud. In 1924, the plants were exhibited in Paris and were the basis for the description by GUILLAUMIN. Plants of this species were reportedly discovered again in 1922, by the collector POILANE in the Vietnamese province of Annam.

Although most authors consider *Paphiopedilum delenatii* to be a Vietnamese endemic, I consider it quite probable that this species has a larger distribution within Indochina. This hypothesis is also supported by the appearance of the related species *Paphiopedilum armeniacum* and *Paphiopedilum emersonii* in China. Unfortunately, up to date, only little information about the distribution of these plants is to be found in the literature as it is still very difficult for western scientists to study Chinese orchids in the field, an unfortunate state of affairs which seems to improve only very slowly.

Paphiopedilum delenatii is an herbaceous plant growing as a humus epiphyte. The stem is short, erect, up to 3 cm long, bearing up to about 7 closely-placed leaves. The leaves are placed in two rows, oblong-lanceolate, obtuse, up to 10 cm long

Diese zweireihig angeordneten Blätter sind länglich lanzettlich, stumpf endend, leicht fleischig, bis 10 cm lang und bis ca. 3 cm breit. Sie sind blaßgrün mit vielen kleinen roten Punkten auf der Unterseite und blaßgrün mit dunklerer Marmorierung auf der oberen Seite.

Die aufrecht stehende, bis zu 15 cm hohe und ca. 4 mm dicke Infloreszenz ist rötlich, rund und deutlich behaart. Sie trägt eine oder zwei Blüten.

Die Braktee ist breit eiförmig, ungefähr 12 mm lang, blaßgrün mit kleinen roten Punkten.

Die Blüten erreichen einen Durchmesser von etwa 8 cm.

Der Fruchtknoten ist ca. 5 cm lang, tief rotgrün bis braungrün und von steifen Härchen überzogen.

Die Fahne ist eiförmig, fein zugespitzt, ca. 3 cm lang und bis zu 2,5 cm breit.

Das Synsepalum ist rundlich oval, spitz endend, ca. 3 cm lang, und fast ebenso breit.

Sowohl die Fahne als auch das Synsepalum sind an der Innenseite weiß und außen rosa mit dunkleren rötlichen Flecken.

Die Petalen sind verkehrt eiförmig, stumpf oder gerundet an den Enden, ca. 4 cm lang und 3 cm breit. Sie sind ähnlich wie die Sepalen (Fahne und Synsepalum) gefärbt, aber sie weisen weniger deutliche Flecken auf der Außenseite auf.

Petalen und Sepalen sind von kurzen, abstehenden Härchen überzogen.

Die Lippe ist breit pantoffelförmig, umgekehrt eiförmig, vorne sehr deutlich gerundet, mit deutlich nach innen geschlagenen Rändern; insgesamt bis 3,5 cm lang und bis 3 cm breit, unten weiß, oben rosa und mit einer, vor allem an der Vorderseite, deutlich ausgeprägten Behaarung.

Das breit eiförmige Staminodium ist kurz gestielt. Die Basis ist mehr oder weniger herzförmig. Das Staminodium ist ungefähr 1,5 cm im Durchmesser, tiefrosa mit gelber Mitte und einem gelben Fleck an der Basis.

Paphiopedilum delenatii war sehr lange Zeit eine Seltenheit in den europäischen und amerikanischen Orchideensammlungen. Der Grund dafür ist sicherlich nicht zuletzt in den politischen Gegebenheiten des Heimatgebietes zu suchen, die es unmöglich machen, Wildpflanzen zu bekommen. Inzwischen ist diese Art aber über Samen vermehrt worden und im Handel zu erhalten.

Das Klima im nordvietnamesischen Heimatgebiet besteht aus einer kühlen Periode von Ende September bis Ende März, die von den trockenen nordöstlichen Monsunen geprägt wird und von

and up to about 3 cm broad, slightly fleshy. They are pale green with numerous fine red spots beneath. The upper surface is tesselate, darker green on a pale green background.

The peduncle is erect, up to 15 cm high and about 4 mm in diameter, terete, reddish, with spreading red hairs, bearing one or two flowers apically.

The bract is broadly ovate, about 12 mm long, pale green with fine red spots.

The flowers are up to about 8 cm in diameter.

The ovary is about 5 cm long, deep reddish to brownish green, covered with stiff, spreading hairs.

The dorsal sepal is ovate, acute and apiculate, about 3 cm long and up to 2.5 cm across.

The synsepalum is orbicular ovate, apiculate, about 3 cm long and equally broad.

Both sepals are white inside, pink outside, with dark reddish spots.

The petals are obovate-orbicular, obtuse to rounded at the apex, about 4 cm long and up to 3.5 cm broad. They are similarly-coloured as the sepals but the spots on the external surface are not so distinct.

Petals and sepals are covered with short, spreading hairs.

The lip is pouch-like, obovoid, distinctly rounded in front, overall about 3.5 cm long and 3 cm broad, white beneath, pink above and, especially in front, shortly pubescent, the margins of the mouth much inflexed.

The staminodal plate is broadly orbicular-ovate, shortly stalked, more or less cordate at the base, about 1.5 cm in diameter, deep pink with a yellow middle part and a yellow spot at the base.

For a considerable period, *Paphiopedilum delenatii* was scarcely represented in European and American orchid collections. The reason for this certainly being the political situation in the area where this species has its distribution, thus making it impossible to search for wild growing plants.

In the mean time, however, *Paphiopedilum delenatii* has been propagated from seed and plants are available from a number of orchid dealers.

The climate in the natural habitat of these orchids consists of a cool dry period between the end of September and the end of March, characterized by a dry northeastern monsoon and a foggy, wet period between December and March. During this period, the nocturnal temperatures are as low as 9 °C, with temperatures during the daytime averaging about 20 °C.

Dezember bis März durch Nebel, Schauer und Taubildung gekennzeichnet ist. In dieser Zeit sinken die Nachttemperaturen zeitweilig auf 9 °C ab, wobei die Tagestemperaturen um 20 °C schwanken.

Nach einer etwas wärmeren Übergangsperiode beginnen Ende Mai die starken Sommerregenfälle, die bis September anhalten. In dieser „Sommerzeit" fallen ungefähr 60% der jährlichen Niederschlagsmenge, die ungefähr 2000 mm beträgt, und es herrschen Temperaturen um 18 °C des Nachts und bis 28 °C am Tage.

Das Areal ist niemals trocken, und die relative Luftfeuchtigkeit schwankt zwischen 60 und 85%.

Was die Kultur anbelangt, so wird *Paphiopedilum delenatii* oft als schwierig betrachtet. Die Pflanzen brauchen eine andauernd hohe Luftfeuchtigkeit und sollten niemals austrocknen.

Die Art gedeiht am besten im warmen Gewächshaus bei 18 bis 21 °C im Sommer und 12 bis 15 °C im Winter.

Getopft wird in ein gut durchlässiges Pflanzsubstrat mit etwas Moos als Beimischung.

Während unserer Sommermonate sollte oft und ausgiebig gegossen werden. Im Winter werden die Wassergaben stark reduziert, ohne die Pflanzen austrocknen zu lassen, was sich durch ein Schrumpfen der Blätter bemerkbar macht.

Nach der Blüte (März bis April) kann *Paphiopedilum delenatii* ohne weiteres umgetopft werden.

Varietäten:

Obwohl *Paphiopedilum armeniacum* geraume Zeit als Variante von *Paphiopedilum delenatii* betrachtet wurde, sind nunmehr alle Fachleute, trotz einiger Vorbehalte, der Meinung, daß man es mit zwei unabhängigen Taxa zu tun hat. (Siehe auch unter *Paphiopedilum armeniacum* Chen & Liu.)

After a somewhat warmer interim period lasting until the end of May, strong summer rains dominate until September. In this period about 60% of the yearly rainfall (averaging 2000 mm) occurs. The temperatures are about 18 °C at night and up to 28 °C during daytime.

The area is never really dry and the relative humidity is between 60 and 85%.

The cultivation of *Paphiopedilum delenatii* is generally considered to be difficult. The plants need a constant high humidity and should never be allowed to dry out.

This species does best in the warm house at 18 to 21 °C in the summer and 12 to 15 °C during the winter months.

Paphiopedilum delenatii should be potted in an open substrate containing some moss.

During our summer months, waterings should be abundant. In the winter months only little water should be given without allowing the vegetative organs to shrink.

After flowering (usually March-April), *Paphiopedilum delenatii* can be repotted without problem.

Varieties:

Although *Paphiopedilum armeniacum* was considered to be a variant of *Paphiopedilum delenatii* for quite some time, and in spite of a number of questions yet to be answered, all experts now consider both species to be autonomous taxa (cfr. under *Paphiopedilum armeniacum* Chen & Liu).

Paphiopedilum emersonii Koopowitz & Cribb

The Orchid Advocate 12 (3): 84–86 (1986)

Paphiopedilum emersonii Koopowitz & Cribb

Paphiopedilum emersonii ist die neueste Entdekkung, was die chinesischen *Paphiopedilum*-Arten anbelangt. KOOPOWITZ und CRIBB schreiben:

„. . . da *Paphiopedilum emersonii* vielleicht die größte Überraschung darstellt. Die Art ist ohne Zweifel mit den anderen Arten der Untergattung PARVISEPALUM verwandt und sollte, unserer Meinung nach, dort eingereiht werden . . . Die Laubblätter aber sind einzigartig in der Untergattung, da sie oberseitig jegliche Marmorierung und unterseitig eine purpurne Befleckung vermissen lassen . . ."

Das ist eindeutig falsch. Die Laubblätter von *Paphiopedilum emersonii* sind oberseitig leicht, aber deutlich marmoriert, und auf der Basis der Unterseite ist, zumindest bei einem Teil der Pflanzen, eine eindeutige purpurne Befleckung zu erkennen.

Dieser Fund beweist, daß nach der Neuentdekkung von *Paphiopedilum armeniacum* und *Paphiopedilum micranthum* noch einiges an Orchideenfunden aus der Volksrepublik China erwartet werden kann.

Die genaue Herkunft dieser Art ist ungeklärt. Nicht bestätigten Meldungen zufolge, wurde *Paphiopedilum emersonii* bei Hai Num in der südöstlichen chinesischen Provinz Ghuizhou gesammelt.

Paphiopedilum emersonii is the latest addition to the Chinese slipper orchids. KOOPOWITZ and CRIBB wrote:

". . . as *Paphiopedilum emersonii* is perhaps the most surprising of all. It is undoubtedly allied to the other species of PARVISEPALUM and belongs there, in our opinion . . . Yet its leaves are unique in the sub-genus, lacking entirely any tesselation on the upper surface or purple markings beneath . . ."

That is very simply incorrect. The leaves of *Paphiopedilum emersonii* have a faint, but distinct tesselation on the upper side, and at least in part of the plants they are distinctly spotted purple in the basal part of the under side.

In any case, the finding of *Paphiopedilum emersonii* proves, that after the discovery of *Paphiopedilum armeniacum* and *Paphiopedilum micranthum*, even more can be expected in terms of orchid discoveries from the Chinese Peoples Republic. The detailed provenance of this species is still doubtful. On the basis of non-confirmed information it can be said that the natural habitat of *Paphiopedilum emersonii* is in the south-eastern Chinese province Ghuizhou.

This species was named in honour of Emerson "Doc" CHARLES, a renowned Californian orchid

Die Art wurde zu Ehren Emerson „Doc" CHARLES, einem bekannten kalifornischen *Paphiopedilum*-Sammler, in dessen Sammlung die Art erstmals im April 1986 blühte, benannt.

Paphiopedilum emersonii ist eine als Humusepiphyt wachsende krautige Pflanze, die insgesamt bis zu ungefähr 18 cm hoch werden kann. Die Laubblätter sind bandförmig, stumpf, bis 23 cm lang und bis ungefähr 4 cm breit. Sie sind oberseitig leicht, aber deutlich marmoriert und erscheinen nur auf den ersten Blick als einfarbig grün. Die untere Seite ist grasgrün, der basale Teil zum Teil purpur gefleckt, deutlich gekielt.

Die aufrecht stehende Infloreszenz ist im allgemeinen einblütig. Es werden sich jedoch, genau wie bei den anderen Arten der Untergattung **PARVISEPALUM**, sicherlich auch zweiblütige Exemplare finden lassen. Der Blütenstandsschaft ist bis ungefähr 12 cm hoch und 4 mm im Durchmesser, cremefarben, samtartig strukturiert, weiß behaart. Die Braktee ist elliptisch, spitz, gefaltet, bis 2,8 cm lang und ca. 2 cm breit, weiß, papierartig in der Struktur. Die Blüte ist im Vergleich zu den anderen Pflanzenteilen groß. Sie ist 8 bis 9,5 cm im Durchmesser.

Die Sepalen und Petalen sind fleischig, weiß, die Petalen mit einem leichten lila Hauch an der Basis. Die sackförmige Lippe ist cremefarben bis gelb, mit einem lila Rand, innen tief purpurn gefleckt. Das Staminodium ist hellgelb mit rötlicher Zeichnung. Der gestielte Fruchtknoten ist bis ca. 3 cm lang, gelblich weiß, samtartig behaart.

collector, who was the first to flower a plant of the species in cultivation (April 1986).

Paphiopedilum emersonii is an herbaceous plant growing epiphytically in leaf mould. The plants can attain an overall size of 18 cm in height. The leaves are coriaceous, ligulate, the upper side faintly but distinctly tesselate, appearing uniformly green only at first sight, under side a bright grass green, the basal portions more or less spotted purple, prominently keeled.

The inflorescence is erect and generally one-flowered. Most certainly two-flowered specimens will be found analogous to the other species of the subgenus **PARVISEPALUM.** The scape is about 12 cm high and ca. 4 mm in diameter, cream-coloured, velvety, white pilose.

The bract is elliptic, acute, conduplicate, up to 2.8 cm long and about 2 cm broad, white, papery-structured. The flower is very large for the size of the plant, 8.5 to 9.5 cm across.

The sepals and the petals are fleshy, white. The petals are faintly-flushed pink at the base. The lip is distinctly calceolate, cream-coloured to yellow with a pink flush around the rim, the inside spotted with deep purple. The staminodal plate is bright yellow with reddish markings. The pedicel and ovary are overall up to 3 cm long, pale yellowish-white, velvety-pilose.

The dorsal sepal is somewhat hooded, curving forward over the lip, 4 to 4.5 cm long and about 3.3 cm broad, elliptic-ovate, obtuse, pubescent on both surfaces and keeled on the outer side.

Paphiopedilum emersonii
Koopowitz & Cribb

Das dorsale Sepalum ist etwas kapuzenförmig und ein wenig in Richtung der Lippe gebogen. Es ist 4 bis 4,5 cm lang und ungefähr 3,3 cm breit, eiförmig, stumpf, beidseitig behaart und außen gekielt. Die Seitenränder sind zurückgebogen.

Das Synsepalum ist fast kreisrund, stumpf, 3,5 cm lang und ebenso breit, beidseitig behaart und mit zwei Kielen auf der Außenseite. Die Petalen sind nach innen gebogen, breit elliptisch bis fast rund mit stumpfer bis gerundeter Spitze, bis 4,5 cm lang und ungefähr ebenso breit, beidseitig fein behaart, an der Basis zottig.

Die Lippe ist tief sackförmig, bis 3,5 cm lang und ungefähr 3 cm breit. Die Seitenränder sind nach innen gefaltet und entlang den Adern gerillt.

Die Säule ist kurz, das Staminodium konvex, lang dreieckig, gestutzt, bis 2 cm lang und 1 cm breit, mit einer der Länge nach verlaufenden tiefen Furche, die zu der Spitze hin breiter wird. Die Narbe ist spatelförmig, kurz papillös.

Wie bereits erwähnt, ist über die Verbreitung der Art wenig bekannt. Wahrscheinlich ist *Paphiopedilum emersonii* auf Südost-China begrenzt.

Paphiopedilum emersonii wächst in Humus und Moos auf feuchten bis nassen Felsen.

Die Pflanzen sollten im warmen Gewächshaus bei konstant hoher Luftfeuchtigkeit kultiviert werden. Das ganze Jahr über sollte reichlich gewässert werden. Diese Art kennt keine Ruheperiode und blüht über das ganze Jahr.

The synsepalum is elliptic-suborbicular, obtuse, 3.5 cm long and equally broad, pubescent on both surfaces and with two keels on the outer side. The petals are incurved, broadly elliptic to subcircular, obtuse to rounded at the apex, up to 4.5 cm long and wide, finely pubescent on both sides, villose at the base.

The lip is deeply urceolate, up to 3.5 cm long and about 3.0 cm broad. The apical margins are incurved and grooved along the veins.

The column is short, the staminodal shield is convex, trullate, truncate, about 2 cm long and up to 1 cm wide, deeply longitudinal sulcate with a groove dilating towards the apex. The stigma is spathulate, shortly papillose.

Little is known about the natural distribution of this species. It is rather probable that *Paphiopedilum ermersonii* is restricted to the south-eastern parts of China.

Paphiopedilum emersonii grows in leaf-moulder and moss on moist to wet rocks.

The plants should be kept in the warm house by constant high humidity and water should be given abundantly over the whole year. This species has no resting period and flowers over the whole year.

Paphiopedilum malipoense Chen & Tsi

Acta Phytotaxonomica Sinica 22 (2): 119–124 (1984)

Obwohl Pflanzen dieser Art bereits am 11. November 1947 durch den chinesischen Botaniker K. M. FENG in der Provinz Yunnan gesammelt wurden und zumindest ein Exemplar seinen Weg zum Herbarium der Universität Peking fand, wurde die Orchidee erst 1984 durch CHEN & TSI beschrieben *(loc. cit.)*.

Bis blühende Pflanzen vorhanden waren, wurde diese Art aufgrund der vegetativen Ähnlichkeit als eine mögliche chinesische Variante von *Paphiopedilum × wardii* angesehen. Die Blüte der Art belegt jedoch eindeutig, daß *Paphiopedilum malipoense* in der Untergattung PARVISEPALUM anzusiedeln ist. Die Feststellung, ob hier eine Naturhybride, möglicherweise mit einer Art aus der *Paphiopedilum venustum*-Verwandtschaft vorliegt, wird erst durch cytogenetische und/oder Kreuzungsversuche möglich sein.

Although the Chinese botanist K. M. FENG collected plants of this species in the province of Yunnan as far back as the 11. November 1947, and although at least one specimen found its way into the herbarium of the University of Peking, it was not until 1984 that this slipper orchid was described *(loc. cit.)*.

Until flowering plants of this species were available, *Paphiopedilum malipoense* was supposed to be the Chinese variant of *Paphiopedilum × wardii*, a misinterpretation based mainly on vegetative similarity. The flower of *Paphiopedilum malipoense*, however, proves that the species belongs to the subgenus PARVISEPALUM. The hypothesis that *Paphiopedilum malipoense* is a natural hybrid, possibly related to a species from the *Paphiopedilum venustum*-complex awaits to be proven.

Paphiopedilum malipoense Chen & Tsi

Paphiopedilum malipoense ist eine als Humusepiphyt wachsende krautige Pflanze mit einem verlängerten, mehr oder weniger kriechenden, knotigen Rhizom, welches 2 bis 3,5 mm im Durchmesser erreicht. Die 7 bis 8 Laubblätter sind oval bis länglich oval, lederig. Sie stehen schräg aufrecht und werden 10 bis 23 cm lang und 2,5 bis 4 cm breit. Die mehr oder weniger scharfe Spitze endet asymmetrisch. Die Laubblätter sind dunkelgrün, oberseitig ziemlich deutlich marmoriert, unterseitig mit purpurnen Flecken überzogen und deutlich gekielt.

Die Infloreszenz ist bis zu 30 cm hoch, rötlich braun, zottig behaart. Die gefaltete Braktee ist 4,5 cm lang, grünlich, purpurn gefleckt, außen behaart und mit deutlich bewimperten Rändern. Die Infloreszenz trägt meistens nur eine Blüte, die 9 cm im Durchmesser erreicht.

Das dorsale Sepalum (die Fahne) ist elliptisch-lanzettförmig, ungefähr 4,5 cm lang und bis zu 1,8 cm breit, langgespitzt, außen zottig, innen spärlich behaart, mit bewimperten Rändern.

Das Synsepalum ist in etwa so groß wie die Fahne, länglich-lanzettförmig, mit zwei Spitzen und hinten mehr oder weniger deutlich doppelt gekielt. Außen zottig, innen spärlich behaart, an den Rändern bewimpert. Die lateralen Petalen sind ebenso lang und breit wie das dorsale Sepalum, eiförmig mit spitzem Ende, beidseitig schwach flaumig behaart mit Ausnahme der zottig behaarten Basis. Die Ränder sind wie die der Sepalen bewimpert. Die deutlich sackförmige, an der Basis schmal keilförmige Lippe ist ca. 4 bis 4,5 cm lang, innen an der Basis zottig behaart. Der Rand des Hauptlappens ist, typisch für diese Gruppe, stark nach innen geschlagen.

Die Staminodiumplatte ist breit eiförmig, fast rautenförmig, ungefähr 1,3 cm lang und bis zu 1,1 cm breit, unten gestutzt, die basalen Ränder bewimpert, außen gekielt, innen mit vier herausragenden Wölbungen (zwei nahe der Spitze und zwei nahe der Basis) versehen.

Die Sepalen und Petalen sind grünlich in der Grundfärbung, durchzogen von purpurroten Adern, an der Basis zusätzlich dunkelrot bis purpurn gefleckt. Der Schuh ist leicht grünlich, mit etwas dunkleren Adern, innen intensiv dunkelrot bis purpurn gepunktet, wobei die Punkte zur Basis hin kleiner werden. Die Staminodiumplatte ist oben weißlich, unten dunkelrot-braun.

Wie bei allen aus China stammenden *Paphiopedilum*-Arten ist auch für *Paphiopedilum malipoense* die Verbreitung ziemlich unbekannt. Aufgrund des

Paphiopedilum malipoense is an herbaceous plant, growing epiphytically in leaf mould. The species has a prolonged, more or less repent, nodose rhizome which is 2 to 3.5 mm in diameter. There are 7 or 8 oblong to elliptic-oblong leaves. They have a leathery texture, are semi-erect, up to 23 cm long and 2.5 to 4 cm broad. The apex is more or less acute and asymmetric. The leaves are dark green, the upper surface distinctly tesselate, the under surface spotted purple and distinctly keeled.

The inflorescence is up to 30 cm high, reddish brown, villose. The bract is folded, up to 4.5 cm long, greenish, purple spotted, pilose on the outer surface and with distinctly ciliate margins. The Inflorescence usually bears a single flower of up to 9 cm in diameter.

The dorsal sepal is elliptic-lanceolate, about 4.5 cm long and up to 1.8 cm wide, with a long-acuminate apex.

The synsepalum is about the same size, ovate-lanceolate, with a twofold apex and indistinctly double-keeled on the outside. The sepals have a villose outer surface and ciliate margins, on the inside they are only sparsely pilose. The lateral petals are about equally long and broad as the dorsal sepal, obovate, subacute, both surfaces villose. Their margins are just as ciliate as those of the sepals. The lip is distinctly calceolate with a small cuneiform base. The lip is about 4 to 4.5 cm long, on the inside with a villose base, the margins of the mouth are distinctly inflexed as is typical for this group, the rim to the front of the pouch is indented.

The staminodal plate is oblong-ovate, nearly rhomboid, about 1.3 cm long and up to 1.1 cm broad with a subsessile base and ciliate margins, keeled on the outside and on the inside with 4 protrusions (two near the apex and two near the base).

The petals and the sepals are greenish with purple veins, near the base spotted deep red to purple. The calceolate lip is greenish with darker veins, on the inside intensively spotted dark red to purple, the spots getting smaller towards the base. The staminodal plate is white at the base, deep reddish brown near the apex.

As for all *Paphiopedilum*-species originating from China, little can be said about the geographical distribution of *Paphiopedilum malipoense*. It is however rather probable that the species is restricted to the Malipo-area in the province of Yunnan. The holotype was discovered in this area between angegebenen Fundortes kann man davon ausgehen, daß die Art vor allem im Malipogebiet der Provinz Yunnan beheimatet ist. Hier wurde die Typuspflanze in einer Höhenlage von 1300 bis 1600 m über NN in gemischten Laubwäldern auf humusbedecktem Felsen gefunden.

Die Blütezeit ist im allgemeinen von Oktober bis Dezember, in unseren Breiten etwa drei Monate später.

1300 and 1600 meter above sea-level, in mixed deciduous woods in leaf mould on rocks.

The flowering period in the natural habitat is normally from October to December, in Middle-Europe about 3 months later.

Paphiopedilum micranthum Tang & Wang

Acta Phytotaxonomica Sinica 1: 56 (1951)

Paphiopedilum micranthum Tang & Wang

Paphiopedilum micranthum[1)] Tang & Wang war sicherlich eine der aufregendsten Entdeckungen, die Orchideen Chinas betreffend. Die Art wurde durch TANG und WANG im Jahre 1951 beschrieben. Das Konzept basierte auf Pflanzen, die WANG im Malipogebiet, im Südosten der Provinz Yunnan, fand.

Die als Humusepiphyt wachsende, krautige Pflanze hat ein deutlich verlängertes Rhizom, fleischige Wurzeln und pro Trieb vier bis fünf länglich bandförmige, stumpfe, lederartige Laubblätter, die beidseitig unbehaart, unterseitig dunkelpurpur ge-

[1)] Der Name dieser Pflanze bedeutet „kleinblütig". In der Tat spricht die Originalbeschreibung von einer Pflanze mit sehr kleinen Blüten, wobei die Lippe nur 7 mm lang sein soll. Es wird aber im allgemeinen davon ausgegangen, daß die Originalbeschreibung aufgrund einer knospigen Pflanze ohne ausgewachsene Blüten geschah (siehe P. J. CRIBB & C. Z. TANG (1982): The Genus *Paphiopedilum* (Orchidaceae) in China, **The Plantsman 4** (3): 165–170).

Paphiopedilum micranthum[1)] Tang & Wang was most certainly one of the most exciting discoveries in respect to Chinese orchids. The species was described by TANG and WANG in 1951. The concept was based on plants which WANG had collected in the Malipoarea in the southeast of the province of Yunnan.

Paphiopedilum micranthum is an herbaceous plant and grows as an epiphyte in leaf mould. The plant has a distinctly prolonged rhizome, fleshy roots and four to five leaves per growth. The leaves are ligulate, obtuse, leathery; upper and under surface

[1)] This name means "with a small flower". In fact, the original description of the species states that the flowers are very small, with a lip not longer than 7 mm. It is however generally accepted that the original description was based on a plant in bud and not in flower. (See P. J. CRIBB & C. Z. TANG (1982) The Genus *Paphiopedilum* (Orchidaceae) in China. **The Plantsman 4** (3): 165–170).

fleckt und oberseitig dunkel und hellgrün marmoriert sind. Sie werden bis zu 14 cm lang und bis ca. 2,5 cm breit.

Die aufrecht stehende Infloreszenz ist bis ca. 26 cm lang, dicht behaart und fast immer einblütig.

Die Blütenbraktee ist gefaltet, grün, purpur gefleckt, bis 1,2 cm lang und ungefähr 7 mm breit.

Der Fruchtknoten ist ungefähr 4 cm lang, hellgrün, an den Rändern tief purpur gefleckt.

Die, im Vergleich zum Zwergwuchs der Pflanze, große Blüte zeigt eine riesig erscheinende, aufgeblasene, rosa bis weißlichrosa Lippe, die bis 8,5 cm lang und 6,5 cm breit werden kann. Die Petalen und Sepalen sind klein im Vergleich zur Lippe. Die Fahne ist oval, stumpf, bis ca. 3 cm lang und 2,5 mm breit, innen kahl, außen kurz behaart, mit bewimperten Rändern.

Das Synsepalum ist eiförmig bis fast rund, mehr oder weniger spitz, bis 3 cm lang und bis ca. 2,8 cm breit.

Die Petalen sind kurz eiförmig, ca. 3 mm lang und bis 2,5 cm breit. Die Petalen und Sepalen sind weiß, der basale Teil grünlich, mehr oder weniger rosa-lila überhaucht und mit dunkleren Längsstreifen.

Die Lippe ist sackförmig, außen kahl, bis 6,5 cm lang und bis 5,5 cm breit. Die Lippenbasis zeigt eine charakteristische dreieckige Aussparung auf jeder Seite.

Vor diesen Aussparungen liegt eine kellenförmige, weiße, leicht kastanienbraun getüpfelte Staminodiumplatte mit einem deutlich dreieckigen gelben Fleck an der Spitze der apikalen Seite. Die Platte ist ca. 1,3 cm lang und ca. 7 mm breit.

Die Blüten dieser Art erinnern an Pflanzen der Gattung *Cypripedium* L.

Paphiopedilum micranthum ist nur aus dem Südosten Yunnans bekannt. Das Verbreitungsgebiet dieser Art liegt an der Grenze zu Vietnam zwischen den Städten Marlipo und Shi Chou. Die Pflanzen wachsen terrestrisch am Waldboden oder an schattigen Stellen der Flußufer in Höhenlagen um 1400 m über NN. Die Standorte sind immer sehr feucht.

Paphiopedilum micranthum sollte halbschattig bis schattig im warmen Gewächshaus kultiviert werden. Die Pflanzen sollten in ein gut durchlässiges Substrat getopft werden und niemals austrocknen. Die relative Luftfeuchtigkeit muß jederzeit hoch gehalten werden.

Die Hauptblütezeit dieser Art ist zwischen Januar und April, aber die Pflanzen können das ganze Jahr hindurch blühen.

glabrous. The upper surface is tesselate dark and pale green, the under side is spotted deep purple. They are up to about 14 cm in length and up to about 2.5 cm broad.

The inflorescence is erect, up to about 26 cm high, densely pilose, in general bearing a single flower.

The bract is folded, green, spotted purple, up to 1.2 cm long and about 7 mm in width.

The ovary is up to about 4 cm long, bright green, the ridges spotted deep purple.

The flower is large for the plant, primarily due to the enormously inflated, pink to whitish-pink lip, up to 8.5 cm long and up to about 6.5 cm broad.

The petals and the sepals are small in comparison to the lip. The dorsal sepal is broadly ovate, more or less acute, up to about 3 cm long and about 2.5 cm across, the inside surface glabrous, the outside shortly pilose, the inflexed margins ciliate.

The synsepalum is ovate to suborbicular, more or less acute, up to 3 cm long and up to 2.8 across.

The petals are ovate, more or less obtuse, up to about 3 cm long and up to 3.5 cm across. The petals, as well as the sepals, are white, in the basal section greenish, covered with a more or less intense pink and with longitudinal dark stripes.

The lip is clearly calceolate, glabrous on the outside, up to 6.5 cm long and up to about 5.5 cm across, with a characteristic triangular indentation on each side of the base.

In front of these indentations, there is a folded staminodal plate which is up to about 1.5 cm high and about 7 mm across; white to light pink, the front spotted maroon, with a distinct triangular yellow area at the top of the apical part.

Paphiopedilum micranthum is known only from the southeastern part of the Chinese province of Yunnan between the villages of Marlipo and Shi Chou. This area borders Vietnam. The plants are said to grow on the floor of the woods or in shady areas of the river banks about 1400 meter above sea-level. The habitats are always very wet.

Paphiopedilum micranthum should be cultivated shaded to semi-shaded in the warm house, potted in a very open substrate. The plants should never be allowed to dry out and the relative humidity should be kept as high as possible all the time.

The main flowering period of this species in our latitudes is from January until April, but plants can flower all year long.

Untergattung BRACHYPETALUM Hallier

Diese Untergattung gehört ebenso wie die Untergattung PARVISEPALUM zu den primitiveren Gruppen innerhalb der Gattung. Sie hat mit der letztgenannten Untergattung mehrere gemeinsame morphologische Charakteristika, wie deutlich nach innen gefaltete Seitenränder des Lippenhauptlappens, wenig differenzierte Aurikel, deutliche Ähnlichkeit zwischen Petalen und Sepalen sowohl in Farbe als auch in der Form und einer Färbung der Lippe, die der der anderen Blütenblätter ähnlich ist, sowie die Chromosomenzahl von 2n = 26.

Trotzdem ist sie deutlich von der vorhergehenden Untergattung durch die dicke, längliche, elliptische, außen glatte Lippe, die viskösen Pollinien und die Abwesenheit von Stolonenbildung verschieden.

Obwohl die einzelnen Taxa innerhalb dieser Untergattung mehr oder weniger cytogenetisch stabil sind, so zeigen sich doch deutliche morphologische Variationen innerhalb jeder Art, vor allem was die Blütencharaktere anbelangt. Diese großen Variationen deuten an, daß die phylogenetische Entwicklung dieser Gruppe der Frauenschuhorchideen noch nicht abgeschlossen ist, so daß es ohne weiteres möglich, ja sogar wahrscheinlich ist, daß sich noch weitere Arten innerhalb der Untergattung BRACHYPETALUM bilden werden.

Soweit heute bekannt, erstreckt sich diese Gruppe von Birma über Thailand, Laos, Kambodscha und Vietnam bis zu der Malaysischen Halbinsel und nördlich bis zu dem südlichen Teil Chinas.

Subgenus BRACHYPETALUM Hallier

This subgenus, together with the subgenus PARVISEPALUM, belongs to the more primitive groups of the genus *Paphiopedilum*. Both subgenera share a number of morphological characteristics such as the distinctly inflexed margins of the main labellum lobe, poorly differentiated lateral ears of the lip, close similarity of petals and sepals in shape as well as in colouration, similar colour pattern of the lip, the petals and the sepals, and a mitotic chromosome number of 2n = 26.

In spite of all this, the subgenus BRACHYPETALUM differs clearly from the subgenus PARVISEPALUM by its thick, ovoid labellum with a plain surface, its viscous pollinia and by the lack of stolons.

Although the taxa belonging to this subgenus are, more or less, cytogenetically stable, each species does show distinct intraspecific morphological variation – especially in respect to floral characteristics. These considerable variations indicate that the phylogenetic development of this group of slipper orchids has not as yet been completed. It is therefore possible, even probable, that other species within the subgenus BRACHYPETALUM will develop.

As far as we know today, this orchid group has a distribution from Burma, Thailand, Laos, Cambodia and Vietnam to the Malaysian Archipelago and as far north as the southern part of China.

Schlüssel zu den Arten der Untergattung BRACHYPETALUM

1. Blütenblätter fahl gelb, Fahne und Petalen spärlich mit sehr kleinen tief-purpurnen Flekken überzogen. Lippe schmal länglich, mehr oder weniger spitz auslaufend . . . . *Paphiopedilum concolor*

1a. Blütenblätter schwach purpurn behaucht, Lippe elliptisch bis breit elliptisch mit gerundeter Spitze 2

2. Petalen und Sepalen fast einfarbig, nur an der Basis leicht mit sehr kleinen purpurnen Flecken überzogen. Blütenstandsachse länger als 5 cm . . . . *Paphiopedilum niveum*

Key to the species of the subgenus BRACHYPETALUM

1. Perianth pale yellow, upper sepal and petals finely and sparsely speckled with minute, deep purple spots; lip narrow ovoid, more or less acute . . . . . . *Paphiopedilum concolor*

1a. Perianth white or faintly-tinged with pale purple, lip ovoid to broad-ovoid, more or less obtuse 2

2. Petals and sepals nearly uni-coloured, faintly speckled with pale purple spots only in the basal portions; inflorescence longer than 5 cm . . . . *Paphiopedilum niveum*

2a. Petalen und Sepalen ganz mit großen tief-purpurnen Flecken überzogen. Blütenstandsachse sehr kurz (meistens 2 bis 5 cm) . . . 3

3. Lippe spärlich und unregelmäßig mit größeren tief-purpurnen Flecken überzogen. Befleckung der übrigen Blütenblätter auch unregelmäßig. Blütenstandsachse sehr kurz (2 bis 3 cm) . . . . *Paphiopedilum bellatulum*

3a. Lippe kaum befleckt, fast einfarbig weiß. Kleine Flecken auf den Petalen und Sepalen in einem reihenartigen Muster, Blütenstandsschaft 3 bis 5 cm lang . . . . *Paphiopedilum* × *godefroyae*

2a. Petals and sepals entirely speckled with large, deep purple spots; inflorescence very short, usually 2–5 cm long 3

3. Lip sparsely and irregularly speckled with large, deep purple spots; speckling of the petals and sepals also irregular; inflorescence very short, usually 2–3 cm . . . . *Paphiopedilum bellatulum*

3a. Lip hardly speckled, almost uniformly whitish; small speckles on the petals and the sepals arranged in a linear pattern; inflorescence usually 3–5 cm long . . . . *Paphiopedilum* × *godefroyae*

Paphiopedilum bellatulum (Rchb. f.) Stein

Stein's Orchideenbuch: 456 (1892)

Basionym: *Cypripedium bellatulum* Rchb. f.
Gardeners' Chronicle, 3. Ser., **3:** 648 (1888)

Synonym: *Cordula bellatula* (Rchb. f.) Rolfe
Orchid Review 20 (1): 2 (1912)

Paphiopedilum bellatulum (Rchb. f.) Stein

Paphiopedilum bellatulum wurde erstmals im Frühjahr 1888 durch die englische Firma LOW aus Birma eingeführt und noch im selben Jahr durch REICHENBACH im **Gardeners' Chronicle** als *Cypripedium bellatulum* beschrieben.

Paphiopedilum bellatulum ist eng mit *Paphiopedilum concolor* und *Paphiopedilum niveum* verwandt, und es findet sich eine Reihe von Pflanzen in Kultur, von denen man aufgrund der intermediären Farbgestaltung der Blüte und der Staminodiummorphologie annehmen könnte, daß sie Hybriden darstellen.

Paphiopedilum bellatulum ist eine kleinwüchsige, selten über 10 cm hohe, krautige Pflanze mit elliptischen, 15 bis 25 cm langen und 6 bis 8,5 cm breiten Blättern. Sie sind oberseitig sattgrün und schwach marmoriert, unterseitig tiefpurpurn getüpfelt, stumpf und ungleich zweilappig endend.

Paphiopedilum bellatulum reached Europe in the spring of 1888, when the first plants of this species were imported from Burma by the English firm of LOW. The species was described in the same year by H. G. REICHENBACH in the **Gardeners' Chronicle** as *Cypripedium bellatulum.*

Paphiopedilum bellatulum is closely related to *Paphiopedilum concolor* and *Paphiopedilum niveum.* A number of plants showing intermediate characteristics, especially in respect to flower colouration and the shape of the staminodal plate, are in cultivation, indicating that they may possibly be of hybrid origin.

Paphiopedilum bellatulum is a small-growing plant, seldom higher than 10 cm. The herbaceous epiphyte has elliptic leaves, 15 to 25 cm long and 6 to 8.5 cm broad; obtuse with an asymmetrically split apex. Their upper surface is a vivid green,

Die Infloreszenz ist sehr kurz, so daß die Blüten die Laubblätter berühren, tiefpurpur, flaumig behaart, meist einblütig.

Die Braktee ist länger als die Hälfte des Fruchtknotens.

Die Blüte ist bis zu 7 cm breit, weiß bis hellgelb mit violettbraunen Flecken.

Die Fahne ist ungefähr 2,5 cm lang und 4 cm breit, elliptisch bis fast kreisrund, hinten gekielt, kurz behaart.

Das Synsepalum ist fast so lang wie die Fahne, aber schmaler, weiß, kurz behaart.

Die Petalen sind bis zu 5 cm lang und 2,8–3,5 cm breit, breit eiförmig, etwas herabgebogen, am Ende eingekerbt, kurz bewimpert.

Die sackförmige Lippe ist ca. 3,5 cm lang und ungefähr 2 cm breit. Sie ist außen etwas weniger getüpfelt als der Rest der Blüte, aber der nach innen gefaltete Seitenlappen ist sehr intensiv gefleckt. Das Staminodium ist rautenförmig bis fast kreisrund, an der Spitze mit drei Zähnen, ca. 9 mm lang und 8 bis 11 mm breit, weiß mit violetten Tüpfeln und einem gelben Mittelfleck.

slightly tesselate; the under surface is spotted deep purple.

The inflorescence is villose, generally uniflorous, very short, the flower therefore touching the leaves.

The bract exceeds half the length of the ovary.

The flower is up to 7 cm across, white, rarely with a yellow hue, spotted irregularly deep purple.

The dorsal sepal is about 2.5 cm long and up to 4 cm broad, elliptic to suborbicular, on the outside distinctly keeled, shortly pilose.

The synsepalum is nearly as long as the dorsal sepal but somewhat narrower, white, shortly pilose.

The petals are up to 5 cm long and about 2.8 to 3.5 cm broad, broadly ovate, curved somewhat downwards, the apex somewhat cut, shortly ciliate.

The lip is narrowly ovoid to sub-ellipsoid, saccate, up to about 3.5 cm long and about 2 cm across, with fewer and smaller spots externally than the other segments but densely spotted on the infolded lobes.

Paphiopedilum bellatulum (Rchb. f.) Stein

Paphiopedilum bellatulum wurde, wie bereits erwähnt, ursprünglich in Birma entdeckt. Die meisten der heute in Kultur befindlichen Pflanzen dieser Art stammen aus den an Birma grenzenden thailändischen Landesteilen. Seit ungefähr zwei Jahren sind Pflanzen auf dem Markt, die über Hongkong aus China gekommen sind. Sie unterscheiden sich von den bisher bekannten Varietäten durch eine schwefelgelbe Grundfärbung und reichliche Befleckung. Im Botanischen Garten von Beijing befinden sich einige Pflanzen von *Paphiopedilum bellatulum* in Kultur, die angeblich in der südchinesischen Provinz Yunnan gesammelt wurden[1].

Im Jahre 1974 wurde auch von einem *Paphiopedilum bellatulum*-Fund aus dem indischen Bundesstaat Meghalaya berichtet[2]. Dieser „Fundort" liegt weitab vom übrigen Verbreitungsgebiet der Art, könnte aber zutreffend sein, obwohl er bisher nicht bestätigt wurde und sehr unwahrscheinlich ist.

In der Tat sprechen einige Anzeichen dafür (Vorkommen der Naturhybriden siehe auch unter *Paphiopedilum × godefroyae*), daß die Verbreitung von *Paphiopedilum bellatulum* viel größer ist (oder war) als bisher im allgemeinen angenommen wird.

Wie seine engen Verwandten ist *Paphiopedilum bellatulum* sehr kalkliebend und wächst in Spalten oder auf hervorstehenden Stellen von Kalkfelsformationen, meist in Blatthumus und Moos.

Paphiopedilum bellatulum ist etwas kälteresistenter als seine Verwandten, was auf die Höhenlage seiner Standorte (1000 bis 1500 m ü. NN) zurückzuführen ist.

Das Klima im birmanischen Heimatareal kann in vier monsunbedingte Perioden aufgeteilt werden:

1. Mitte Mai bis September mit warmen sommerlichen Monsunwinden,
2. eine etwas kühlere Periode zwischen Oktober und November,
3. Wintermonsun von Dezember bis Februar und
4. wieder mildere Bedingungen von März bis Mai.

Fast 90 % des jährlichen Niederschlags fallen von Ende Mai bis September. Die Periode von Oktober bis Anfang Mai ist ausgesprochen trocken.

Die kältesten Monate sind Dezember und Januar mit bis zu 24 °C am Tage und Nachttemperaturen, die bis auf den Gefrierpunkt (manchmal auch darunter) absinken können. Im April und Mai ist es am

[1] Siehe P. J. CRIBB, **Curtis's Botanical Magazine 184** (4): 186 (1983).

[2] PEMPAHISHEY, **American Orchid Society Bulletin 43:** 716 (1974).

The staminodal plate is sub-rhomboidal to suborbicular, at the apex with three teeth, about 8 mm long and up to 11 mm across, white, spotted purple and with a yellow middle part.

As noted above, *Paphiopedilum bellatulum* was discovered in Burma. Most of the specimens which are in culture today however, originate from Thailand.

For about 2 years, plants very much resembling the typical *Paphiopedilum bellatulum* from China have reached us via Hong Kong. They differ from the typical forms of the species in that the ground colouration of the perianth is distinctly yellow and more intensely spotted.

Moreover, in the Botanical Garden of Beijing, some plants, said to originate from the southern Chinese province Yunnan, which cannot be differentiated from the typical form of *Paphiopedilum bellatulum* are being cultivated.[1]

In 1974, it was reported that *Paphiopedilum bellatulum* was found in the Indian state of Meghalaya.[2] This "habitat" is far away from the other areas where the species is known to occur. Although this report has not yet been confirmed, one should note that dislocated habitats of one species are possible, although improbable.

However, there is some indication (distribution of the natural hybrids, see under *Paphiopedilum × godefroyae*) that the natural distribution of *Paphiopedilum bellatulum* is (or was) much larger than the distribution generally accepted today.

Paphiopedilum bellatulum is, just like its close relatives, strongly calcicolous, growing in cracks and crevices of limestone outcrops usually in a layer of leaf-mould and moss.

Paphiopedilum bellatulum is somewhat more cold-tolerant than most of the other species, growing at between 1000 m and 1500 m altitude.

The climate of the areas where *Paphiopedilum bellatulum* occurs can be divided into 4 monsoon controlled periods:

1. mid-May to September with a warm summer monsoon,
2. a somewhat cooler period between October and November,
3. winter monsoon from December to February and,
4. again milder conditions from March until May.

[1] See P. J. CRIBB, **Curtis's Botanical Magazine 184** (4): 186 (1983).

[2] PEMPAHISHEY, **American Orchid Society Bulletin 43** (?): 716 (1974).

wärmsten – mit Temperaturen von 24 bis 34 °C am Tage und 17 bis 20 °C in der Nacht.

Paphiopedilum bellatulum sollte temperiert bis kühl kultiviert werden.

Obwohl die Pflanzen am Tage ziemlich hohe Temperaturen vertragen können, gedeihen sie nur bei einer ausgeprägten Nachtkühle gut. Im Frühjahr sollte man die Pflanzen kühl und trocken stellen. Pflanzen, die zu warm kultiviert werden, blühen nicht. *Paphiopedilum bellatulum* mag hohe Luftfeuchtigkeit und einen durchlässigen Pflanzstoff mit Moosbeimischung.

Paphiopedilum bellatulum ist im Wurzelbereich sehr salzempfindlich. Es sollte daher nur geringfügig und selten gedüngt werden.

Kranke Pflanzen erholen sich am besten durch Umtopfen in reines, frisches Waldmoos (eventuell ergänzt durch Gießen mit destilliertem Wasser).

Paphiopedilum bellatulum blüht normalerweise von Mai bis Juli. Je nach Kulturbedingungen sind auch andere Blühzeiten möglich.

Varietäten:

Eine ganze Reihe von „Varietäten" von *Paphiopedilum bellatulum* sind beschrieben worden. Mit einer Ausnahme (var. *album*) sind diese Pflanzen innerhalb der natürlichen Bandbreite der Art anzusiedeln und gewährleisten keineswegs eine Aufrechterhaltung als eigenständige taxonomische Varietät. Sie basierten auf einzelnen Exemplaren, die von der Nominalform unwesentlich in Blütengröße, Form und Intensität der Befleckung oder durch die Bildung einer zweiblütigen Infloreszenz abweichen. In einigen Fällen wurden sogar lediglich „sehr schöne Exemplare" in den Varietätsstand erhoben *(sic)*. Der interessierte Leser sei somit auf die einschlägige Fachliteratur verwiesen.

Paphiopedilum bellatulum var. *album* O'Brien
Gardeners' Chronicle, 3. Ser., **17:** 748, 1895

Blüte ganz weiß ohne jegliche Tupfung.

Nearly 90 % of the yearly rain falls between the end of May and September. The period between October and the beginning of May is very dry.

December and January are the coldest months with daytime temperatures of about 24 °C and nighttime temperatures dropping to or below the freezing point.

The warmest period with temperatures of 24 to 34 °C during daytime and 17 to 20 °C at night is between April and May.

Consequently, *Paphiopedilum bellatulum* needs to be grown cooler than most species of the genus.

Although the plants can stand rather high temperatures during daytime, they will only do well when kept cool at night. In the springtime *Paphiopedilum bellatulum* should be kept cool and dry. Plants grown warmer will not flower. *Paphiopedilum bellatulum* likes high humidity and should be potted in an open moss containing substrate.

This species is very sensitive to salts, especially in the root zone. One should therefore only fertilize infrequently and use weak fertilizer.

Sickly plants will quickly recover when repotted in fresh moss (possibly supported by waterings with distilled water).

The normal flowering period of *Paphiopedilum bellatulum* is between May and July. Depending on the state of cultivation, it is possible that the flowers will develop at other times of the year.

Varieties:

A whole series of "varieties" of *Paphiopedilum bellatulum* have been reported. With the exception of var. *album* all these plants are to be seen within the natural variation of a species and do not warrant their confirmation as a autonomous taxon. They are based on single specimens differing more or less from the nominal form by the shape, colour and speckling of the flower or by the generation of a bi-flowered inflorescence. In some cases merely "very beautiful specimens" have been elevated to separate "varieties" *(sic)*. The interested reader is referred to the pertinent literature.

Paphiopedilum bellatulum var. *album* O'Brien
Gardeners' Chronicle, 3. ser., **17:** 748 (1895)

Flower uniformly white, without any spots.

Paphiopedilum concolor (Bateman) Pfitzer

in ENGLER & PRANTL: **Nat. Pflanzenfamilien. 2,** (6): 84 (1888)

Basionym: *Cypripedium concolor* Bateman
Curtis's Botanical Magazine 91, t. 5513 (1865)

Synonym: *Cordula concolor* (Bateman) Rolfe
Orchid Review 20 (1): 2 (1912)

Paphiopedilum concolor (Bateman) Pfitzer

Diese durch Blütenfarbe und Blütenform unverwechselbare Art wurde erstmals 1859 durch den englischen Anglikanerpfarrer PARISH auf Kalksteinfelsen in der Nähe von Tenarassim (Moulmein) in Birma gefunden.

Der Geistliche sandte eine Skizze seiner Entdekkung an William Jackson HOOKER, den damaligen Direktor des Königlichen Botanischen Gartens in Kew.

Die ersten lebenden Pflanzen erreichten England aber erst 1864, als PARISH eine Sammlung an die Firma LOW sandte.

Fast zur gleichen Zeit erhielt der Botanische Garten in Kew Pflanzen derselben Art durch den englischen Offizier BENSON, die dieser am selben Standort wie PARISH gefunden hatte.

Die Beschreibung durch BATEMAN im **Botanical Magazine** basierte auf einer Pflanze, die in der

This unmistakable species is unique in its flower shape and colour and was discovered in 1859 by the English anglican minister PARISH on the limestone formations near Tenarassim (Moulmein) in Burma.

PARISH send a jotting of his discovery to William Jackson HOOKER who, at that time, was the director of the Royal Botanic Gardens at Kew, England.

The first living plants recorded in England were send by PARISH to the orchid dealer LOW. At just about the same time, plants of this species were send to Kew by an English army officer named BENSON. Those plants originated from the same site as the PARISH ones.

The original description by Bateman was based on a plant which flowered in the collection of RÜCKER in the spring of 1865.

Sammlung von RÜCKER im Frühjahr 1865 zur Blüte kam.

Paphiopedilum concolor ist eine als Humusepiphyt wachsende krautige Pflanze, die bis ca. 20 cm hoch wird.

Die Laubblätter, bis fünf an der Zahl, werden 7 bis 15 cm lang und 3 bis 5 cm breit. Sie sind vorne eingeschnitten, elliptisch, auf der Oberseite dunkelgrün mit hellerer Marmorierung, unterseitig mehr oder weniger intensiv purpurn gefleckt.

Die bis zu 12 cm lange Infloreszenz ist dunkelgrün, flaumig behaart und ein- bis zweiblütig (selten mehrblütig).

Die Brakteen sind nur 1 bis 2 cm lang und somit viel kürzer als der Fruchtknoten, eiförmig.

Die 5 bis 8 cm großen Blüten sind blaß- bis schwefelgelb, mit mehr oder weniger intensiver, sehr feiner purpurner Betüpfelung. Die Betüpfelung ist meistens an der Basis der Blütenblätter am intensivsten.

Die Fahne ist ungefähr 2,8 cm hoch und 2 bis 2,4 cm breit, eiförmig, mehr oder weniger spitz endend.

Paphiopedilum concolor is an herbaceous plant, growing as an humus epiphyte in leaf-mould. The plants are up to 20 cm high and have about 5 leaves per growth.

The leaves are 7 to 15 cm long and 3 to 5 cm broad. They are cut at the apex, elliptic-oblong, the upper surface deep green, tesselated with silvery green, the under surface more or less intensely spotted purple.

The inflorescence is up to about 12 cm long, dark brown, villose, usually bearing one to two flowers (rarely more).

The ovate floral bracts are only 1 to 2 cm long and, therefore, much shorter than the ovary.

The flowers are 5 to 8 cm across, pale yellow to sulphureous, speckled, more or less, intensively purple, whereby the spots are more numerous at the base of the perianthal parts.

The dorsal sepal is about 2.8 cm long and 2 to 2.4 cm broad, ovate, more or less acute.

The synsepal is similar but only about 1.8 to 2 cm long and narrower than the dorsal one.

Paphiopedilum concolor (Bateman) Pfitzer

Das Synsepalum ist ähnlich, aber nur ungefähr 1,8 bis 2 cm lang und schmaler als die Fahne.

Die länglich- bis breitelliptischen Petalen werden ca. 3 cm lang und ungefähr 1,6 bis 1,8 cm breit. Sie sind stumpf oder abgerundet. Die Ränder der Petalen und Sepalen sind fein behaart.

Das Labellum (die Lippe) ist ca. 2,5 cm lang und 1,5 cm breit, schmal-kegelig bis sackförmig und mehr oder weniger zusammengedrückt.

Das sehr variable Staminodium ist mehr oder weniger herzförmig, gelb mit dunkelgelbem Mittelfleck, purpurrot getüpfelt.

Paphiopedilum concolor hat ein ziemlich großes Verbreitungsgebiet. Die Art besiedelt meistens Küstengegenden oder Flachlandhügel und Kalkfelsformationen in Birma, Thailand, Kambodscha, Vietnam und Südchina, im allgemeinen in Höhenlagen bis 300 m (mindestens ein Standort über 1000 m ü. NN ist bekannt), in Felsnischen und -spalten, wo sich Humus angesammelt hat.

Innerhalb des großen Heimatareals gibt es viele verschiedene Klimazonen, die aber alle durch kühle Monsunwinde von November bis März beeinflußt werden. Diese Jahreszeit ist gekennzeichnet durch Nebelbildung, andauernde niedrige Wolken, Nieselregen sowie nächtliche Taubildung.

Im Dezember ist es am kältesten, mit Tagestemperaturen um 22 bis 24 °C und einem nächtlichen Absinken auf 9 bis 7 °C. Ab Mai wird es etwas wärmer, und während des Aprils, normalerweise dem wärmsten Monat des Jahres, herrschen Temperaturen um 32 °C am Tage und um 16 °C in der Nacht.

Die Regenperiode beginnt normalerweise im Mai und hält bis Oktober an. Die regenreichsten Monate sind Juni und Juli.

Die jährliche Niederschlagsmenge variiert von Areal zu Areal zwischen 2300 und 3100 mm.

Die relative Luftfeuchtigkeit ist immer sehr hoch.

Was die Kultur anbelangt, ist zu bemerken, daß *Paphiopedilum concolor* sehr empfindlich auf hohe Salzkonzentrationen im Substrat reagiert. Man sollte daher ein moosreiches Substrat wählen und während der Sommermonate ausgiebig gießen. Nur ganz geringe Düngergaben sind zu empfehlen.

Ein sonniger Standort fördert eine reiche Blütenbildung, während ein schattiger Platz die Blattmarmorierung deutlicher macht.

The petals are broad-elliptical to elongated-oval, about 3 cm long and about 1.6 to 1.8 cm across. They are sub-obtuse to rounded. The margins of the sepals and the petals are finely pilose.

The lip is about 2.5 cm long and more or less 1.5 cm across, conical with a rounded bottom to saccate and sometimes compressed laterally.

The shape of the staminodal plate is very variable, more or less cordate, yellow with a dark yellow area in the middle, spotted purple.

Paphiopedilum concolor has a widespread occurrence. The species usually inhabits coastal areas, low hillsides and limestone rocks in Burma, Thailand, Cambodia, Vietnam and the southern part of China. The plants usually grow in rock cracks and crevices at altitudes of up to 300 m above sea-level. (One site has been reported at an elevation of more than 1000 meter.)

In the enormous area where *Paphiopedilum concolor* occurs, a number of different microclimates may be found. All of them are influenced by cool monsoon winds from November till March. That period of the year is characterized by fog, prevailing low clouds, drizzles and nighttime dews.

December is the coldest month with daytime temperatures around 22 °C and nighttime temperatures as low as 7 to 9 °C. Beginning in March, the weather warms and in April, usually the warmest month of the year, temperatures are around 32 °C in the daytime and 16 °C at night.

The rainy season usually starts in May and prevails until October. June and July are the wettest months.

The average amount of precipitation differs from area to area and lies between 2300 and 3100 mm.

The relative humidity is very high at all times.

Concerning the cultivation of this species, it can be said that *Paphiopedilum concolor* is very intolerant to high concentration of salts in the substrate. One should therefore use plenty of moss in the potting medium and water extensively during the summer months. Fertilization should be kept very low.

Growing *Paphiopedilum concolor* in a sunny place will cause many flowers to be produced, whereas the cultivation in a shady spot results in nicely tesselate leaves.

Varietäten:

Paphiopedilum concolor ist, wie auch alle anderen Arten der Unterfamilie **BRACHYPETALUM**, eine sehr variable Art. Eine Reihe von Varietäten sind beschrieben worden[1], und FOWLIE[2] hielt es sogar für angebracht, die Art in vier Unterarten aufzuteilen. Mit Ausnahme der reinweißen Form, die hier als var. *album* beschrieben wird, weicht keine der beschriebenen „Varietäten" so entscheidend von der Nominalform ab, daß sie aufrechterhalten werden kann.

Paphiopedilum concolor var. ***album*** Braem, *var. nov.*

Typus similis, sed differt petala, sepala et labellum alba immaculata.

Blüte einfarbig weiß, ohne jegliche rötlichen Pigmente.

[1] als *Cypripedium concolor* var.
[2] **Orchid Digest 41** (5): 190–197 (1977).

Varieties:

Paphiopedilum concolor is, like all other members of the subgenus **BRACHYPETALUM** very variable. A series of varieties has been described[1] and FOWLIE considered it useful to divide the species into four subspecies.[2] With the exception of the pure white form, here described as var. *album*, none of these "varieties" differs to such an extent from the nominal form as to warrant them varietal status.

Paphiopedilum concolor var. ***album*** Braem, *var. nov.*

Typus similis, sed differt petala, sepala et labellum alba immaculata.

Flowers uniformly white without any red pigmentation.

[1] as *Cypripedium concolor* var.
[2] **Orchid Digest 41** (5): 190–197 (1977)

Paphiopedilum niveum (Rchb. f.) Stein

Stein's Orchideenbuch: 478 (1892)

Basionym: *Cypripedium niveum* Rchb. f.
Gardeners' Chronicle, 2 Ser., **9:** 1038 (1869)

Synonyma: *Cordula nivea* (Rchb. f.) Rolfe
Orchid Review 20 (1): 2 (1912)

Cypripedium concolor var. *niveum* Rchb. f.
Gardeners' Chronicle, 2 Ser., **19:** 16 (1883)

Paphiopedilum niveum (Rchb. f.) Stein

Die Art wurde erstmals 1868 durch die Firma LOW nach England importiert, in der Annahme, es handele sich um *Paphiopedilum concolor.*

Obwohl die Sendung aus Moulmein kam, stand bereits kurz darauf fest, daß die Pflanzen von den dem Festland vorgelagerten Inseln stammten.

SEIDENFADEN[1] aber berichtet, sowohl *Paphiopedilum niveum* als auch *Paphiopedilum niveum* var. *album* auch auf dem thailändischen Festland gefunden zu haben.

Paphiopedilum niveum ist eine bis zu 26 cm hoch werdende, kleinwüchsige, krautige, als Humusepiphyt wachsende Pflanze. Die elliptischen Laubblätter werden bis zu 15 cm lang und 2,7 bis 3,5 cm breit. Die Oberseite der Blätter ist dunkelgrün mit

[1] **Botanisk Tidsskrift 67:** 78–80 (1972)

The species was introduced to Europe in 1868 by the English orchid dealer LOW which imported the plants assuming that the shipment contained *Paphiopedilum concolor* from the Moulmein district.

As there was no evidence of *Paphiopedilum niveum* growing in the vicinity of Moulmein it was soon considered that the plants had to originate from the Langkawi Islands, situated a few miles in the sea.

However, SEIDENFADEN[1] reported to have found *Paphiopedilum niveum* as well as *Paphiopedilum niveum* var. *album* on mainland Thailand.

Paphiopedilum niveum is a dwarf, up to 26 cm high, herbaceous plant growing in leaf-mould. The

[1] **Botanisk Tidsskrift 67:** 78–80 (1972)

grauer Zeichnung, die Unterseite ist purpurn gefleckt.

Die aufrecht stehende Infloreszenz wird bis zu 17 cm hoch und trägt 1 bis 2 Blüten, die einen Durchmesser bis ca. 8 cm erreichen.

Die Blüten sind weiß mit feinen purpurnen Tupfen an der Basis der Petalen und der Fahne.

Die Fahne wird bis zu 4 cm hoch und 2 cm breit. Sie ist mehr oder weniger länglich und abstehend, konkav, hinten gekielt und rötlich getupft.

Das Synsepalum ist länglich eiförmig, schmaler als die Fahne.

Die abstehenden Petalen sind leicht nach unten gebogen, länglich oval bis breit oval, stumpf, manchmal fast rund.

Die Lippe ist bis zu 3 cm lang, im Umriß elliptisch, mit schmaler Öffnung, die durch Säule und Staminodium fast verschlossen ist.

Das Staminodium ist sehr variabel, mehr oder weniger elliptisch bis fast nierenförmig und in der Mitte gekielt, gelb mit weißen Rändern.

Fundorte von *Paphiopedilum niveum* sind bekannt von den Langkawi Inseln (West Malaysia), Thailand und von den zwischen Borneo und Singapur gelegenen Tambelan Inseln.

Die Pflanzen werden auf Kalkböden in Seenähe bis in Höhenlagen von ungefähr 80 m ü. NN gefunden.

Das Klima im Verbreitungsgebiet von *Paphiopedilum niveum* ist eindeutig vom naheliegenden Meer geprägt.

Die Luftfeuchtigkeit ist naturgemäß das ganze Jahr über sehr hoch. Während der Monate Juni bis Oktober gibt es viel Regen, im Dezember ist es relativ trocken.

Die Temperaturen im Sommer bewegen sich um 31 °C am Tage und um 20 °C in der Nacht. Im Winter sinken die Temperaturen auf ca. 25 °C bzw. 15 °C ab.

Paphiopedilum niveum ist eine sehr beliebte Pflanze, nicht zuletzt des relativ kleinen Wuchses wegen. Die Art ist recht einfach zu kultivieren und blühwillig.

Man sollte *Paphiopedilum niveum* im warmen Haus bei hoher Luftfeuchtigkeit halten und während der Sommermonate reichlich mit salzarmem Wasser gießen.

Ein heller Standort bewirkt reiche Blütenbildung zwischen Juni und August.

leaves are elliptic, up to 15 cm long and about 2.7 to 3.5 cm broad. The upper surface is dark green, tesselate with grayish green, the under surface is spotted purple.

The inflorescence is erect, up to 17 cm high, usually bearing one or two flowers. These attain a diameter of about 8 cm.

The flowers are white with small purple spots at the basal parts of the petals and dorsal sepal.

The dorsal sepal is up to 4 cm long and about 2 cm broad, more or less elongate, spreading, concave, keeled and stained with reddish purple behind.

The synsepalum is ovate-oblong, narrower than the dorsal one.

The petals are spreading, slightly deflexed, broadly oblong, obtuse, sometimes nearly sub-orbicular.

The lip is ovoid, up to 3 cm long with a contracted mouth, which is nearly covered by the column and the staminode.

The staminodal plate is very variable, more or less elliptic to sub-reniform, keeled in the middle, yellow, bordered with white.

Paphiopedilum niveum has been reported from the Langkawi Islands (West Malaysia), Thailand and from the Tambelan Islands, situated between Singapore and Borneo.

The plants grow on limestone near the sea up to elevations of about 80 m above sea-level.

The climate of the distribution area of this species is obviously influenced by the nearby ocean.

The relative humidity is constantly very high the whole year over. The region is only influenced by a single monsoon during the months of June to October bringing plenty of rain. The wettest month is August and the driest is December.

Summer temperatures are about 31 °C in the daytime and about 20 °C during the nights. During the winter, temperatures drop to about 25 °C during the day and to 15 °C at night.

Paphiopedilum niveum is a very popular plant among orchid growers, mainly because of its dwarf habit. The cultivation of the species is fairly easy and the plants flower willingly.

Paphiopedilum niveum should be kept in the warm house. One should make sure that the humidity remains high at all times. During the summer months abundant waterings with saltfree water are indicated.

Cultivating *Paphiopedilum niveum* under intense light will be rewarded by rich flowering between June and August.

Varietäten:

Paphiopedilum niveum var. *album*
Hort. ex Linden
Lindenia 3: 90 (1887)

Pflanze mit reinweißen Blüten ohne jegliche Fleckung.

Alle anderen beschriebenen „Varietäten" beziehen sich auf die mehr oder weniger intensive Fleckung der Blüte und auf die Farbvariationen dieser Flekken. Sie sind ohne Ausnahme innerhalb der natürlichen Bandbreite der Art anzusiedeln und können als eigenständige Varietäten im Sinne der botanischen Taxonomie kaum aufrechterhalten werden. Auch hier verweise ich den interessierten Leser auf die einschlägige Fachliteratur.

Varieties:

Paphiopedilum niveum var. *album*
Hort. ex Linden
Lindenia 3: 90 (1887)

Plants with snow-white flowers lacking any spots.

All other "Varieties" described are based on single specimens differing from the nominal form solely by, more or less, intensive spotting of the flower and by the different shades of these spots. They are, without any exception, to be considered part of the natural variation of a species in the wild and do not warrant separate taxonomic status. Here again, I refer the interested reader to the pertinent literature.

Paphiopedilum × *godefroyae* (Godefroy) Stein (pro sp.) stat. nov.

Basionym: *Cypripedium godefroyae* Godefroy-Lebeuf
L'Orchidophile: 830 (1883)

Synonyma: *Cordula godefroyae* (Godefroy) Rolfe
Orchid Review 20 (1): 2 (1912)
Cypripedium concolor var. *godefroyae* (Godefroy) Collett & Hemsley
Journ. Linn. Soc. 28: 135 (1891)
Paphiopedilum godefroyae Stein
Stein's Orchideenbuch: 468 (1892)

Paphiopedilum × *godefroyae* (Godefroy) Stein

Paphiopedilum × *godefroyae* wurde von Herrn MURTON in Thailand entdeckt. Der genaue Fundort ist unbekannt geblieben.

GODEFROY-LEBEUF bekam die Pflanze von einem Engländer namens ALABASTER, einem Bekannten MURTONs. Er brachte sie 1876 nach Europa und benannte die Art *(loc. cit.)* nach seiner Frau als *Cypripedium godefroyae.*

Paphiopedilum × *godefroyae* ist eng verwandt mit *Paphiopedilum niveum, Paphiopedilum bellatulum* und *Paphiopedilum concolor.*

Viele namhafte Autoren[1] betrachten diese Pflanze als Naturhybride zwischen *Paphiopedilum bellatulum* und *Paphiopedilum concolor.*

Einige gärtnerische Kreuzungen zwischen den oben erwähnten Arten brachten Pflanzen hervor, die von den in der freien Natur gefundenen Exemplaren von *Paphiopedilum* × *godefroyae* kaum zu unterscheiden waren.

[1] KARASAWA, K., **Bull. of the Hiroshima Botanical Garden** Nr. 2: 1–149 (1979)
ROLFE, R. A., **Orchid Review 6:** 231 (1898)
ROLFE, R. A., **Orchid Review 11:** 281, 359, 361 (1903)
HOLTTUM, R. E., **Flora of Malaya 1:** 71 (1957) und andere

Paphiopedilum × *godefroyae* was originally discovered by an Englishman named MURTON in Thailand. The exact location remained unknown.

GODEFROY-LEBEUF obtained the plants from a Mr. ALABASTER, an acquaintance MURTONS. GODEFROY brought the plants with him to Europe in 1876, and described them *(loc. cit.)* as a new species, supposedly, in honour of his wife, *Cypripedium godefroyae.*

Paphiopedilum × *godefroyae* is a close relative of *Paphiopedilum niveum, Paphiopedilum bellatulum* and *Paphiopedilum concolor.* Many distinguished authors[1] consider this plant to be a natural hybrid between *Paphiopedilum bellatulum* and *Paphiopedilum concolor.*

Some artificial crosses between the putative parent species as cited above have resulted in plants that cannot be differentiated from the plants of *Paphiopedilum* × *godefroyae* found in the wild.

[1] KARASAWA, K., **Bull. of the Hiroshima Botanical Garden** Nr. 2: 1–149 (1979)
ROLFE, R. A., **Orchid Review 6:** 231 (1898)
ROLFE, R. A., **Orchid Review 11:** 281, 359, 361 (1903)
HOLTTUM, R. E., **Flora of Malaya 1:** 71 (1957) and others

Das Argument, das am häufigsten gegen die These des Naturhybridentums vorgebracht wird, ist die Verbreitung von *Paphiopedilum bellatulum*, dessen bekannte Standorte mindestens 800 km vom *Paphiopedilum* × *godefroyae*-Standort entfernt sind.

Dieses Argument aber hält keiner kritischen Prüfung stand. Es ist viel wahrscheinlicher, daß es *Paphiopedilum bellatum*-Standorte gibt, die uns nicht bekannt sind, oder, daß es solche Standorte gab, die inzwischen (durch welchen Umstand auch immer) vernichtet worden sind.

Die immense Variabilität des Staminodiums ist ein weiterer Beweis dafür, daß *Paphiopedilum* × *godefroyae* als Hybride anzusehen ist und zeigt außerdem an, daß diese, was ihren genetischen Pool anbelangt, noch nicht stabilisiert ist. Daher ist *Paphiopedilum* × *godefroyae* weniger als selbständige Art, sondern eher als Naturhybride zu betrachten. Das Taxon sei hiermit entsprechend umkombiniert.

Paphiopedilum × *godefroyae* ist eine meist auf Felsen als Humusepiphyt wachsende, krautige Pflanze, die bis ca. 15 cm hoch wird.

Die Laubblätter sind bis 18 cm lang und bis 4 cm breit. Sie sind länglich-linealisch, dunkelgrün, marmoriert mit hellgrünen Flecken und unterseitig intensiv purpurbraun getupft.

Die 3 bis 10 cm hohe Infloreszenz ist ein- bis zweiblütig, hellgrün, braun getupft, filzig behaart.

Die Braktee hat ungefähr ein Drittel der Länge des Fruchtknotens. Die Blüten erreichen einen Durchmesser von 5 bis 7 cm. Sie sind schwach flaumig behaart, weiß bis fahlgelb, mehr oder weniger intensiv mit magentaroten Tupfen überzogen.

Die Fahne ist ca. 2,5 cm hoch und bis 3 cm breit, breit eiförmig bis fast kreisrund, hinten gekielt.

Das Synsepalum ist viel schmaler, länglich elliptisch, bis 2 cm lang.

Die Petalen sind bis zu 3,5 cm lang und 3 cm breit, länglich elliptisch, mehr oder weniger herabgeschlagen wie bei *Paphiopedilum concolor.*

Die fast zylindrische Lippe ist bis zu 3 cm lang und ca. 1,5 cm breit, oft mit winzigen magentaroten Tupfen.

Das längliche Staminodium ist auf beiden Seiten abgerundet, sehr variabel, unten mit ein bis drei Zähnchen versehen, in der Mitte gelb und mehr oder weniger vollständig mit winzigen purpurnen Tupfen überzogen.

Paphiopedilum × *godefroyae* wächst epiphytisch in Blattmoder auf moosbewachsenen Felsen oder

The argument used most to discredit the hypothesis that *Paphiopedilum* × *godefroyae* is a natural hybrid is the known distribution of *Paphiopedilum bellatulum.* Indeed there are about 800 km between the populations of *Paphiopedilum* × *godefroyae* and the nearest population of *Paphiopedilum bellatulum.*

This argumentation however, does not withstand a critical examination. It is much more probable that there are populations of these species we do not know. Or, maybe, the plants did grow together in populations that have been destroyed since.

The immense variation in the shape of the staminodal plate is another indication pointing to *Paphiopedilum* × *godefroyae* being of hybrid origin and, furthermore, shows, that the taxon is not yet stabilized as far as the gene pool is concerned.

I believe that the arguments for the treatment of *Paphiopedilum* × *godefroyae* as a natural hybrid are more conclusive than those against it. Therefore, the taxon is considered accordingly.

Paphiopedilum godefroyae is a dwarf, up to 15 cm tall, herbaceous plant, usually growing in leaf-mould on the top of limestone rocks.

The leaves are linear-oblong, up to about 18 cm long and up to about 4 cm wide, deep green, more or less tesselate with pale green above, densely spotted with deep purple beneath.

The Inflorescence is 3 to 10 cm tall, pale green spotted with purple, tomentose, 1–2 flowered.

The bract is about one third as long as the ovary.

The flowers are about 5–7 cm across, white to pale yellow, spotted with magenta-purple and obscurely pubescent.

The dorsal sepal is broadly ovate to sub-orbicular, 2.5 cm long and about 3 cm across, keeled behind.

The synsepalum is much smaller, elliptic-oblong, about 2 cm long.

The petals are elliptic-oblong, broad and reflexed as in *Paphiopedilum concolor,* up to about 3.5 cm long and about 3 cm broad.

The lip is sub-cylindric, up to about 3 cm long and about 1.5 cm across, often minutely spotted magenta-purple.

The staminodal plate is very variable, more or less roundish-oblong, with 1 to 3 small teeth at the lower part, minutely spotted like the lip, with a yellow stain in the center.

Paphiopedilum × *godefroyae* grows on moss-covered limestone rocks or more seldom on moss-covered trees on a limestone underground. Someseltener auf moosbewachsenen Bäumen, in Gegenden mit Kalksteinböden. Manchmal wurden auch Pflanzen als Humusepiphyten am Boden gefunden. Die Standorte sind meist sehr hell.

Die Pflanzen wachsen direkt am Meer an steilen Felshängen der sogenannten „Vogelnestinseln“ gegenüber der westmalaysischen Küste sowie in Thailand und Birma.

Paphiopedilum × *godefroyae* hat ein ziemlich ausgedehntes Verbreitungsgebiet, so daß einheitliche Klimaangaben kaum möglich sind. Die Pflanzen wachsen aber im allgemeinen unmittelbar am Meeresrand und sind ausgesprochen kalkliebend.

Wie alle Pflanzen der BRACHYPETALUM-Gruppe sind sie sehr empfindlich gegen Salze im Wurzelbereich.

Paphiopedilum × *godefroyae* gedeiht gut im warmen Gewächshaus bei sehr hoher Luftfeuchtigkeit und viel Licht. Man sollte aber für eine gute Luftumwälzung und Frischluftzufuhr sorgen.

Die Blütezeit ist von März bis Juni nach einer Ruheperiode von Dezember bis Januar, während der die Pflanzen kühler stehen und weniger Wasser bekommen sollten.

times the plants are found in leaf-mould on the earth. The habitats are, in general, very bright and mainly in cracks and crevices of the limestone rocks of the so-called "bird-nest islands" near the westmalaysian coast, in Thailand and in Burma.

The distribution of *Paphiopedilum* × *godefroyae* is fairly impressive. Therefore it is not easy to give generally valid information about the climate. However, the plants grow generally in close vicinity of the sea and are very calcicolous.

Like all plants of the BRACHYPETALUM-group they are extremely sensitive to high salt concentration in the root area.

Paphiopedilum × *godefroyae* grows well in the warm greenhouse under constant high humidity and intensive light. Good air circulation and adequate fresh air are imperative.

The flowering period is from March to June after a resting period from December to January in which the plants should be kept dryer and cooler.

Varietäten:

Von *Paphiopedilum* × *godefroyae* sind eine Reihe von „Varietäten“ beschrieben worden[1]. Keine davon ist als eindeutig abweichend von der Nominalform anzusehen und nur zwei werden hier eingehend besprochen. Die erste ist FOWLIEs Konzept von *Paphiopedilum* × *ang-thong.* Diese ist ohne Zweifel als Varietät von *Paphiopedilum* × *godefroyae* zu betrachten und dementsprechend umzukombinieren. Die zweite Varietät wird von einigen Autoren als eigenständige Art angesehen, aber hier, ihrem richtigen taxonomischen Status entsprechend, als *Paphiopedilum* × *godefroyae* var. *leucochilum* behandelt.

Varieties:

A number of "Varieties" of *Paphiopedilum* × *godefroyae*[1] have been published. None of them is clearly different from the nominal form and only two will be discussed in detail.

The first is FOWLIEs concept of *Paphiopedilum* × *ang-thong.* This is without any doubt to be considered a variety of *Paphiopedilum* × *godefroyae* and is redescribed accordingly.

The second one has been considered a autonomous species by some authors and will be treated as *Paphiopedilum* × *godefroyae* var. *leucochilum.*

Paphiopedilum × *godefroyae* var. *ang-thong* (Fowlie) Braem, *comb. et stat. nov.*

Basionym: *Paphiopedilum* × *ang-thong* Fowlie
Orchid Digest 41 (3): 117 (1977)

Vor einigen Jahren wurden durch den thailändischen Orchideensammler und -händler SUKHAKUL die von den Ang Thong-Inseln stammenden *Paphiopedilum*-Pflanzen als *Paphiopedilum ang-thong* auf den Markt gebracht.

Paphiopedilum × *godefroyae* var. *ang-thong* (Fowlie) Braem, comb. et stat. nov.

Basionym: *Paphiopedilum* × *ang-thong* Fowlie
Orchid Digest 41 (3): 117 (1977)

Some years ago, the Thai orchid dealer SUKHAKUL offered some *Paphiopedilum*-plants from the Ang-thong Islands for sale as *Paphiopedilum ang-thong.*

[1] Meist als *Cypripedium godefroyae* var.

[1] usually as *Cypripedium godefroyae* var.

FOWLIE hielt es für angebracht, die Pflanze als eine Naturhybride zwischen *Paphiopedilum niveum* und *Paphiopedilum godefroyae* als *Paphiopedilum* × *ang-thong* zu beschreiben.

Die Pflanzen von den Ang Thong-Inseln sind kaum von Pflanzen des Taxons *Paphiopedilum* × *godefroyae* zu unterscheiden. Sie stammen lediglich von einem gesonderten Standort. Sie sind daher ohne Zweifel als geographische Varietät von *Paph.* × *godefroyae* zu betrachten.

Natürlich werden einige Autoren beweisen wollen, daß *Paphiopedilum* × *godefroyae* var. *ang-thong* im Sinne FOWLIEs eine Hybride zwischen *Paphiopedilum niveum* und *Paphiopedilum* × *godefroyae* ist. Dazu werden sie eine gärtnerische Kreuzung zwischen diesen Pflanzen heranziehen, die von den Naturformen der Ang Thong-Inseln (und auch von *Paph.* × *godefroyae*) nicht zu unterscheiden ist.

Die Verfechter dieser Theorie scheinen sich, vielleicht aufgrund ihrer Unkenntnis der cytotechnischen Gegebenheiten nicht im klaren darüber zu sein, daß *Paph. niveum* eben nur eine weiße Blütenfärbung mit mehr oder weniger intensiven, Magenta-Tupfen vererben kann. Aus dem selben Grunde wird *Paph.* × *godefroyae* auch nur weiße bis gelblich weiße Grundfärbung mit magentaroten Tupfen vererben. Die Nachkommenschaft wird also zwangsläufig aus Pflanzen bestehen, die eine weiße Grundfärbung (eventuell mit einem leichten Gelbstich) haben und magentarot getupft sind.

FOWLIE considered the plants to be a natural hybrid between *Paphiopedilum niveum* and *Paphiopedilum godefroyae* and described them as *Paphiopedilum* × *ang-thong*.

However, those plants from the Ang-thong Island are by no means different from plants belonging to *Paphiopedilum* × *godefroyae*. The only difference is that they originate from another habitat. They are, therefore, without the slightest doubt, to be considered as a geographic variant of *Paphiopedilum* × *godefroyae*.

Of course, some authors will want to prove that *Paphiopedilum* × *ang-thong* is a hybrid between *Paphiopedilum niveum* and *Paphiopedilum* × *godefroyae* (*sensu* FOWLIE). To substantiate their claim, they will refer to a artificial cross between those two taxa, and indeed, the progeny of such a cross can not be differentiated from the plants found on the Ang-thong Islands (nor from *Paphiopedilum* × *godefroyae*).

Those propagating that theory seem not to understand basic genetics: *Paphiopedilum niveum* cannot but transmit the genes for a white flower more or less spotted intensively magenta-purple; and *Paphiopedilum* × *godefroyae* cannot but transmit the genes for a white to yellowish flower, more or less intensively spotted magenta-purple. Therefore, the progeny cannot but have flowers of a white to yellow background colour, spotted more or less intensively magenta-purple.

Paphiopedilum* × *godefroyae* var. *leucochilum
(Masters) Hallier pro sp. *godefroyae*
Ann. Jard. Bot. Buitenzorg 14: 35 (1897)

Synonym: *Cypripedium godefroyae* var. *leucochilum* Rolfe
Orchid Review 2: 145 (1894)
Paphiopedilum leucochilum (Rolfe) Fowlie
Orchid Digest 39 (3): 110 (1975)

Paphiopedilum × *godefroyae* var. *leucochilum* unterscheidet sich von den übrigen *Paphiopedilum* × *godefroyae*-Formen durch ein ungeflecktes Labellum. Dieser Unterschied und ein Unterschied in der geographischen Verbreitung (diese Varietät stammt von den Inseln an der malaysischen Westküste) reichen keineswegs aus, um diese Pflanzen als selbständige Art zu betrachten.

Paphiopedilum* × *godefroyae* var. *leucochilum
(Masters) Hallier pro sp. *godefroyae*
Ann. Jard. Bot. Buitenzorg 14: 35 (1897)

Synonym: *Cypripedium godefroyae* var. *leucochilum* Rolfe
Orchid Review 2: 145 (1894)
Paphiopedilum leucochilum (Rolfe) Fowlie
Orchid Digest 39 (3): 110 (1975)

Paphiopedilum × *godefroyae* var. *leucochilum* differs from all other forms of the species by a immaculate, unspotted labellum. This difference and the different habitat (this variety originates from the island along the Malaysian West-coast) do not suffice to warrant the treatment of the plants as an autonomous species.

Untergattung POLYANTHA (Pfitzer) Brieger

Die Untergattung **POLYANTHA** wurde durch BRIEGER beschrieben,[1] basiert auf der PFITZERschen Sektion **Polyantha,**[2] und entspricht größtenteils der Untergattung **ANOTOPETALUM** wie sie durch PFITZER 1903 beschrieben wurde.[3]

Während PFITZER nur die Pflanzen, die keine seitlichen Aurikeln an den Lippenhauptlappen aufwiesen, in dieser Gruppe zusammenfaßte, die er in drei Sektionen aufteilte,[4] beinhaltet die Untergattung **POLYANTHA** *sensu* Brieger alle Arten mit einfarbig grünen, nicht marmorierten, dicken, fleischigen Laubblättern und einer mehrblütigen Infloreszenz.

Subgenus POLYANTHA (Pfitzer) Brieger

The subgenus **POLYANTHA** was described by BRIEGER[1] on the basis of the section **Polyantha** *sensu* Pfitzer[2] and corresponds largely to the subgenus **ANOTOPEDILUM** as described by PFITZER in 1903.[3]

Whereas PFITZER included only those plants having a non-auriculate labellum and divided his subgenus into three sections,[4] the subgenus **POLYANTHA** *sensu* BRIEGER acommodates all species having uni-coloured, non-tesselate, thick fleshy green leaves, and a pluriflorous inflorescence.

Schlüssel zu den Sektionen der Untergattung POLYANTHA

1. Fahne und Synsepalum fast gleich in Form, Größe und Farbgebung; Lippenhauptlappen mit stark nach innen gefaltetem oberem Rand und ohne Aurikeln (Ausnahme *Paphiopedilum randsii*) . . . Sektion **MASTIGOPETALUM**

1a. Fahne und Synsepalum deutlich verschieden in Form, Größe und/oder Farbgebung; oberer Rand des Lippenhauptlappens nicht nach innen gebogen, sondern gerade oder leicht nach außen gerichtet und mit gut entwickelten Aurikeln 2

2. Seitliche Petalen zur Spitze hin allmählich breiter werdend, schmal löffelförmig, leicht nach unten gerichtet, kaum gedreht . . . . Sektion **POLYANTHA**

2a. Seitliche Petalen zur Spitze hin allmählich schmaler werdend, stark gedreht und deutlich nach unten gerichtet . . . . Sektion **MYSTROPETALUM**

Key to the sections of the subgenus POLYANTHA

1. Upper and lower sepals almost the same in shape, size and colour patterns; upper margin of the main lobe of the lip strongly inflexed, no lateral ears (except *Paphiopedilum randsii*) . . . . section **MASTIGOPETALUM**

1a. Upper and lower sepals clearly differentiated in shape, size and colour patterns; upper margin of the main lobe of the lip not incurved but straight or directed slightly outward; lateral ears well-developed 2

2. Lateral petals gradually widened towards the apex, narrowly spoon-shaped; spreading slightly downward, hardly twisted . . . . section **POLYANTHA**

2a. Lateral petals gradually narrowed towards the apex, strongly twisted and drooping . . . . section **MYSTROPETALUM**

---

[1] BRIEGER, F. G. in SCHLECHTER, R.: **Die Orchideen,** 3. Ausgabe: 175 (1971).

[2] PFITZER, E. in ENGLER, A.: **Bot. Jahrb. 19:** 41 (1894).

[3] PFITZER, E. in ENGLER, A.: **Das Pflanzenr. IV** (50) Heft 12: 58 (1903).

[4] a) Sektion **Gonatopedilum,** mit *Paphiopedilum rothschildianum.*
b) Sektion **Coryopedilum,** mit *Paphiopedilum glanduliferum, Paphiopedilum praestans, Paphiopedilum philippinense, Paphiopedilum roebbelinii* und *Paphiopedilum sanderianum.*
c) Sektion **Prenipedilum** mit *Paphiopedilum stonei.*

---

[1] BRIEGER, F. G. in SCHLECHTER, R. **Die Orchideen,** 3. Edition: 175 (1971)

[2] PFITZER, E. in ENGLER, A.: **Bot. Jahrb. 19:** 41 (1894)

[3] PFITZER, E. in ENGLER, A.: **Das Pflanzenr. IV** (50) Heft 12: 58 (1903)

[4] a: section **Gonatopedilum** with only *Paphiopedilum rothschildianum;*
b: section **Coryopedilum** with *Paphiopedilum glanduliferum, Paphiopedilum praestens, Paphiopedilum philippinense, Paphiopedilum roebbelinii* and *Paphiopedilum sanderianum;*
c: section **Prenipedilum** with only *Paphiopedilum stonei*

Sektion MASTIGOPETALUM
Hallier

Diese Sektion wurde durch HALLIER[1] für *Paphiopedilum elliottianum, Paphiopedilum rothschildianum, Paphiopedilum sanderianum* und *Paphiopedilum stonei* begründet.

Ich setze alle Arten der Untergattung **POLYANTHA**, deren Fahne und Synsepalum in Form, Größe und Farbgebung gleichgestaltet sind, in diese Sektion. Alle diese Arten (mit Ausnahme von *Paphiopedilum randsii*) haben Lippenhauptlappen **ohne** seitliche Aurikeln.

Ich betrachte es daher als möglich, daß *Paphiopedilum randsii* eine (Natur)Hybride darstellt, wahrscheinlich mit *Paphiopedilum philippinense* als einem Elternteil.

[1] HALLIER, H. **Annales du Jardin Botanique de Buitenzorg 14:** 18–52 (1897)

Section MASTIGOPETALUM
Hallier

This section was described by HALLIER[1] who included *Paphiopedilum elliottianum, Paphiopedilum rothschildianum, Paphiopedilum sanderianum* and *Paphiopedilum stonei.*

The section, as treated in this book includes those species of the subgenus **POLYANTHA** having uniformity of the upper and lower sepals as far as the shape, size and colour patterns are concerned. All but one species – *Paphiopedilum randsii* – lack lateral ears to the lip.

I therefore suggest, that *Paphiopedilum randsii* is a (natural) hybrid, probable with *Paphiopedilum philippinense* as one of the parent species.

[1] HALLIER, H. **Annales du Jardin Botanique de Buitenzorg 14:** 18–52 (1897)

Schlüssel zu den Arten der Sektion MASTIGOPETALUM

1. Lippe ohne seitliche Aurikeln 2

1a. Lippe mit seitlichen Aurikeln; seitliche Petalen kurz und mit warzenlosen Rändern; Blüte grünlich . . . . *Paphiopedilum randsii*

2. Seitliche Petalen nicht länger als 20 cm . 3

2a. Seitliche Petalen sehr lang, meist 50 bis 70 cm, aber möglicherweise bis 120 cm lang . . . . *Paphiopedilum sanderianum*

3. Lippenrand mehr oder weniger zurückgebogen; Staminodium nach oben geneigt . . . . *Paphiopedilum elliottianum*

3a. Lippenrand deutlich nach innen gebogen, Staminodium nicht nach oben geneigt . . . 4

4. Seitliche Petalen horizontal oder fast horizontal abstehend 5

4a. Seitliche Petalen deutlich nach unten gerichtet 6

5. Blüten größtenteils rötlich braun . . . . *Paphiopedilum rothschildianum*

5a. Blüten größtenteils gelb bis grünlich gelb . . . . *Paphiopedilum philippinense* var.

6. Seitliche Petalen ohne Schulter, von der Basis aus nach unten gebogen, mehr oder weniger intensiv gedreht 7

6a. Seitliche Petalen bogenförmig, nicht intensiv gedreht 9

7. Seitliche Petalen über die ganze Länge mehr oder weniger stark gedreht 8

7a. Seitliche Petalen nur in der apikalen Hälfte gedreht . . . . *Paphiopedilum supardii*

8. Staminodiumplatte ein gerundetes Viereck, mit kleinen hornartigen Auswüchsen an beiden oberen Enden *Paphiopedilum glanduliferum*

8a. Staminodiumplatte mehr oder weniger eiförmig, ohne Auswüchse . . . . *Paphiopedilum philippinense*

9. Seitliche Petalen bogenförmig, größtenteils braun, nicht grün, schmal linealisch sichelförmig; Staminodiumplatte länglich, deutlich gekerbt am unteren Ende . . . . *Paphiopedilum stonei*

9a. Seitliche Petalen einfarbig grün, bandförmig, schräg nach unten gerichtet; Staminodiumplatte rundlich *Paphiopedilum kolopakingii*

Key to the species of section MASTIGOPETALUM

1. Lip exauriculate at mouth 2

1a. Lip auriculate at mouth; lateral petals short, margins of lateral petals without any wards; flower greenish . . *Paphiopedilum randsii*

2. Lateral petals less than 20 cm long . . . 3

2a. Lateral petals very long, usually 50–70 cm but possibly up to 120 cm long . . . . *Paphiopedilum sanderianum*

3. Margin of lip more or less recurved; staminode inclined upwards . . . . *Paphiopedilum elliottianum*

3a. Margin of lip distinctly incurved; staminode not inclined upwards 4

4. Lateral petals spreading horizontally or nearly so 5

4a. Lateral petals distinctly directed downwards 6

5. Flower dominantly reddish-brown . . . . *Paphiopedilum rothschildianum*

5a. Flowers dominantly yellow to greenish-yellow . . . . . *Paphiopedilum philippinense* var.

6. Lateral petals straightly bending, without a shoulder, more or less strongly twisted 7

6a. Lateral petals bending arcuate; not strongly twisted 9

7. Lateral petals more or less strongly twisted over the whole distance 8

7a. Lateral petals twisted in the apical half only . . . . *Paphiopedilum supardii*

8. Staminodal shield a rounded quadrangle, with minute hornshaped protuberances at both upper ends . *Paphiopedilum glanduliferum*

8a. Staminodal shield obovoid, without protuberance . . . . *Paphiopedilum philippinense*

9. Lateral petals bending arcuate, mainly brown, not green, narrow linear-falcate; staminodal shield oblong, distally distinctly notched . . . . *Paphiopedilum stonei*

9a. Lateral petals pure green, strap shaped, bending obliquely downwards; staminodal shield orbicular . . . . *Paphiopedilum kolopakingii*

Paphiopedilum elliottianum (O'Brien) Stein

Stein's Orchideenbuch: 466 (1892)

Basionym: *Cypripedium elliottianum* O'Brien **Gardeners' Chronicle,** 3. Ser., **4:** 501 (1888)

Synonyma: *Paphiopedilum adductum* Asher **Orchid Digest 47** (6): 213–236 (1983)

Selenipedium elliottianum Gower **The Garden 34:** 408 (1888)

Am 3. November 1888 wurde im **Gardeners' Chronicle** ein Artikel des Herausgebers James O'BRIEN abgedruckt, worin dieser eine neu entdeckte Pflanze, die durch die Fa. SANDER eingeführt wurde, als *Cypripedium elliottianum n. sp.* beschrieb.

Die Pflanze sollte von den Philippinen stammen und laut O'BRIEN deutlich von *Cypripedium rothschildianum* zu unterscheiden sein.[1]

Eine Woche später wurde durch REICHENBACH in derselben Zeitschrift[2] nochmals eine Beschreibung dieser Art veröffentlicht. Auch REICHENBACHs Artikel trägt den Titel „*Cypripedium elliottianum n. sp.*" und er vermerkt lapidar, daß O'BRIEN die Art eine Woche zuvor „erwähnt" hätte.

Das Chaos wurde vervollständigt durch die nochmalige Veröffentlichung desselben Taxons durch GOWER als *Selenipedium elliottianum.*[3]

Viele Jahre war diese Art dann verschollen, bis FOWLIE die Wiederentdeckung feierte[4] und (in Unkenntnis von **Stein's Orchideenbuch**) das Taxon überflüssigerweise nochmals als *Paphiopedilum elliottianum* umkombinierte.

Damit hätte (abgesehen von FOWLIEs „Fauxpas") die „*elliottianum*-Story" an und für sich ein Happy-End, wäre da nicht der amerikanische Zoologe ASHER, der behauptete,[5] die Pflanze, welche nunmehr als *Paphiopedilum elliottianum* (O'Brien) Stein in den Gewächshäusern der Orchideenlieb-

[1] **Gardeners' Chronicle,** 3. Ser., **4:** 501 (1888): „... which call to mind the beautiful *Cypripedium rothschildianum* and *Cypripedium praestans,* from both of which, however, it is widely distinguished. ..."

[2] **Gardeners' Chronicle,** 3. Ser., **4:** 532 (1888)

[3] **The Garden 34:** 408 (1888)

[4] **Orchid Digest 44** (2): 72–76 (1980)

[5] **Orchid Digest 47** (6): 224 (1983)

The first description of this species was rendered by James O'BRIEN under the title of *Cypripedium elliottianum nov. sp.* in an article published on November 3, 1888, in the **Gardeners' Chronicle.**

The plants were imported to England by the SANDER Co. and noted to originate from the Philippines.

O'BRIEN wrote about his *Cypripedium elliottianum:*

"... which call to mind the beautiful *Cypripedium rothschildianum* and *Cypripedium praestans,* from both of which, however, it is widely distinguished ..."

One week later, another description of this same species by REICHENBACH appeared in the same journal.[1] REICHENBACHs article also was entitled "*Cypripedium elliottianum nov. sp.*" and REICHENBACH lapidarily notes that O'BRIEN had "mentioned" the species a week before.

The chaos was completed by a third description of the same species under the name of *Selenipedium elliottianum* by GOWER.[2]

For many years this species disappeared. It was FOWLIE who celebrated the rediscovery and redescribed the taxon (unaware of **Stein's Orchideenbuch)** as *Paphiopedilum elliottianum.*[3]

In that sense (overlooking FOWLIEs "faux pas") the "*elliottianum*-story" could have been considered as having an happy end if it were not for the Ameri-

[1] **Gardeners' Chronicle,** 3. Ser., **4:** 532 (1888)

[2] **The Garden 34:** 408 (1888), erroneously cited as *Cypripedium elliottianum* by Sander in his advertisement the **Gardeners' Chronicle,** 3. Ser., **4:** 556 (1888)

[3] **Orchid Digest 44** (2): 72–76 (1980)

Paphiopedilum elliottianum (O'Brien) Stein

Paphiopedilum elliottianum (O'Brien) Stein

haber steht, sei gar nicht *Paphiopedilum elliottianum*, sondern eine neu zu beschreibende Art. Er beschrieb dann auch das Taxon neu als *Paphiopedilum adductum* und stellte *Paphiopedilum elliottianum* (O'Brien) Stein als Synonym zu *Paphiopedilum rothschildianum* (Rchb. f.) Stein.

Nun, auch dies könnte man noch als Happy-End betrachten, wären da nicht einige Ungereimtheiten!

Die Argumentation ASHERs ist nämlich gar nicht so schlüssig wie es auf den ersten Blick erscheint!

Paphiopedilum rothschildianum wurde nur 5½ Monate vor *Paphiopedilum elliottianum* beschrieben und die Pflanzen von *Paphiopedilum rothschildianum* waren sowohl O'BRIEN und GOWER als auch REICHENBACH wohlbekannt. O'BRIEN spricht ja auch von einem deutlichen Unterschied. Ferner wurden *Paphiopedilum rothschildianum* und *Paphiopedilum elliottianum* von derselben Firma importiert, deren Inhaber SANDER als Orchideenfachmann zu betrachten ist. Als Herkunft von *Paphiopedilum rothschildianum* wurde der Malai-

can zoologist ASHER claiming[4] that the taxon labelled *Paphiopedilum elliottianum* (O'Brien) Stein is in fact not at all *Paphiopedilum elliottianum* but a species new to science. Consequently, he described the "new" taxon as *Paphiopedilum adductum* and set *Paphiopedilum elliottianum* (O'Brien) Stein as synonym to *Paphiopedilum rothschildianum* (Rchb. f.) Stein.

Now, even that could be considered an "Happy End" if there weren't any obscurities.

In fact, ASHERs argumentation is not at all as conclusive as it seems to be on first sight.

Paphiopedilum elliottianum was published only 5½ month after the original description of *Paphiopedilum rothschildianum* and plants of the latter species were well known to O'BRIEN and GOWER as well as to REICHENBACH. And indeed, O'BRIEN mentions a clear difference between the species.

Furthermore, *Paphiopedilum elliottianum* and *Paphiopedilum rothschildianum* were imported to

[4] **Orchid Digest 47** (6): 213–236 (1983)

ische Archipel angegeben. Dies war üblich für Pflanzen aus Borneo. Bei *Paphiopedilum elliottianum* wurden die Philippinen genannt.

Das wichtigste Argument für die Betrachtung von *Paphiopedilum elliottianum* als selbständige Art wird allgemein übersehen. Eine Woche nach der Beschreibung des Taxons als *Cypripedium elliottianum* durch O'BRIEN und am Tag der „Neubeschreibung" durch REICHENBACH wurde ein Inserat der Fa. SANDER veröffentlicht[1] **mit einer Zeichnung** von *Paphiopedilum elliottianum.*

Es ist wohl unbestritten, daß die damaligen Illustratoren sehr genau arbeiteten.

Die Zeichnung zeigt eine Pflanze, die mit Sicherheit nicht mit *Paphiopedilum rothschildianum* identisch ist, da die Blüte deutlich herabhängende Petalen zeigt.

Diese herabhängenden („drooping") Petalen werden auch in einer Notiz erwähnt, die Lewis CASTLE (der vierte, der aus dem Hause SANDER eine Pflanze von *Paphiopedilum elliottianum* erhielt) im **Journal of Horticulture and Cottage Gardener** am 1. November 1888 veröffentlichte.

Der des öfteren als Beweis für die Interpretation ASHERs herangezogene Typusbogen aus dem LINDLEY Herbar in Kew, der in der Tat eine Blüte von *Paphiopedilum rothschildianum* zeigt, ist nicht schlüssig. Die Typusbezeichnung erfolgte viele Jahre nach der Beschreibung, und was noch aufschlußreicher ist, das Typusexemplar stimmt nicht mit dem Herbarbeleg REICHENBACHs überein.

Obwohl die Interpretation ASHERs zur Zeit von vielen Autoren anerkannt wird, bin ich der Meinung, daß die eigentlichen Beweise dagegen sprechen.

Paphiopedilum elliottianum ist eine selbständige Art, deren Namen irrtümlicherweise als Synonym für *Paphiopedilum rothschildianum* benutzt wurde. *Paphiopedilum adductum* Asher ist eindeutig als Synonym zu *Paphiopedilum elliottianum* zu betrachten.

Paphiopedilum elliottianum ist eine als Humusepiphyt, meist am Waldboden wachsende, krautige Pflanze mit ca. sechs Laubblättern pro Trieb. Diese länglich bandförmigen Laubblätter erreichen eine Länge von ca. 25 cm bei einer Breite von 3,7 bis 4,5 cm. Sie stehen zweireihig, am jungen Trieb aufrecht, später horizontal. Sie sind abgerundet, am Ende mit einer Einkerbung. Die Blätter sind gefaltet. Somit ist die dunkelgrüne Oberseite rinnenförmig und die heller grüne Unterseite deutlich gekielt.

[1] **Gardeners' Chronicle,** 3. Ser., **4:** 556 (1888)

England by the same firm, whose owner, SANDER, was a distinguished orchid expert of his time. The origin of *Paphiopedilum rothschildianum* was cited as the Malaysian Archipelago, as it was usable for plants originating from Borneo. The origin of *Paphiopedilum elliottianum* however was said to be the Philippines.

The best reason to consider *Paphiopedilum elliottianum* as a good autonomous species is generally overlooked. One week after the description of the taxon as *Cypripedium elliottianum* by O'BRIEN and in the same issue as the "new description" by REICHENBACH *(loc. cit.),* SANDER advertised *Paphiopedilum elliottianum* and his advertisement **included a drawing** of the species.[1]

It is unchallenged, that the illustrators of that time were very naturalistic and exact.

The drawing shows a plant, certainly not identical to *Paphiopedilum rothschildianum*, and the flower depicts clearly drooping lateral petals.

These drooping petals were also mentioned in a note by Lewis CASTLE (the fourth person having received a plant of *Paphiopedilum elliottianum* from SANDER) in the **Journal of Horticulture and Cottage Gardener** for 1. November 1888.

Often, the type specimen – deposited in the Lindley Herbarium at Kew – is cited to prove ASHERs thesis. Indeed, that herbarium sheet depicts a flower of *Paphiopedilum rothschildianum*, but this is not at all conclusive. The type specimen was designated many years after the description, and, on top of that, the type specimen does not correspond to the herbarium specimen deposited in the Reichenbach Herbarium at Vienna.

Although at this time, ASHERs interpretation is generally accepted, it is my opinion that it is not all substantiated by the facts.

Therefore, I consider *Paphiopedilum elliottianum* to be a good autonomous species often erroneously classified to be conspecific with *Paphiopedilum rothschildianum. Paphiopedilum adductum* Asher therefore is, without any doubt to be considered a synonym of *Paphiopedilum elliottianum.*

Paphiopedilum elliottianum is a herbaceous plant, generally growing epiphytically in leaf mould.

There are about 6 leaves per growth. The leaves are oblong-ligulate to slightly oblanceolate or lanceolate, distichous, emerging vertically, becoming horizontally in age, about 25 cm long and about 3.7 to 4.5 cm wide, margins narrowly hyaline, eciliate, slightly uneven, apices equal or subequal, obtuse,

[1] **Gardeners' Chronicle,** 3. Ser., **4:** 556 (1888)

Die Infloreszenz ist bis zu 40 cm hoch, rund, ca. 3 mm im Durchmesser, purpurbraun, manchmal mit gelben Streifen, von kurzen, ca. 1 mm langen, purpurnen Härchen überzogen. Sie ist meist zweiblütig, manchmal auch mehrblütig.

Die Braktee ist 3 bis 4 cm lang und 0,6 bis 1,3 cm breit, länglich, deutlich zusammengefaltet, grün mit purpur-braunem Überzug an der Basis, und deutlich dunkel purpurn geadert.

Der Fruchtknoten ist 3,5 bis 4,5 cm lang und 6 bis 10 mm dick, grün mit leicht purpurbraunem Überzug und mit sechs, bis zu 2 mm breiten, dunkelpurpurnen Linien versehen; im Querschnitt breit oval.

Die Fahne ist 5 bis 6,5 cm lang und ca. 3 cm breit, eiförmig, spitz, konkav, nach vorne über den Schuh gebeugt, hinten deutlich gekielt, an der Basis hellgrün, zur Spitze hin in dunkleres Grün überlaufend, außen kahl, mit Nektartropfen auf der oberen äußeren Hälfte, innen teilweise von kurzen purpurnen Härchen überzogen, innen und außen mit unterschiedlichen, bis 1,5 mm breiten purpurnen Adern.

Das Synsepalum ist meistens etwas länger als die Fahne, 5 bis 7 cm lang und 2,3 bis 3 cm breit, eiförmig, spitz, leicht konkav im basalen Teil und deutlich konkav im apikalen Teil. Die Mitteladern der verwachsenen Sepalen sind sichtbar, nicht gekielt. Das Synsepalum ist ähnlich wie die Fahne gefärbt und auch innen teilweise behaart.

Die Petalen sind 8 bis 15 cm lang, an der Basis 7 bis 8 mm, an der Spitze nur ca. 1 mm breit. Sie sind nach unten gebogen, dunkel behaart und mit purpurnen Flecken überzogen. Diese Flecken sind an der Basis der Petalen kleiner und deutlicher voneinander getrennt als an der Spitze. Die Petalen sind oft von feinen purpurnen Adern überzogen, wovon einige über die ganze Petalenlänge führen.

Die Lippe ist umgekehrt helmförmig und steht fast senkrecht zum Fruchtknoten. Sie ist 3,6 bis 4,8 cm lang und 1,5 bis 2,2 cm breit. Es sind keine seitlichen Aurikeln vorhanden. Die basalen Ränder sind nach innen gerollt, im Bereich der Pollinien außen etwas behaart. Die Rückseite der Lippe ist eingedellt. Die Rückseite und der basale Teil der Vorderseite sind hellgrün. Der obere Teil der Vorderseite ist netzförmig purpurbraun überzogen. Die Innenwand der Lippe zeigt viele kleine purpurne Flecken, die nach unten größer und weniger differenziert werden.

Das Staminodium ist eine gestielte rechteckige Platte, im Querschnitt T-förmig. Die Platte ist ca.

notched, a persistent midvein frequently extending to the apex, adaxially canaliculate and dark green, abaxially keeled and lighter green.

The inflorescence is terete, up to about 40 cm high, about 3 mm in diameter, uniformly purple-brown or purple-brown with short yellow streaks, pilose with short, about 1 mm long purple hairs, generally two-flowered, sometimes multi-flowered.

The floral bracts are about 3–4 cm long and about 0.6 to 1.3 cm wide, lanceolate, purplish to greenish at the apex, green suffused with purple-brown to the base, lined with darker purple-brown veins.

The ovary is 3.5 to 4.5 cm long and 6 to 10 mm in diameter, broadly obovate in cross section, base colour green suffused with light purple-brown, lined with 6, up to 2 mm wide, darker purple-brown stripes.

The dorsal sepal is 5 to 6.5 cm long and about 3 cm broad, ovate, acuminate, concave, arching forward over the labellum, with a prominent keel, light green at the base, becoming darker green in the distal three-fourths, glabrous outside, with irregular nectar droplets on the apical exterior two-thirds, inside basal one-fourth pilose with short purple hairs, inside and outside variably striped with purple veins to 1.5 mm wide.

The synsepalum is usually slightly longer than the dorsal sepal, 5 to 7 cm long and 2.3 to 3 cm broad, ovate, acuminate, moderately concave at the base and strongly concave in the apical half. The midveins from the fused sepals are apparent but not keeled. The synsepalum is coloured similar to the dorsal sepal, the inside also sparsely pilose.

The petals are 8 to 15 cm long, 7 to 8 mm wide basally, tapering to about 1 mm wide apically. They are bent downward, the apical part with short, dark, glandular hairs, denser adaxially, glabrous elsewhere; the basal one-fourth light green, the apical three-fourths suffused with purple; basally with purple-brown spots of variable size and shape becoming larger and less distinct apically, variably striped with fine purple veins, some running over the entire length.

The labellum is invertedly helmet-shaped, jutting forward, nearly perpendicular to the ovary, 3.6 to 4.8 cm long and 1.5 to 2.2 cm broad, without lateral ear-like projections. The basal edges are rolled inwards, sparsely pilose externally in the region near the pollinia, bottom grooved along the distal one-half, laterally and basally light green, upper distal portion purplish-brown reticulate; interior of

9 mm lang und 4 bis 7 mm breit, seitlich dicht behaart, in der Mitte unbehaart.

In Kultur befindliche Pflanzen von *Paphiopedilum elliottianum* stammen von der philippinischen Insel Mindanao. Ein anderer Standort ist zur Zeit nicht bekannt.

Paphiopedilum elliottianum ist relativ einfach zu kultivieren. Die Pflanzen brauchen viel Feuchtigkeit und viel Licht. Sie mögen ein durchlässiges Substrat, möglichst mit Moosbeimischung.

Wie die meisten ihrer Verwandten blüht *Paphiopedilum elliottianum* an 2 bis 3 Jahre alten Trieben, wobei nur mehrtriebige Pflanzen regelmäßig blühen.

Die Blütezeit ist im allgemeinen von April bis Mai, jedoch sind andere Blütezeiten in Kultur keine Seltenheit.

labellum with small purple spots becoming larger and less distinct distally.

The staminode is a stalked rectangular shield, T-shaped in cross section. The shield is about 9 mm long and 4 to 7 mm broad, densely hirsute laterally, glabrous centrally.

The specimens of this species now in cultivation originate from the Philippine island of Mindanao. No other habitats are known.

The growing of *Paphiopedilum elliottianum* is rather easy. The plants require high humidity and plenty of light. They do well in an open potting medium with the addition of moss.

Paphiopedilum elliottianum, as most of its relatives, flowers from 2–3 year old growths. Only strong multi-growth plants will flower regularly.

Paphiopedilum elliottianum generally flowers between April and May, not seldom however also at other times of the year.

Paphiopedilum glanduliferum (Blume) Stein

Stein's Orchideenbuch: 468 (1892)

Basionym: *Cypripedium glanduliferum* Blume
Rumphia 4: 56, t. 195 (1848)

Synonyma: *Cordula glanduligera* (Blume) Rolfe
(lapsus linguae)
Orchid Review 20 (1): 2 (1912)

Cordula praestans (Rchb. f.) Rolfe
Orchid Review 20 (1): 2 (1912)

Cypripedium praestans Rchb. f.
Gardeners' Chronicle, 2. Ser., **26:** 776 (1886)

Paphiopedilum praestans (Rchb. f.) Pfitzer
in ENGLER, A.: **Bot. Jahrb. 19:** 41 (1894)

Paphiopedilum wilhelminiae L. O. Williams
American Orchid Society Bulletin 10: 373–375 (1942)

Nomina nuda: *Cypripedium gardineri*
Erwähnt von GUILLEMARD im Buch **„The Cruise of the Marchesa to Kamschatka and New Guinea" 2:** 309 (1886)

Paphiopedilum bodegomensis
Paphiopedilum bodegomii

Oft erwähnt aber nirgendwo beschrieben.

often referred to but never officially described

Nomen illegitimum: *Paphiopedilum gardineri* (Guillemard) Pfitzer
in ENGLER, A.: **Bot. Jahrb. 19:** 41 (1894)

Obwohl durch PFITZER umkombiniert, ist auch dieses ungültig, da es auf einem ungültigen Taxon basiert.

Based on an *nominum nudum* and therefore just as invalid.

Paphiopedilum glanduliferum (Blume) Stein

Die Liste der Synonyme, *nomina nuda* und das *nomen illegitimum* deuten an, daß auch diese Pflanzenart in taxonomischer Hinsicht für einige Verwirrung gesorgt hat. Verwirrung kann aber entworren werden, wenn man genau recherchiert.

The list of *synonyma*, *nomina nuda* and the *nomen illegitimum* suggest that this species also has been the cause of some confusion in respect to taxonomy. However, confusion can very often be clarified with sound investigation.

Die Art wurde durch BLUME in der **Rumphia** *(loc. cit.)*. anhand von Materialien beschrieben, die er von ZIPPEL, einem der frühesten Neu-Guinea-Reisenden bekam. Die Beschreibung BLUMEs ist ausführlich, genau und von einer Farbtafel begleitet. Fachleute, die diese Veröffentlichung eingehend studieren und das unter vielen verschiedenen Namen reichlich vorhandene Pflanzenmaterial untersuchen, müssen unzweifelhaft zu der Auffassung kommen, daß alle anderen oben aufgeführten Konzepte höchstens als Varianten von *Paphiopedilum glanduliferum* anzusehen sind.

Paphiopedilum glanduliferum ist eine als Humusepiphyt wachsende, krautige Pflanze mit zweireihig stehenden, schmal länglichen bis ungefähr 60 cm (je nach geographischer Variante) langen Laubblättern.

Diese sind vorn abgerundet, bis ca. 3 cm breit, ziemlich fleischig, einfarbig grün und unterseits deutlich gekielt.

Die im allgemeinen bis ca. 50 cm hohe (manchmal noch höhere) Infloreszenz ist dunkel purpurn, teilweise grün gefleckt und flaumig graubraun behaart. Sie trägt normalerweise bis zu 5 Blüten.

Die Braktee ist 2,5 bis 4,5 cm lang, scharf gekielt, braun mit dunkelbraunen Adern, lanzettlich geformt.

Die Blüten werden bis 12,5 cm hoch.

Die Fahne ist etwa 5,5 cm hoch und bis 2,5 cm breit, eiförmig, spitz, gelblich weiß, manchmal mit dunkelgelber Mitte und dunkelgelben bis rotbraunen Adern, hinten gekielt.

Das Synsepalum ist ähnlich, aber kleiner.

Die linealischen bis bandförmigen Petalen sind 10 bis 13 cm lang und ungefähr 0,5 cm breit, in einem Winkel von etwa 45° nach unten gerichtet, an den Rändern gewellt und mit schwarzen, behaarten Wärzchen besetzt. Sie sind mehr oder weniger gedreht, gelbgrün mit dunkleren gelbgrünen bis rotbraunen Längsadern.

Die deutlich sackförmige Lippe ist bis zu 4,5 cm lang und ca. 2 cm breit, hellgrün bis gelblich grün oder leicht rosa, mit dunklerer Aderung.

Das Staminodium ist länglich, kantig und fleischig. An den Seiten ist es mit Knötchen und unten mit kurzen Spitzen versehen. Es ist gelb bis rötlich gefärbt. Die Seitenränder zeigen eine rotbraune, stoppelige Behaarung.

Paphiopedilum glanduliferum ist auf Neu Guinea weit verbreitet und sowohl in Küstennähe als auch in Gebirgsgegenden bis ca. 1800 m ü. NN anzutreffen.

The species was described by the Dutch botanist BLUME in the journal **Rumphia** *(loc. cit.)*. His concept was based on material given to him by ZIPPEL, one of the earliest New Guinea travellers. BLUMEs description is extensive, precise and accompanied by a coloured plate. Botanists carefully studying this publication and the abundant plant material available under a multitude of names, will have no difficulty in coming to the conclusion that the different concepts listed above can merely be considered varieties of *Paphiopedilum glanduliferum*.

Paphiopedilum glanduliferum is an herbaceous plant, generally growing epiphytically in leaf-mould. Leaves narrow-linear, up to 60 cm long, up to about 3 cm wide, rounded at the apex, fairly fleshy, uniformly green, distinctly keeled underneath (the size of the leaves varies considerably according to the geographic variant).

Inflorescence in general up to 50 cm high (sometimes higher still), deep-purple, partly spotted green and covered with grey-brown hairs; usually bearing up to 5 flowers.

The floral bracts are 2.5 to 4.5 cm long, sharply keeled, brown with darker veins, lanceolate.

The flowers are up to 12,5 cm high.

The dorsal sepal is about 5.5 cm high and up to about 2.5 cm wide, ovate, acute, pale yellow to whitish, sometimes with a dark-yellow middle part, striped deep yellow to reddish-brown, keeled externally.

The synsepalum is similar but smaller.

The petals are linear to strap-shaped, about 10 to 13 cm long and about 0.5 cm broad, standing in an angle of approximately 45° downwards, margins undulate, covered with blackish hairy warts, more or less helically twisted, pale yellow-green with darker yellow-green to reddish-purple longitudinal veins.

The lip is distinctly saccate, up to about 4.5 cm long and about 2 cm across, green to yellowish-green or light pink, with darker venation.

The staminodal shield is elongate, angular, fleshy, cucullate with two umbonate projections near the base, sides recurved, apex truncate, obtusely macronate, middle glabrous, buff-yellow to reddish, the broadly inflexed lateral margins studded with short red-brown bristles.

Paphiopedilum glanduliferum is widely distributed on New Guinea. It can be found near the coast as well as in altitudes of up to 1800 meters above sea-level.

Diese Art ist schwer zu etablieren, aber einmal eingewöhnt, sind die Pflanzen relativ einfach in der Kultur.

Man sollte für viel Licht und extrem hohe Luftfeuchtigkeit sorgen. *Paphiopedilum glanduliferum* mag ein durchlässiges Substrat mit Moosbeimischung und viel Wasser während der Wachstumsperiode. Gut etablierte Pflanzen sind blühwillig, wenn sie im Winter eine kurze Ruheperiode bekommen.

Die Blütezeit ist im allgemeinen von Mai bis Juni. Andere Blütezeiten sind jedoch nicht selten. Gut kultuvierte Exemplare blühen oft nochmals im Spätherbst.

Varietäten:

Wie bereits besprochen, geistern eine Reihe von Beschreibungen durch die Literatur, die eigentlich zu *Paphiopedilum glanduliferum* gehören. Diese basieren oft auf Unterschieden in der Blütenfärbung. Ich halte es daher für die sauberste Lösung, diese Beschreibungen hier als Varianten von *Paphiopedilum glanduliferum* abzuhandeln.

Paphiopedilum glanduliferum var. ***gardineri*** Braem, *var. nov.*

Differt forma nominalis in sepalum dorsalum album, vittatum rubrobrunneum; labellum luteo viridis, reticulatis atrobrunneis. Staminodium rubescens.

Synonyma: *Cypripedium gardineri, nomen nudum*
Paphiopedilum gardineri, (Guillemard) Pfitzer *nomen illegitimum*

Diese Varietät beruht auf dem von GUILLEMARD erwähnten[1], aber niemals gültig beschriebenen *Cypripedium gardineri.* Sie ist gekennzeichnet durch eine weiße Fahne mit rotbrauner Streifung sowie durch eine gelbe Lippe mit dunkelbrauner Aderung und rötlichem Überzug. Die Staminodiumplatte ist rötlich.

Paphiopedilum glanduliferum var. ***praestans*** (Rchb. f.) Braem, *comb. et stat. nov.*

Basionym: *Cypripedium praestans* Rchb. f. **Gardeners' Chronicle,** 2. Ser., **26:** 776 (1886)

Synonyme: *Cypripedium praestans* var. *illustre* Linden **Lindenia 3:** 90 (1886)

[1] **The Cruise of the Marchesa to Kamschatka and New Guinea 2:** 309 (1886)

This species is very difficult to establish in culture, but once the plants have been established successfully, they are relatively easy to grow.

Intensive light (but not direct sun) and a constantly high humidity is a must. *Paphiopedilum glanduliferum* does well in an open substrate with moss in the root area, abundant water during the growing season. A brief rest in the winter will enhance flowering.

Paphiopedilum glanduliferum generally flowers from May to June, but this varies quite distinctly according to the provenance of the variety. Well grown specimens often flower a second time in late autumn.

Varieties:

As already mentioned, a lot of different descriptions obviously belonging to the concept of *Paphiopedilum glanduliferum* can be found in the literature. Those descriptions are often based on differences in the colouration of the flowers. I consider it best to treat those concepts as varieties of *Paphiopedilum glanduliferum.*

Paphiopedilum glanduliferum var. ***gardineri*** Braem, *var. nov.*

Differt forma nominalis in sepalum dorsalum album, vittatum rubro-brunneum; labellum luteo viridis, reticulatis atrobrunneis. Staminodium rubescens.

Synonyma: *Cypripedium gardineri, nomen nudum*
Paphiopedilum gardineri (Guillemard) Pfitzer *nomen illegitimum*

This variety is based on *Cypripedium gardineri,* a taxon mentioned by GUILLEMARD[1] that has never been described validly. This variety is characterized by a white lateral sepal striped reddish-brown as well as by a yellowish coloured lip reticulated reddish-brown. The staminodal shield is reddish.

Paphiopedilum glanduliferum var. ***praestans*** (Rchb. f.) Braem, *comb. et stat. nov.*

Basionym: *Cypripedium praestans* Rchb. f. **Gardeners' Chronicle,** 2. Ser., **26:** 776 (1886)

Synonyma: *Cypripedium praestans* var. *illustre* Linden **Lindenia 3:** 90 (1886)

[1] **The Cruise of the Marchesa to Kamschatka and New Guinea 2:** 309 (1886)

Cypripedium praestans var. *spectabile* Linden
Lindenia 3: 90 (1886)

Cypripedium praestans var. *splendens* Linden
Lindenia 4: 100 (1887)

Cypripedium praestans var. *warocqueanum* Linden ex Desbois
Cypripedium, Selenipedium & Uropedium – Monographie: 401 (1898)

Pflanzen mit weißer Fahne mit dunkelbrauner Aderung. Die Lippe ist gelb mit gelbbraunem Überzug und Aderung. Die Staminodiumplatte ist gelb.

Paphiopedilum glanduliferum var. *wilhelminiae* (L. O. Williams) Cribb
The Genus Paphiopedilum: 111 (1987)

Synonym: *Paphiopedilum praestans* ssp. *wilhelminiae* (L. O. Williams) M. W. Wood
Curtis's Bot. Mag. n. s. **6,** t. 743 (1977)

Dies ist die „Gebirgsrasse" von *Paphiopedilum glanduliferum.* Die Pflanzen unterscheiden sich vom Nominaltyp durch einen kleineren Wuchs und durch kleinere Blüten. Die Laubblätter sind im allgemeinen bis zu 15 cm lang und bis zu 2 cm breit. Die Blüten sind 6 bis 10 cm hoch.

Die Blütenfärbung entspricht weitgehend jener der var. *praestans* (Rchb. f.) Braem.

Anmerkung:

Das von LINDEN beschriebene *Cypripedium praestans* var. *kimballianum*[1)] sowie die als *Paphiopedilum bodegomensis* oder *Paphiopedilum bodegomii* bezeichneten Pflanzen entsprechen dem Nominaltyp von *Paphiopedilum glanduliferum* (Blume) Stein.

[1)] **Lindenia 6,** t. 249 (1890)

Cypripedium praestans var. *spectabile* Linden
Lindenia 3: 90 (1886)

Cypripedium praestans var. *splendens* Linden
Lindenia 4: 100 (1887)

Cypripedium praestans var. *warocqueanum* Linden ex Desbois
Cypripedium, Selenipedium & Uropedium – Monographie: 401 (1898)

Plants with a white dorsal sepal veined dark-brown. The lip is yellow to yellowish-brown with darker venation. The staminodal shield is yellow.

Paphiopedilum glanduliferum var. *wilhelminiae* (L. O. Williams) Cribb
The Genus Paphiopedilum: 111 (1987)

Synonym: *Paphiopedilum praestans* ssp. *wilhelminiae* (L. O. Williams) M. W. Wood
Curtis's Bot. Mag. n. s. **6,** t. 743 (1977)

This is the "mountain race" of *Paphiopedilum glanduliferum.* The plants of this variety differ from the nominal type by a smaller habit and smaller flowers. The leaves are in general up to 15 cm long and up to about 2 cm wide. The flowers are 6 to 10 cm high.

The colour of the flower is very similar to that of var. *praestans* (Rchb. f.) Braem.

Note:

The "variety" described as *Cypripedium praestans* var. *kimballianum* by LINDEN[1)] as well as the plants designated as *Paphiopedilum bodegomensis* or *Paphiopedilum bodegomii* do not differ from the nominal type of *Paphiopedilum glanduliferum* (Blume) Stein.

[1)] **Lindenia 6,** t. 249 (1890)

Paphiopedilum kolopakingii Fowlie

Orchid Digest 48 (1): 41 (1984)

Paphiopedilum kolopakingii Fowlie

Paphiopedilum kolopakingii wurde durch einen indonesischen Pflanzensammler Anfang 1982 nach Ostjava gebracht und von dem Orchideenhändler LIEM KHE WIE aufgekauft.

Die Pflanzen befanden sich zwischen anderen *Paphiopedilum*-Arten und fielen durch ihre Blattlänge von bis zu 60 cm auf.

Aufgrund dieser Blattgröße und der Tatsache, daß die Pflanzen zusammen mit *Phalaenopsis maculata* – einer Begleitpflanze von *Paphiopedilum rothschildianum* – in Borneo gesammelt wurden, nahm LIEM KHE WIE an, daß es sich hierbei um *Paphiopedilum rothschildianum* oder eine neue „Varietät" dieser Art handeln würde. Er kaufte alle gesammelten Pflanzen auf und kultivierte sie. Seine Überraschung wird groß gewesen sein, als sich die Blüten öffneten und es sich herausstellte, daß es sich keineswegs um *Paphiopedilum rothschildianum* handelte.

1983 besuchte FOWLIE Ostjava und besuchte LIEM KHE WIE, der ihm erzählte, daß er die Pflanzen selber entdeckt hatte, was FOWLIE veranlaßte, die neue Art nach ihm zu benennen.

Immer noch gibt es Zweifel über die Identität und Herkunft dieser Art. Es ist sogar gemutmaßt worden, daß *Paphiopedilum kolopakingii* eine gärtne-

Paphiopedilum kolopakingii was brought to east Java in 1982 by an indigenous orchid hunter and bought by the orchid dealer LIEM KHE WIE.

The plants were among other *Paphiopedilum* species and were characterized by a leaf length of up to 60 cm.

Because of this leaf size and because of the fact that the plants were collected in Borneo, together with *Phalaenopsis maculata* – a species often growing sympatrically with *Paphiopedilum rothschildianum* – LIEM KHE WIE thought that he was buying *Paphiopedilum rothschildianum* or a variety thereof. He purchased the whole collection thinking to make the deal of his life. He must have been rather surprised when he saw the first flowers open and realized that the plants were by no means *Paphiopedilum rothschildianum*.

Still now, there is some discussion and doubt about the origin of these plants. It has even been suggested that *Paphiopedilum kolopakingii* is an artificial hybrid between *Paphiopedilum haynaldianum* and *Paphiopedilum philippinense*.

During my voyage to Indonesia in the spring of 1985 I convinced myself that *Paphiopedilum kolopakingii* is a real species from the wild, and I was informed that the natural site of collection, kept

Paphiopedilum kolopakingii Fowlie

rische Hybride zwischen *Paphiopedilum haynaldianum* und *Paphiopedilum philippinense* darstellt.

Während meiner Reise nach Indonesien im Frühjahr 1985 konnte ich feststellen, daß es sich bei *Paphiopedilum kolopakingii* um eine echte Wildart handelt und daß der von KOLOPAKING (LIEM KHE WIE) geheimgehaltene Standort inzwischen bekannt sei.

Paphiopedilum kolopakingii gehört zweifellos in die Verwandtschaft von *Paphiopedilum rothschildianum* und *Paphiopedilum stonei.* Die Pflanzen wachsen als Humusepiphyten an steilen Hängen. Die Rhizome sind kurz, so daß Gruppen von eng stehenden Blättern hervorgebracht werden.

Die Pflanzen haben fleischige, bis ca. 1 cm dicke und im allgemeinen bis zu 45 cm lange Wurzeln.

Jedes Stolon trägt bis zu 12 Laubblätter, die 40 bis 60 cm lang und bis zu 8 cm breit werden.

Die Blätter sind schmal bandförmig und meist einfarbig sattgrün, manchmal auch mit einer leichten Marmorierung.

Die Infloreszenz ist aufrecht, 40 bis 70 cm hoch, oben aufgrund des Blütengewichtes herabgebogen, bis ca. 1 cm dick.

Die Pflanzen bringen bis zu 15 Blüten hervor, die sich am Naturstandort gleichzeitig öffnen. Die einzelnen Blüten haben einen Durchmesser von bis zu 10 cm.

Die Braktee ist 4,5 bis 5 cm lang und ca. 1,5 cm breit, spitz endend.

secret up to that time by LIEM KHE WIE is now known to other people.

Paphiopedilum kolopakingii is, without any doubt, to be regarded related to *Paphiopedilum rothschildianum* and *Paphiopedilum stonei.* The plants grow as epiphytes in leafmould in rock cracks and crevices on steep cliffs. The rhizomes are short, the growth habit is therefore tufted.

The roots are fleshy, generally up to 45 cm long and up to about 1 cm in diameter.

Each stolon bears up to about 12 leaves. They are 40 to 60 cm long and up to about 8 cm broad, narrowly strap-shaped, usually uniformly deep green, sometimes with a slight marbling.

The inflorescence is suberect, 40 to 70 cm long, drooping under the weight of the flowers, up to about 1 cm in diameter, carrying up to 15 flowers usually all open at the same time. The single flowers are up to 10 cm in diameter.

The floral bract is 4.5 to 5 cm long and about 1.5 cm broad, with a triangular acuminate apex.

The dorsal sepal is narrowly-elliptic, acuminate, 5 to 6.5 cm high and 2.2 to 3.5 cm broad and provided with 9 to 15 longitudinal stripes; whitish, sometimes suffused green, and in some clones with dark red-brown longitudinal stripes. Some of these are prominent and nearly full-length. They are intersperced with secondary ones which end halfway up the dorsal sepal.

The synsepalum is similar to the dorsal sepal but smaller, 3 to 4 cm long and 2 to 2.6 cm broad, in general less intensively striped.

Paphiopedilum kolopakingii Fowlie

Die Fahne ist schmal elliptisch, spitz, 5 bis 6,5 cm hoch und 2,2 bis 3,5 cm breit; weißlich bis grünlich weiß, mit mehr oder weniger ausgeprägten dunkelbraunen Längsstreifen, meist 9 bis 15 an der Zahl. Streifen, die über die ganze Länge der Fahne gezogen sind, wechseln ab mit kürzeren, die ungefähr in der Mitte der Fahne enden.

Das Synsepalum ähnelt der Fahne, ist aber kleiner, 3,5 bis 4 cm lang und 2 bis 2,6 cm breit und im allgemeinen weniger intensiv gestreift.

Die Petalen sind länglich schmal, bandförmig. Sie stehen in einem Winkel von ca. 45° zur Lippe. Sie sind bis 7 cm lang und bis 8 mm breit, grün mit 7 bis 9 rötlichen Längsstreifen.

Die Lippe ist umgekehrt helmförmig, 4,5 bis 6 cm lang und 2 bis 3 cm breit, mit einer 1,5 bis 2,5 cm langen Klaue. Die Lippe ist grün mit intensiver brauner Aderung. Die Basis ist fast weiß.

Die Staminodiumplatte ist rundlich, 1 bis 1,5 cm hoch und ungefähr 1 cm breit, an den Seiten deutlich kurz behaart. Das untere Ende der Staminodiumplatte ist abgeflacht und weist eine kleine, leicht nach oben gebogene Spitze auf.

Paphiopedilum kolopakingii wurde entlang des Baritoflusses im inneren Kalimantan auf Felsen an Steilhängen um 700 m ü. NN entdeckt. Dieser Standort ist inzwischen bestätigt worden, und andere Standorte in vergleichbarer Lage sind bekannt.

Paphiopedilum kolopakingii blüht im allgemeinen zwischen Juli und September.

Wenn die Pflanzen einmal etabliert sind, sind sie relativ einfach zu kultivieren. Diese Art wächst gut im temperierten oder warmen Haus in einem durchlässigen Pflanzstoff bei regelmäßigen Wassergaben und konstant hoher Luftfeuchtigkeit.

The lateral petals are strap-shaped, divaricate at about 45°. They are up to 7 cm in length and up to about 8 mm wide, pure green with 7 to 9 reddish longitudinal stripes.

The lip is inverted helmet-shaped, 4.5 to 6 cm long and 2 to 3 cm broad, with a 1.5 to 25 cm long claw; greenish, distally with brownish venation. The base of the labellum is pure white.

The staminodal shield is orbicular, 1 to 1.5 cm high and about 1 cm across, along the lateral and superior sides provided with copious amounts of twisted hairs. The apex of the staminode is truncate and minutely subacute, distally with a light upwards turn.

Paphiopedilum kolopakingii was discovered along the Barito-river in central Kalimantan growing on rocks on steep cliffs at about 700 m above sea-level. This habitat has been confirmed and other habitats with comparable conditions have been reported.

Paphiopedilum kolopakingii generally flowers between July and September.

Once the plants have been established, they are fairly easy to grow. This species grows well in the intermediate or warm house, potted in an open compost. Waterings should be regular and a constant high humidity is indicated.

A

B

C

D

Paphiopedilum philippinense (Rchb. f.) Stein

Stein's Orchideenbuch: 480 (1892)

Basionym: *Cypripedium philippinense* Rchb. f.
Bonplandia 10: 335 (1862)

Synonyma: *Cordula philippinensis* (Rchb. f.) Rolfe
Orchid Review 20 (1): 2 (1912)

Cypripedium cannaertianum Linden
Lindenia 3: 93, t. 141 (1887)

Cypripedium laevigatum Bateman
Curtis's Botanical Magazine 91: t. 5508 (1865)

Cypripedium roebbelinii Rchb. f.
Gardeners' Chronicle, 2. Ser., **20:** 684 (1883)

Paphiopedilum laevigatum (Bateman) Pfitzer
in ENGLER, A.: **Natürliche Pflanzenfamilien 2** (6): 84 (1888)

Paphiopedilum roebbelinii (Rchb. f.) Pfitzer
in ENGLER, A.: **Bot. Jahrb. 19:** 41 (1894)

Selenipedium laevigatum (Bateman) May
Rev. Hort.: 301 (1885)

REICHENBACH beschrieb dieses Taxon anhand eines Herbarexemplars dessen Ursprung ungeklärt ist.

Zwei Jahre später (1864) fand J. G. VEITCH die Art auf der philippinischen Insel Guimares. Die Pflanzen wuchsen auf den Wurzeln von *Vanda batemanii.*

Aufgrund der von VEITCH nach England gebrachten Pflanzen veröffentlichte BATEMAN *(loc. cit.)* die Art als *Cypripedium laevigatum.* Fast 20 Jahre später erhielt REICHENBACH von dem im Dienste SANDERs stehenden Pflanzensammler RÖBBELEN drei getrocknete Einzelblüten einer *Paphiopedilum*-Art. Obwohl REICHENBACH die enge Affinität zu seinem *Cypripedium philippinense* bzw. zu dem *Cypripedium laevigatum* BATEMANs erkannte, beschrieb er die Pflanze als neue Art mit dem Namen *Cypripedium roebbelinii (loc. cit.).*

REICHENBACH described this taxon *(loc. cit.)* on the basis of an herbarium specimen of unknown origin.

Two years later (1864) plants of this species were found growing on the roots of *Vanda batemanii* on the Philippine island of Guimares by J. G. VEITCH. Some of that material was passed on to BATEMAN resulting in the description of *Paphiopedilum laevigatum.*

About another 20 years later, RÖBBELEN, an orchid collector in the services of SANDER, send REICHENBACH three dried flowers of a *Paphiopedilum*-species. Although REICHENBACH recognized the close affinity of these flowers to those of his *Cypripedium philippinense* and to BATEMANs *Cypripedium laevigatum,* he used them to publish a third concept as *Cypripedium roebbelinii.*

Tafel Seite 74 / Plate page 74

A/B verschiedene Formen von / different forms of *Paphiopedilum philippinense*
C *Paphiopedilum philippinense* var. *roebbelinii* (Rchb. f.) Cribb
D *Paphiopedilum philippinense* – Nahaufnahme / close-up

Es besteht keinerlei Zweifel daran, daß diese drei Konzepte sich auf die gleiche Art beziehen. Somit ist dem ältesten Konzept – also *Cypripedium philippinense* Rchb. f. – zu folgen. Die Umkombination zu *Paphiopedilum* erfolgte dann 1892 durch STEIN in seinem **Orchideenbuch.**[1)]

Die Blätter dieser bis 50 cm hohen, als Humusepiphyten wachsenden, krautigen Pflanzen sind länglich elliptisch, glänzend einfarbig grün, bis 30 cm lang und 4 cm breit, unten deutlich gekielt, gerundet oder ungleich zweilappig an der Spitze.

Die Infloreszenz wird bis zu 50 cm hoch. Sie ist flaumig behaart, grün mit braunvioletten Schlieren und trägt meistens 3 bis 5 Blüten.

Die ca. 2,2 cm lange Braktee ist kahnförmig, rötlich braun, behaart, und erreicht ein Drittel der Länge des Fruchtknotens.

Die Blüten werden bis 20 cm hoch und 8 cm breit.

Die Fahne wird ca. 4 cm hoch und 2 bis 3 cm breit. Sie ist breit eiförmig, spitz, an den Rändern flaumig behaart, weiß mit braunvioletten Streifen.

Das Synsepalum ist, wie die Fahne, eiförmig aber etwas schmaler, bis 3,8 cm lang, weiß mit grünen Adern.

Die Petalen sind 12 bis 15 cm lang, linealisch, bandförmig, verdreht, herabhängend, die Ränder des basalen Teils purpurrot überzogen und gehen an der Spitze in ein bleiches Grün über.

Die umgekehrt helmförmige Lippe ist bis 3 cm lang und ca. 1,7 cm breit. Sie ist ledergelb mit hellgrüner bis brauner Beaderung.

Die Staminodiumplatte ist ei- bis herzförmig, an den Seitenrändern flaumig behaart.

Paphiopedilum philippinense wächst epiphytisch im Humus am Waldboden, auf Bäumen, in Moos (auch auf moosbewachsenen Felsen), immer an hellen Stellen.

Das Klima des Heimatareals besteht hauptsächlich aus einer heißen „Sommer"-Periode während der Monate April bis Juni und einer Ende Juni anfangenden und bis zum Oktober andauernden Regenperiode. Es gibt keine ausgeprägte Trokkenzeit.

[1)] Seit einiger Zeit wird eine angeblich neue *Paphiopedilum*-Art von der Insel PALAWAN als *Paphiopedilum palawense* angeboten. Es handelt sich hier eindeutig um eine geographische Variante von *Paphiopedilum philippinense* mit „*exul*-ähnlichen" Blättern, aber mit typischen *Paphiopedilum philippinense*-Blüten.

There can be no doubt whatsoever that these three concepts refer to one single species. Therefore, the eldest description – *Cypripedium philippinense* – is the valid on. The combination to *Paphiopedilum* was published 1892 by STEIN in his **Orchideenbuch**[1)].

The leaves of this herbaceous, epiphytic plant, usually growing in leafmould, are ligulate-oblong, up to about 30 cm long and 4 cm broad, coriaceous with a polished surface, uniformly green, distinctly keeled underneath, obtuse or unequally bilobed at the apex.

The inflorescence is up to about 50 cm high, downy pubescent, green with purple-brown streaks, generally bearing 3 to 5 flowers.

The floral bracts are boat shaped, reddish-brown, hairy, about one third as long as the ovary.

The flowers are up to about 20 cm high and 8 cm across.

The dorsal sepal is up to about 4 cm long and 2 to 3 cm broad, broadly ovate, pointed, whitish with purple-brown veins, the margins downy pubescent.

The synsepalum is, just like the dorsal petal, ovate, but somewhat narrower, up to about 3.8 cm long, whitish with green veins.

The petals are ribbon-like, 12 to 15 cm long, pendulous, twisted, the margins of the basal part with hairy warts, yellowish at the base, passing into dull reddish purple along the major part of their length, pale green at the apex.

The lip is inverted helmet-shaped, up to 3 cm long and about 1.7 cm wide, buff-yellow, faintly striated with brown.

The staminodal shield is sub-cordiform, emarginate, with black hairs on each side.

Paphiopedilum philippinense grows epiphytically in leaf-mould on the ground in forests, on trees and in moss (also on moss-covered rocks), always in bright places.

The climate of the natural distribution area is characterized mainly by a hot "summer period" during the months of April to June followed by a rainy period lasting until October. There is no distinct period of dryness.

[1)] For some time now, a *Paphiopedilum*-species originating from the Philippine island of PALAWAN is being sold under the name of *Paphiopedilum palawense*. This plant, again, is merely a geographic variant of *Paphiopedilum philippinense* differing from the nominal form solely by its leaves "reminiscent to those of *Paphiopedilum exul*".

Die Luftfeuchtigkeit bleibt das ganze Jahr über sehr hoch.

Die Sommertemperaturen schwanken zwischen 32 °C am Tage und 20 °C in der Nacht. In der Regenzeit liegen sie zwischen 25 °C am Tage und 15 °C nachts.

Paphiopedilum philippinense blüht normalerweise von März bis Juli.

Diese Art ist in unseren Breiten einfach zu kultivieren. Entsprechend seines Herkunftareals braucht *Paphiopedilum philippinense* sehr viel Licht. Die Pflanzen gedeihen gut in einem lockeren Substrat und mögen Moos im Wurzelbereich. Die Pflanzen werden bei entsprechender Kultur sehr massig, und es empfiehlt sich, schwere Töpfe zu verwenden.

Beim Teilen und Umpflanzen sollte man vorsichtig verfahren, da die Neutriebe bis zu 2 Jahre brauchen, um Wurzeln zu bilden. Man sollte daher immer mindestens drei Triebe zusammenlassen.

Varietäten:

Auch von *Paphiopedilum philippinense* sind eine Reihe von Varietäten beschrieben worden. Wie so oft variieren die meisten kaum von dem Typusexemplar. Nur die als *Paphiopedilum roebbelinii* bekannte Abart kann als solche anerkannt werden.

Paphiopedilum philippinense var. *roebbelinii* (Rchb. f.) Cribb
The Genus Paphiopedilum 111 (1987)

Synonym: *Paphiopedilum roebbelinii* (Rchb. f.) Pfitzer
in ENGLER, A.: **Bot. Jahrb. 19:** 41 (1894)

PFITZER[1] führt diese Varietät mit dem Hinweis auf, daß sie von VEITCH[2] beschrieben sei. Dies ist jedoch nicht der Fall.

Die Pflanzen, welche im allgemeinen als *Paphiopedilum philippinense* var. *roebbelinii* betrachtet werden, unterscheiden sich von der Nominalform durch schmalere Laubblätter und intensiver gedrehte Petalen.

[1] in ENGLER, A.: **Das Pflanzenr. 4** (50): 62 (1903).
[2] **A Manual of Orchidaceous Plants 4:** 42 (1889)

The relative humidity is very high over the whole year.

Summer temperatures are between 32 °C during the day and 20 °C at night. During the rainy period, temperatures are about 25 °C during the day and 15 °C at nighttime.

Paphiopedilum philippinense generally flowers between March and July.

This species is easy to grow. According to its natural distribution, *Paphiopedilum philippinense,* should be kept in a bright spot. The plants do well when they are potted in an open compost with plenty of moss in the root area. Well grown plants become very large and top-heavy. They should therefore be kept in heavy pots.

One should be very cautious in dividing plants of this species as the new growths need up to 2 years to develop roots. Divisions should be not smaller than 3 growths.

Varieties:

Paphiopedilum philippinense is not an exception in having a number of "varieties", and, as with so many other species, most of these varieties do not differ much from the type. Only *Paphiopedilum roebbelinii* can be regarded a valid variety of *Paphiopedilum philippinense.*

Paphiopedilum philippinense var. *roebbelinii* (Rchb. f.) Cribb
The Genus Paphiopedilum 111 (1987)

Synonym: *Paphiopedilum roebbelinii* (Rchb. f.) Pfitzer
in ENGLER, A.: **Bot. Jahrb. 19:** 41 (1894)

PFITZER[1] lists this variety, erroneously noting that it has been described by VEITCH[2].

The plants, generally considered as belonging to this variety differ from typical specimens of *Paphiopedilum philippinense* by narrower leaves and more intensily twisted lateral petals.

[1] in ENGLER, A.: **Das Pflanzenr. 4** (50): 62 (1903)
[2] **A Manual of Orchidaceous Plants 4:** 42 (1889)

Paphiopedilum randsii Fowlie

Orchid Digest 33: 321–322 (1969)

Paphiopedilum randsii Fowlie

Paphiopedilum randsii wurde auf der philippinischen Insel Mindanao entdeckt und als „gelbblühendes *Paphiopedilum philippinense* mit besonders kurzen Petalen" bezeichnet. Die Pflanzen wurden als *Paphiopedilum roebbelinii* an die Firma RANDS in den USA verkauft. FOWLIE sah die Pflanzen bei RANDS, erkannte, daß es sich um eine neue Art handelte und beschrieb die Art zu Ehren Ray RANDS *(loc. cit.)*

Obwohl einige amerikanische *Paphiopedilum*-Züchter der Ansicht sind, daß *Paphiopedilum randsii* ein „schlechtes *Paphiopedilum philippinense*" ist, so bin ich doch der Meinung, daß FOWLIE recht hatte, indem er die Pflanze als selbständiges Taxon beschrieb.

Sicherlich haben *Paphiopedilum randsii* und *Paphiopedilum philippinense* eine Reihe von Gemeinsamkeiten in der Blütenstruktur, aber die Unterschiede sind andererseits gravierend.

Paphiopedilum randsii was discovered on the Philippine island of Mindanao and designated as "yellow flowered *Paphiopedilum philippinense* with short petals". The plants were sold as *Paphiopedilum roebbelinii* and purchased by the American orchid firm RANDS.

FOWLIE, who saw the plants in flower at RANDSs recognized that this was a species new to science and described the taxon in honour of Ray RANDS *(loc. cit.)*.

Although some American orchid growers claim *Paphiopedilum randsii* to be merely a "bad *Paphiopedilum philippinense*", it is my opinion that FOWLIE was right in describing these plants as a autonomous taxon.

It is evident that *Paphiopedilum randsii* and *Paphiopedilum philippinense* have a number of common characteristics in respect to the structure of the flower but on the other side, the differences are too obvious.

Ich würde eher postulieren, daß es sich hier um eine Naturhybride von *Paphiopedilum philippinense* handelt, eine These, die durch das Vorhandensein von seitlichen Aurikeln an der Lippe (Unterschied zu allen anderen Taxa dieser Gruppe) sicherlich unterstützt wird.

Paphiopedilum randsii ist eine als Humusepiphyt wachsende, krautige Pflanze mit länglich elliptischen, biś ca. 35 cm langen und 5 bis 6 cm breiten, vorne abgerundeten, weder netzartig gefärbten noch marmorierten, einfarbig hellgrünen Blättern.

Die Infloreszenz wird bis ungefähr 40 cm hoch und trägt im allgemeinen 2 bis 5 Blüten. Sie steht aufrecht und ist dicht behaart.

Die Brakteen sind bis 3 cm lang und ca. 2 cm breit, lanzettförmig, spitz.

Die großen Blüten zeigen weiße bis hell grünlichweiße Sepalen mit deutlichen kastanienbraunen Linien.

Das mittlere Sepalum (die Fahne) ist schmal eiförmig, 3,8 bis 4,3 cm lang und 1,7 bis 2,2 cm breit. Es ist nach vorne gebogen, konkav, an der Spitze schwach behaart.

I would rather postulate that *Paphiopedilum randsii* is a natural hybrid with *Paphiopedilum philippinense* as one of its parent species. The presence of lateral ears on the lip (an exception to all other taxa of this group) does certainly support such an hypothesis.

Paphiopedilum randsii is an herbaceous plant, mainly growing epiphytically in leaf-mould. Its leaves are oblong-elliptic, up to about 35 cm long and 5 to 6 cm wide, rounded at the apex, unicoloured bright green without any tesselation. The inflorescence is up to about 40 cm high, erect, densely pilose, generally bearing 2 to 5 flowers.

The floral bracts are about 3 cm long and about 2 cm wide, lanceolate, acute.

The large flowers have whitish to brilliant greenish-white sepals with a number of distinct, maroon coloured longitudinal stripes.

The upper sepal is narrowly ovate, 3.8 to 4.3 cm long and 1.7 to 2.2 cm broad, somewhat inclined over the lip, concave, the apex sparsely pubescent.

The lateral sepals are united to a synsepalum which is similar to the dorsal sepal but smaller, 3 to

Paphiopedilum randsii Fowlie

Die lateralen Sepalen sind zum Synsepalum verwachsen, welches der Fahne ähnlich ist, aber kleiner. Das Synsepalum ist an der Spitze gestutzt, 3 bis 3,3 cm lang und 1,8 bis 2,2 cm breit.

Die schmal linealischen Petalen sind 3,7 bis 4,6 cm lang und 0,4 bis 0,7 cm breit. Sie sind zur Lippe hin gebogen, abgerundet, an den Rändern kurz bewimpert, hell grünlichweiß, mit kastanienbraunen Längsstreifen.

Die Lippe ist verkehrt helmförmig, hell gelblichgrün, dunkler beadert. Sie ist 2,8 bis 3,2 cm lang und 1,2 bis 1,5 cm breit, dreilappig. Die stumpfen Seitenlappen sind nach innen gebogen. Der Mittellappen ist mit Aurikel versehen und zeigt am Gipfel eine deutliche Rinne.

Das hellgrüne Staminodium ist elliptisch, unten gestutzt, 0,5 cm lang und ca. 0,4 cm breit, an den Seiten flaumig behaart.

Paphiopedilum randsii wächst meist im Humus am Waldboden in Höhenlagen um 500 m ü. NN auf der philippinischen Insel Mindanao. Dieser Teil der Philippinen kennt keine Trockenzeit. Im November und Dezember bringen warme Winde aus dem Südpazifik große Regenmengen mit sich.

Von Januar bis März bläst der nordöstliche Monsun, der dann weniger Niederschlag bringt.

Die trockensten Monate sind April, Mai und Juni.

Von Juli bis Ende Oktober herrscht wieder eine regenträchtige Periode mit südwestlichen Monsunwinden und möglicherweise taifunartigen Stürmen.

Der regenträchtigste Monat ist August.

Die Sommertemperaturen schwanken um 30 °C am Tage und um 18 °C in der Nacht. Im „Winter" sinken sie auf ca. 22 °C am Tage und 14 °C nachts.

Die Luftfeuchtigkeit ist das ganze Jahr über sehr hoch.

Wie bereits erwähnt, wird gemutmaßt, daß *Paphiopedilum randsii*, genauso wie *Paphiopedilum lowii*, auch als Epiphyt auf Bäumen wächst. Es ist daher kaum verwunderlich, daß einige Orchideenzüchter die Pflanzen am besten aufgebunden kultivieren können. *Paphiopedilum randsii* sollte immer bei ziemlich hellem Licht und hoher Luftfeuchtigkeit gehalten werden. Während der Wachstumsperiode sollte nicht mit Wassergaben gespart werden.

Am besten gedeihen die Pflanzen in luftdurchlässigem Substrat im kälteren Teil des temperierten Gewächshauses.

Die Pflanzen blühen im allgemeinen von März bis Mai aus zwei- bis dreijährigen Trieben.

3.3 cm long and 1.8 to 2.2 cm wide, the apex somewhat truncated.

The lateral petals are linear-lanceolate, 3.7 to 4.6 cm long and 0.4 to 0.7 cm wide, bent towards the lip, rounded, the margins short pilose, light green with maroon longitudinal stripes.

The lip is inverted helmet-shaped, more or less bright yellowish green, with darker veins, about 2.8 to 3.2 cm long and 1.2 to 1.5 cm across, trilobate. The sidelobes are blunt, folded inwards. The middle lobe with lateral ears at the base and a distinct grove at the apex.

The staminodal shield is elliptic, truncate at the apex, about 0.5 cm long and about 0.4 cm wide, bright green, the margins distinctly pubescent.

Paphiopedilum randsii usually grows in leaf-mould beneath trees at altitudes of about 500 m above sea-level. The species is only known from the Philippine island of Mindanao. This part of the Philippines does not know any dry season.

From November to December warm winds from the South Pacific bring large amounts of rainfall.

From January until March there is a northeastern monsoon bringing somewhat less precipitation.

It becomes even dryer in April, May and June, but from the end of June until October the southwest monsoon brings more abundant rain again and possibly some typhoons. August is the wettest month.

The summer temperatures are about 30 °C during the day and about 18 °C during the night. In the colder periods, the temperatures sink to about 22 °C during daytime and to about 14 °C during the night.

The relative humidity remains very high all over the year.

It has been suggested, that *Paphiopedilum randsii*, just like *Paphiopedilum lowii*, is also capable of growing epiphytically on trees. In view of this, it is not surprising to hear that some orchid growers can cultivate this species best when mounted on blocks. *Paphiopedilum randsii* should also be kept under bright light and constant high humidity. Periods of 10 to 12 hours of extremely bright light per day are required to get the plants to flower. Abundant waterings are indicated, especially during the vegetative period (between the appearance and maturation of the new growth).

The plants do best when potted in an open compost and kept in the cooler part of the intermediate house.

Paphiopedilum randsii flowers generally between March and May from mature, two- to three-year-old growths.

Paphiopedilum rothschildianum (Rchb. f.) Stein

Stein's Orchideenbuch: 482 (1892)

Basionym: *Cypripedium rothschildianum* Rchb. f. **Gardeners' Chronicle,** 3. Ser., **3:** 457 (1888)

Synonyma: *Cordula rothschildiana* (Rchb. f.) Rolfe **Orchid Review 20** (1): 2 (1912)

Cypripedium neo-guineense Linden **Gardeners' Chronicle,** 3. Ser., **3:** 505 (1888)

Cypripedium nicholsonianum ex Hort. Roy. Bot. Gard. Glasnevin. in Herb. Kew

Im Jahre 1888 sandte die Firma SANDER Pflanzenmaterial einer neu importierten Frauenschuhart an REICHENBACH mit der Bitte um Beschreibung.

REICHENBACH folgte bei der Namensgebung den Wünschen SANDERs und benannte die neue Art zu Ehren von Baron Ferdinand de ROTHSCHILD *(loc. cit.)*.

Jean LINDEN, der damals führende belgische Orchideenzüchter behauptete, er habe die Art zuerst, und zwar im Jahre 1887, nach Europa gebracht. Das Beschreibungskonzept LINDENs (falls es jemals existiert hat) wurde nie publiziert. Er bot aber die Pflanzen als *Cypripedium neo-guineense* in einem Inserat im **Gardeners' Chronicle** *(loc. cit.)* zum Kauf an.

Paphiopedilum rothschildianum ist eine als Humusepiphyt wachsende krautige Pflanze mit grünen, glänzenden, bis 60 cm langen und 4,5 bis 7,5 cm breiten, schmal elliptischen bis riemenförmigen, fast aufrecht stehenden Blättern.

Die Infloreszenz wird bis 45 cm hoch. Sie ist rötlich, schwach kurz behaart und trägt im allgemeinen 3 bis 6 Blüten.

Die Brakteen sind fast so lang wie der kahle Fruchtknoten, am Gipfel dreiteilig, hell gelbgrün mit schwarzen Längsstreifen.

Die Fahne wird ungefähr 6,5 cm hoch und ist 3,5 bis 4,5 cm breit. Sie ist länglich keilförmig, gelblich, mit vielen schwärzlichen Längsstreifen und weißen Rändern.

Das Synsepalum ist ungefähr 5,5 cm lang und fast so breit wie die Fahne.

In 1888, the English orchid dealer SANDER send REICHENBACH some material of a newly imported slipper orchid and requested the species to be named in honour of Baron Ferdinand de ROTHSCHILD.

Jean LINDEN, at that time the major Belgian orchid dealer, claimed to have been first in importing the species to Europe in the year 1887. The concept of the description of the species by LINDEN (supposing it did exist) was never published. He simply offered the plants for sale in the **Gardeners' Chronicle** *(loc. cit.)* naming them *Cypripedium neo-guineense.*

Paphiopedilum rothschildianum is like all tropical slipper orchids i.e., an epiphytic herbaceous plant growing mainly in leaf-mould.

The leaves are suberect, bright green, unicoloured, narrowly elliptic to strap-shaped, up to 60 cm long and about 4.5 to 7.5 cm broad.

The inflorescence is up to about 45 cm high, reddish, minutely pubescent; it generally bears from 3 to 6 flowers.

The floral bracts are nearly as long as the glabrous ovary, apically three-toothed; bright yellow-green with blackish longitudinal stripes.

The dorsal sepal is about 6.5 cm long and 3.5 to 4.5 cm across, ovate, acute, yellowish with numerous blackish longitudinal stripes, the margins white.

The synsepalum is about 5.5 cm long and nearly as broad as the dorsal sepal.

The linear-lanceolate petals are spreading nearly horizontally. They are 10 to 14 cm long and about

Paphiopedilum rothschildianum (Rchb. f.) Stein

Die linealisch lanzettlichen Petalen stehen fast waagerecht. Sie werden 10 bis 14 cm lang und ungefähr 1 cm breit. Sie sind allmählich spitz ausgezogen, an der Basis gewellt, im apikalen Teil gedreht, gelbgrün mit dunklen Längsstreifen, an der Basis dunkel gefleckt. Die Lippe ist 4,5 bis 5,8 cm lang, ungefähr 2,2 cm breit, pantoffelförmig, seitlich etwas zusammengedrückt, hervorstehend; zimtfarben, um die Öffnung zinnoberrot.

Das Staminodium ist schnabelartig (in der Erstbeschreibung wurde es durch REICHENBACH mit Kopf und Hals eines Kranichs verglichen), bestehend aus einer behaarten Basis und einem langen, zweispitzigen Schnabel im rechten Winkel dazu, die untere Oberfläche behaart.

Paphiopedilum rothschildianum kommt auf der Insel Borneo vor. Die bekannten Standorte liegen an den Hängen des Mount Kinabalu.

Die Pflanzen wachsen an steilen Hängen auf Serpentingestein (wobei die Wurzeln den Blatthumus durchdringen) an ziemlich hellen, freien Stellen ohne direkte Sonneneinstrahlung, in Höhenlagen von 500 bis 1000 m ü. NN.

Die jährliche Niederschlagsmenge beläuft sich in diesem Gebiet auf ca. 3750 mm. Die Luftfeuchtigkeit ist über das ganze Jahr sehr hoch.

Die kühlsten Monate sind Juni und Juli mit Temperaturen um 25 °C am Tage und um 15 °C in der Nacht. Im April sind die Temperaturen mit 30 °C beziehungsweise 18 °C am höchsten.

Paphiopedilum rothschildianum ist relativ einfach zu kultivieren, aber extrem schwierig zum Blühen zu bringen. Man sollte die Pflanzen über lange Zeit in Ruhe lassen und vor allem nicht teilen. Die Art gedeiht am besten in einem durchlässigen, aber wasserspeichernden Substrat mit Moosbeimischung. Häufiges Gießen ist neben reichlicher Luftzirkulation wichtig.

Paphiopedilum rothschildianum braucht viel Licht, direkte Sonne ist jedoch zu vermeiden.

Paphiopedilum rothschildianum hat keine bestimmte Blütezeit. Gut kultivierte Pflanzen blühen nach Vollendung des Neutriebes unabhängig von der Jahreszeit.

Varietäten:

Auch von *Paphiopedilum rothschildianum* sind einige Varietäten beschrieben worden. Keine davon weicht in irgendeiner Weise, die das Aufrechterhalten als autonomes Taxon gewährleisten würde, von der „Normalform“ ab.

1 cm broad, attenuate into a more or less acute apex, the basal part undulate, the apical part twisted, greenish-yellow with dark longitudinal stripes and spotted dark near the base.

The lip is distinctly saccate, somewhat laterally compressed, protruding, about 4.5 to 5.8 cm long and about 2.2 cm across, cinnamon coloured, vermillion around the mouth.

The staminode is peculiarly structured, resembling the head and beak of a crane, consisting of a hairy base, with a long twotoothed beak at right angles thereto, hairy on the lower surface.

Paphiopedilum rothschildianum is an endemic of the isle of Borneo. The known habitats are on the slopes of Mount Kinabalu.

The plants grow in steep ravines on eroded serpentine. The roots are completely imbedded in leafy humus. The plants generally grow on upper slopes in exposed locations, mostly in reflected light at altitudes of about 500 to 1000 m above sealevel.

In this area, the yearly amount of precipitation is about 3750 mm; the relative humidity remains very high over the whole year.

June and July are the coolest months, with daily temperatures about 25 °C and nighttime temperatures around 15 °C. April is the hottest month with temperatures about 30 °C during the day and 18 °C at night.

Paphiopedilum rothschildianum is an easy grower but very difficult to flower. Plants should be left undisturbed and undivided over a long period of time. This species does best in a open but moisture retaining compost with abundant moss. Besides a constant high humidity, abundant waterings are just as imperative as is a good circulation of free air.

Paphiopedilum rothschildianum needs plenty of light, however, the plants should not be put into direct sun.

Paphiopedilum rothschildianum has no main flowering period. Wellgrown plants will flower whenever the new growth is completed.

Varieties:

Of *Paphiopedilum rothschildianum* too, a number of "varieties" have been described. None of them, however, differs from the nominal form to such an extent as to be considered an autonomous taxon.

Paphiopedilum sanderianum (Rchb. f.) Stein

Paphiopedilum sanderianum (Rchb. f.) Stein

Stein's Orchideenbuch: 482 (1892)

Basionym: *Cypripedium sanderianum* Rchb. f. **Gardeners' Chronicle,** 2. Ser., **25:** 554 (1886)

Synonyma: *Cordula sanderiana* (Rchb. f.) Rolfe **Orchid Review 20** (1): 2 (1912)

Cypripedium foerstermanni Hort. *nomen nudum*

Paphiopedilum sanderianum (Rchb. f.) Stein

Paphiopedilum sanderianum wurde durch FOERSTERMANN, einem Pflanzensammler im Auftrag SANDERs, im Jahre 1885 entdeckt. Der Fundort wurde traditionsgemäß irreführend als „Malaiisches Archipel" angegeben.

Der echte genaue Fundort der FOERSTERMANNschen Aufsammlung wurde nie bekannt, und bis vor kurzem waren keine Pflanzen dieser Art in Kultur.

1985 gelang es dem österreichischen Ingenieur Paul MATTES, diese seltene Art wiederzuentdekken und einige Pflanzen nach Mitteleuropa zu bringen. Zwei dieser Pflanzen wurden im Frankfurter Palmengarten kultiviert und blühten kurz vor Weihnachten 1985. Wenige Monate später gelangen eine Reihe von Pflanzen dieser Art aus anderer Quelle nach Europa. Einige Exemplare wurden mir zwecks fotografischer Dokumentation zur Verfü-

Paphiopedilum sanderianum was discovered in 1885 by FOERSTERMANN, a plant collector in the services of the English orchid firm SANDER. As for all plants discovered on the isle of Borneo, the place of origin was misleadingly said to be the Malayan Archipelago (as was the tradition at the time!).

The real habitat discovered by FOERSTERMANN was never disclosed and until recently no specimens of this species were in cultivation.

In 1985, the Austrian engineer Paul MATTES rediscovered the species and brought some plants to Europe. Two of these were grown in the botanical gardens of the City of Frankfurt and flowered shortly before Christmas 1985. A few months later, a number of plants of *Paphiopedilum sanderianum* found their way into Europe through other channels. Several specimens were put at my disposal gung gestellt, so daß auch von dieser seltenen Art eine Abbildung veröffentlicht werden kann. Es ist zu hoffen, daß eine Vermehrung über Samen gelingen wird und daß somit sichergestellt werden kann, daß die Art auch in Kultur zu einer größeren Verbreitung gelangt.

Paphiopedilum sanderianum wurde, ohne Illustration, aber mit einer relativ ausführlichen Beschreibung, durch REICHENBACH im **Gardeners' Chronicle** *(loc. cit.)* zu Ehren von Frederick SANDER als *Cypripedium sanderianum* veröffentlicht.

Aufgrund der Seltenheit dieser Art in den Sammlungen wurde sehr viel gemutmaßt. Die Pflanze wurde als gültige autonome Art, als Naturhybride und als Monstrosität gedeutet.

Paphiopedilum sanderianum ist ein krautiger Humusepiphyt mit kurzem Sproß.

Jeder Sproß trägt ungefähr 6 länglich linealische Blätter, die bis ungefähr 45 cm lang und bis 4,5 cm breit werden und asymmetrisch zweiteilig enden.

Die Laubblätter sind unterseitig scharf gekielt, glänzend, oberseitig mehr oder weniger dunkelgrün, im basalen Bereich purpurn gefleckt.

Die aufrechte Infloreszenz wird bis ungefähr 60 cm hoch und 5 mm im Durchmesser. Sie ist purpurn und von kurzen aufrecht stehenden purpurnen Härchen überzogen, zwei- bis fünfblütig.

Die Braktee ist schmal elliptisch, gerundet, bis 8 cm lang und bis 3 cm breit, purpurn mit einer grünen Spitze, mit bewimperten Rändern, dunkelpurpur gestreift.

Der Fruchtknoten ist ungefähr 7 cm lang und 1 cm im Durchmesser, 6kantig, weiß, mit leichter Behaarung.

Die Blüten werden bis 7,5 cm hoch (die langen Petalen nicht mitgerechnet).

Die Fahne ist breit lanzettförmig, spitz, gekielt, ungefähr 5,5 cm lang und 2,5 cm breit; vor allem auf dem Kiel und am spitzen Ende fein behaart. Ihre Farbe ist ein blasses Grüngelb mit ungefähr 10 cm breiten braunpurpurnen Streifen und purpurnen Rändern.

Das Synsepalum ist schmal lanzettförmig, doppelt gekielt, gerundet bis gestutzt, tief konkav, ungefähr 5 cm lang und 2 cm breit. In Farbe und Behaarung ähnelt es der Fahne, aber die purpurnen Streifen sind schmaler.

Die Petalen sind linealisch bis bandförmig, im basalen Teil deutlich gedreht. Sie stehen ungefähr 3 cm horizontal und biegen sich dann abrupt nach

allowing me to study and photograph the plants extensively. In this way, colour photographs of this very rare species can be included in this book. There is some hope that this species will be propagated in one or more horticultural establishments, thus assuring that it will be available to a wider public in the fairly near future.

Paphiopedilum sanderianum was described as *Cypripedium sanderianum* by REICHENBACH in honour of Frederick SANDER. The publication appeared in the **Gardeners' Chronicle** *(loc. cit.)* without an illustration but with an unusually extensive description.

Due to the cultural rarity of this species several hypothesis were articulated. The plant was designated as an autonomous species, as natural hybrid and as a monstrosity.

Paphiopedilum sanderianum is an herbaceous epiphyte, growing generally in leaf-mould.

The stems are short, bearing up to about 6 elongated linear leaves, up to about 45 cm long and up to about 4.5 cm broad, with an asymmetric two-lobed apex. They are shiny, the upper surface uniformly green, the under surface distinctly keeled with a purple spotted basal area.

The inflorescence is erect, up to about 60 cm high and about 5 mm in diameter, purple, covered with short purple hairs and bearing two to five flowers.

The floral bracts are narrowly elliptic, rounded, up to 8 cm long and up to about 3 cm broad, purple, the apex green, ciliate at the edges and striped dark purple.

The ovary is about 7 cm long and 1 cm in diameter, 6 sided, white, sparsely hairy.

The flowers are up to about 7 cm high (not counting the elongated lateral petals).

The dorsal sepal is broadly lanceolate, acute, hooded at the tip, keeled, about 5.5 cm long and about 2.5 cm wide, pubescent, especially on the keel and at the tip, pale greenish-yellow with about 10 brownish-purple stripes and purple margins.

The synsepalum is narrowly lanceolate, 2-keeled, rounded to truncate, deeply concave, about 5 cm long and up to about 2 cm wide, similarly coloured and pubescent as the dorsal sepal but with narrower longitudinal stripes.

The petals are linear and twisted near the base, horizontal for about 3 cm and then abruptly turned downwards and tapering into very long pendulous filaments, generally up to 60 cm long (reported up to 1 m long in literature[1]), about 5 mm wide at the unten. Sie sind bis zu 1 m lang,[1] dabei an der Basis 5 mm breit und ungefähr 9 mm breit an der „Schulter", 1,5 bis 2 mm breit an den fleischig verdickten Spitzen. Sie sind fahl gelbgrün und mit unregelmäßigen purpurbraunen Flecken überzogen, an den verdickten, fleischigen Spitzen vollständig purpurn gefärbt, in der Basisgegend behaart.

Die Lippe ist tief sackförmig mit gerundeter Spitze, 5,5 bis 6,5 cm hoch und 2 bis 2,5 cm breit, gelblichgrün, purpurn überzogen. Die hinteren Ränder sind nach innen gefaltet.

Das Staminodium ist länglich, mit nach hinten gefalteten, stark behaarten Rändern.

Paphiopedilum sanderianum ist höchstwahrscheinlich ein Endemit der Insel Borneo. Die Pflanzen wachsen in Höhenlagen um 200 m ü. NN an schattigen Stellen steiler Dolomitwände. Die Standorte sind teilweise sonnig am frühen Morgen, sonst aber schattig.

Entsprechend des in diesem Gebiet herrschenden Klimas ist die Art warm, feucht und schattig zu halten.

Die Blütezeit fällt im allgemeinen in die Zeit von Juli bis Anfang Dezember, wobei auch von einer Blühperiode zwischen Februar und April berichtet wird.

[1] Über eine Pflanze mit 1,2 m langen Petalen wurde mir vor kurzem berichtet.

base, more or less 9 mm wide at the "shoulder", 1.5 to 2 mm wide above the swollen fleshy tip, pale yellowish-green with irregular spots and blotches of brownish-purple, becoming entirely purple towards the tip, coarsely ciliate near the base.

The lip is deeply saccate, obtusely pointed, 5.5 to 6.5 cm high and about 2 to 2.5 cm wide, yellowish-green flushed with purple, the rear margins inturned.

The staminode is oblong with reflexed and densely pilose margins.

Paphiopedilum sanderianum is most probably endemic to the isle of Borneo. The plants grow at altitude of about 200 meters above sea-level in rock crevices on steep, southeast facing limestone cliffs in places which receive scattered sunlight early in the day but otherwise remain in shade.

According to the climate prevailing in the natural habitat, this species should be grown in warm, humid and shady conditions.

Paphiopedilum sanderianum generally flowers between July and December but flowering plants have also been reported between February and April.

[1] A plant with petals of a length of 1.2 meters was reported to me recently.

Paphiopedilum stonei (Hooker) Stein

Stein's Orchideenbuch: 487 (1892)

Basionym: *Cypripedium stonei* Hooker
Curtis's Botanical Magazine 88: t. 5349 (1862)

Synonym: *Cordula stonei* (Hooker) Rolfe
Orchid Review 20 (1): 2 (1912)

Paphiopedilum stonei (Hooker) Stein

Pflanzen dieser Art wurden erstmals durch Hugh LOW gesammelt und 1860 nach England gebracht, in der Annahme, daß es sich um *Paphiopedilum lowii* handle. Die Pflanzen wurden durch John DAY erstanden, dessen Gärtner Robert STONE sie im darauffolgenden Jahr zur Blüte brachte. Auf Bitten der Firma LOW beschrieb William Jackson HOOKER[1)] die Art als *Cypripedium stonei (loc. cit.)*. HOOKER stellte fest, daß er noch nichts Vergleichbares aus „der alten Welt" gesehen hatte und mutmaßte eine enge Verwandtschaft zum amerikanischen *Phragmipedium* (damals *Cypripedium*) *caudatum* Lindley.

The first plants of this species were collected and brought to England by Hugh LOW assuming that they were specimens of *Paphiopedilum lowii*. The plants were purchased by John DAY whose gardener Robert STONE, managed to bring some into flower the following year. Following a request by LOW, William Jackson HOOKER[1)] described the species as *Cypripedium stonei*. HOOKER noted: "Nothing like this, as far as I know, has ever been received from the Old World; but it is evidently allied to a species gathered by RUÍZ and PAVÓN, in Peru, *Cypripedium caudatum* of Dr. LINDLEY, . . ."

[1)] Nicht dessen Sohn Joseph Dalton HOOKER, wie es des öfteren, vor allem in amerikanischen Arbeiten, behauptet wird.

[1)] And not his son Joseph Dalton HOOKER, as often claimed in American publications.

Paphiopedilum stonei ist ein bis 60 cm hoher, krautiger Humusepiphyt.

Die Laubblätter werden 30 bis 40 cm lang und bis 4 cm breit. Sie sind riemenförmig, ledrig, einfarbig grasgrün.

Die bis zu 60 cm hohe Infloreszenz ist grünlich purpurn, schwach flaumig behaart, 3- bis 5blütig.

Die Braktee ist lanzettförmig, langspitzig. Sie erreicht ungefähr ein Drittel der Länge des Fruchtknotens.

Die Blüten werden bis 10 cm hoch.

Die Fahne ist etwa 6 cm lang, 4,2 cm breit, ei- bis herzförmig, langspitzig, weiß mit 2 bis 3, mehr oder weniger intensiven, purpurnen Längsstreifen, hinten gekielt.

Das Synsepalum ist ähnlich, aber etwas kleiner.

Die linealischen Petalen werden 12 bis 15 cm lang und ungefähr 0,5 cm breit. Sie sind herabhängend, mehr oder weniger gedreht, an der Basis schwach schwarz behaart, gelblich, nach unten zu mit länglichen braunen Flecken überzogen, im Enddrittel karminrot braun.

Paphiopedilum stonei is an herbaceous humus epiphyte attaining an overall height of about 60 cm.

The strap-shaped leaves are about 30 to 40 cm long and up to 4 cm broad, very leathery, uniformly grass-green.

The inflorescence is up to about 60 cm high, greenish-purple, slightly pubescent, bearing 3 to 5 flowers.

The floral bracts are lanceolate, acuminate, covering about a third of the ovary.

The flowers are up to about 10 cm high.

The dorsal sepal is ovate to cordate, acuminate, white, usually with 2–3 or more blackish crimson longitudinal streaks, keeled behind, about 6 cm long and about 4.2 cm wide.

The synsepalum is similar but somewhat smaller.

The petals are linear, 12 to 15 cm long and about 0.5 cm broad, pendent, more or less twisted, with a few black ciliate hairs on each margin towards the base; yellowish, towards the apex suffused with elongated brown spots; the apical third wholly brownish crimson.

Paphiopedilum stonei (Hooker) Stein

Die Lippe ist pantoffelförmig, bis 6 cm lang und ca. 2,5 cm breit, matt rosenrot bis braunrot, mit dunklerer Beaderung, unten und auf den Seitenlappen weißlich.

Das Staminodium ist länglich oval, gelblich weiß, gefranst und mit Ausnahme der vorderen Platte von dichtstehenden, borstigen Härchen überzogen.

Paphiopedilum stonei ist in Westborneo beheimatet. Die Pflanzen wachsen auf Kalkgestein bis zu einer Höhe von 1000 m ü. NN, meist epiphytisch in Humusansammlungen.

Das Klima in diesem Areal ist gekennzeichnet durch eine warme Regenperiode von November bis Januar. Hiernach folgt eine etwas trockenere Zeit, die bis Ende April andauert, mit Temperaturen bis 30 °C am Tage und um 18 °C in der Nacht. Ab Mai gibt es wieder etwas mehr Niederschlag, und die Temperaturen gehen zurück. Juni und Juli sind die kältesten Monate mit 25 °C am Tage und 15 °C in der Nacht. Von August bis Oktober gibt es häufig Gewitter. Diese Zeit geht dann wieder in die ausgeprägte Regenperiode über.

Die Heimatregion von *Paphiopedilum stonei* hat ein ständig feuchtes Klima mit nur kurzen Trockenperioden. Die relative Luftfeuchtigkeit ist demgemäß immer hoch.

Die Blüten kommen meistens im Mai bis Juli zur Entfaltung. Die Pflanzen blühen aus 2- bis 3jährigen Trieben.

Importpflanzen von *Paphiopedilum stonei* brauchen lange Zeit, um sich zu etablieren. Zu diesem Zweck sollte man sie eintopfen und unter einem Vlies in gespannter Atmosphäre halten. Die Pflanzen sollten in dieser Zeit kaum gegossen werden, nur das Vlies ist feucht zu halten.

Etablierte Pflanzen sind relativ problemlos zu kultivieren und blühwillig.

Paphiopedilum stonei wünscht eine gute Luftversorgung der Wurzeln und konstant hohe Luftfeuchtigkeit. Man sollte daher ein gut durchlässiges, aber wasserspeicherndes Substrat verwenden. Die Pflanzen brauchen eine hohe Lichtintensität. Blühende Pflanzen werden kühler gehalten.

Mit dem Teilen von *Paphiopedilum stonei* sollte man eher zurückhaltend sein.

The calceiform lip is projecting, up to about 6 cm long and more or less 2.5 cm wide, dull rose coloured, veined and reticulated with crimson, whitish beneath; the infolded lobes narrow, whitish.

The staminode is oval-oblong, yellowish white, fringed, densely couvered with bristle hairs except for the glabrous front.

The distribution of *Paphiopedilum stonei* is restricted to West Borneo. The plants occur on limestone hills at elevations up to about 1000 m above sea-level, generally growing epiphytically in rock cracks and crevices filled with vegetable soil.

The climate of the habitats of *Paphiopedilum stonei* is characterized by a warm rainy season from November to January followed by a somewhat dryer period lasting until the end of April. During the dryer period, the temperatures are between 30 °C during the day and 18 °C at night. From May on the precipitation intensifies again and it becomes somewhat cooler. June and July are the coolest months with temperatures of 25 °C during the day and about 15 °C during the night. Between August and October there are frequent thunderstorms.

The climate in the regions were *Paphiopedilum stonei* has its natural distribution is constantly wet, with only very short dryer periods. The relative humidity is constantly high.

The flowers generally open from May to July. The plants flower from mature, two to three year old growths.

Plants of *Paphiopedilum stonei*, imported from the wild, need a long time to be established. For that purpose they should be potted and kept in a tense, extremely humid atmosphere. The plants should not be watered during this period.

Well established plant are relatively easy to grow and flower readily.

Paphiopedilum stonei does well when the plants are kept under constant high humidity in an open but moisture retaining substrate that allows for air exchange in the root area. The plants need plenty of light. Flowering plants should be kept somewhat cooler.

One should be cautious when dividing the plants as new growths can be without roots for one to two years.

Varietäten:

Von *Paphiopedilum stonei* sind 11 „Varietäten"[1] beschrieben worden. Eine, *Cypripedium stonei* var. *acrosepalum,* war wahrscheinlich eine Mutante und unterscheidet sich von der Nominalform dadurch, daß die lateralen Sepalen nicht zu einem Synsepalum verwachsen sind. Alle anderen sind mehr oder weniger deutliche Blütenform- bzw. Blütenfarbvarianten, die den Status der Varietät kaum beanspruchen können.

Die durch LINDEN als *Cypripedium stonei* var. *album* beschriebene Pflanze[2] hat mit einer *alba-* oder einer Albinoform überhaupt nichts zu tun. Die Blüte zeigt lediglich eine „weißliche" Grundfärbung.

1863 importierte die Firma LOW eine größere Menge *Paphiopedilum stonei* aus Sarawak. Wiederum erwarb John DAY einige Pflanzen. Eine davon blühte im Juni 1867 und wies für die Art ziemlich breite und „schön befleckte" Petalen auf. Dieser Umstand veranlaßte REICHENBACH dazu, eine neue Varietät unter dem Namen *Cypripedium stonei* var. *platytaenium* zu veröffentlichen.[3] REICHENBACH stellte sogar die These auf, daß es sich bei dieser neuen „Varietät" um eine Naturhybride zwischen *Paphiopedilum stonei* und *Paphiopedilum lowii* handle. FOWLIE postulierte,[4] daß es sich hier um eine Naturhybride zwischen *Paphiopedilum stonei* und *Paphiopedilum hookerae* handle.

Die Pflanzen dieser „Varietät" sind aber eindeutig reine *Paphiopedilum stonei.* Die Breite der Petalen sowie der mehr oder weniger intensive Fleckenüberzug sind innerhalb der normalen Bandbreite einer Art anzusiedeln.

[1] Meistens als *Cypripedium stonei* var.
[2] **Lindenia 4:** 100 (1888).
[3] **Gardeners' Chronicle 27:** 1118 (1867).
[4] **Orchid Digest 40** (1): 15–17 (1976).

Varieties:

11 Varieties of *Paphiopedilum stonei*[1] have been described. One of them, *Cypripedium stonei* var. *acrosepalum* was most probably a mutant differing from the nominal form by free lateral sepals not united to a synsepalum. All other "varieties" differ only by minimal changes in flower colour or flower shape and do not deserve to be treated as separate taxa.

The plant described as *Cypripedium stonei* var. *album* by LINDEN[2] has nothing to do with a *alba-* or *albino*-form but differs from all other clones simply by a "whitish ground colour".

In 1863, a large number of plants of *Paphiopedilum stonei* were imported from SARAWAK. Again, John DAY purchased some of the plants. One of them flowered in June 1867 and showed, in comparison to other specimens, rather broad and "nicely spotted" petals. These two characteristics caused REICHENBACH to publish a new variety named *Cypripedium stonei* var. *platytaenium.*[3] REICHENBACH even suggested that this "variety" represents a natural hybrid between *Paphiopedilum stonei* and *Paphiopedilum lowii.* FOWLIE[4] postulated the variety to be a natural hybrid between *Paphiopedilum stonei* and *Paphiopedilum hookerae.*

Plants of this variety are, however, to be considered as pure *Paphiopedilum stonei.* The wideness of the petals as well as the more or less intensive spotting are within the natural limits of a species.

[1] Generally as *Cypripedium stonei* var.
[2] **Lindenia 4:** 100 (1888).
[3] **Gardeners' Chronicle 27:** 1118 (1867).
[4] **Orchid Digest, 40** (1): 15–17 (1976).

Paphiopedilum supardii Braem & Löb

Die Orchidee, 36 (4): 142–143 (1985)[1)]

Paphiopedilum supardii Braem & Löb

Während einer Forschungsreise nach Java und Sumatra, im Frühjahr 1985, erhielt ich anläßlich eines Gespräches mit Vertretern der indonesischen Orchideengesellschaft Pflanzen einer angeblich neuen *Paphiopedilum*-Art, mit der Bitte, diese zu Ehren des Entdeckers zu beschreiben.

Nach meiner Rückkehr wurden die Pflanzen durch LÖB und mich eingehend untersucht. Wir kamen zu der Auffassung, daß wir es mit einer neuen *Paphiopedilum*-Art der Sektion **Mastigopetalum** Hallier zu tun hatten, und beschrieben die Pflanze dementsprechend und entsprechend den Wünschen meiner Auftraggeber.

Nach unserer Beschreibung ist festgestellt worden, daß unser Konzept mit dem des *Paphio-*

[1)] Veröffentlicht als Separat im voraus am 8. Juni 1985.

Synonym: *Paphiopedilum devogeli* Schoser & Van Deelder, *nomen nudum*

During a research trip to Java and Sumatra in the spring of 1985, representatives of the Indonesian Orchid Society gave me some plant material of a supposedly new *Paphiopedilum*-species with the request to describe the plants in honour of their discoverer.

Upon my return home, the plants were examined carefully by me and my co-author Udo LÖB. We reached the conclusion that the plants represented a new, not yet described, species belonging to the section **Mastigopetalum** Hallier, and published the concept in honour of Mr. SUPARDI who collected the plants in the wild.

After our description, it has become evident, that our concept is identical with that of *Paphiopedilum devogeli* from SCHOSER and VAN DEELDER. However, the latter concept has never been published and is only known from a citation by Dr. SCHOSER at the World Orchid Congress in Bangkok back in 1978.

Paphiopedilum supardii is an herbaceous, epiphytic plant. It has leaves that are up to 50 cm long and about 5.5 cm wide, uniformly coloured dark green, fairly leathery, tongue-shaped.

The multiflowered inflorescence is generally 35 to 40 cm high.

The floral bracts are up to about 5.2 cm long and sowith as long as, or longer than the ovary.

The dorsal sepal is about 5.5 cm high and 2.4 to 2.6 cm wide, ovate, acute, concave, keeled, white, striped distinctly dark-brown, with pale green margins.

The synsepalum is similar but narrower.

The petals are up to about 8–9 cm long and 7 to 9 mm broad, pale green, spotted dark brown and

[1)] Published as an advance excerpt on 8. June 1985.

pedilum devogeli Schoser & Van Deelder identisch ist. Dieses Konzept ist aber niemals veröffentlicht worden und nur aus einer Erwähnung SCHOSERs anläßlich eines Vortrages vor der Weltorchideenkonferenz 1978 in Bangkok bekannt.

Paphiopedilum supardii hat bis ca. 50 cm lange und bis ungefähr 5,5 cm breite, einfarbig dunkelgrüne, ledrig steife, zungenförmige Laubblätter.

Die mehrblütige Infloreszenz ist im allgemeinen 35 bis 40 cm hoch.

Die Brakteen sind mit ca. 5,2 cm Länge ebenso lang oder nur unwesentlich länger als der Fruchtknoten.

Die Fahne ist ungefähr 5,5 cm hoch, bei einer Breite von 2,4 bis 2,6 cm. Sie ist eiförmig, spitz, konkav, mehr oder weniger gekielt, in der Grundfärbung blaßgrün, deutlich dunkelbraun gestreift und mit blaßgrünen Rändern.

Das Synsepalum ist ähnlich, meistens ebensolang, aber schmaler.

Die Petalen sind ungefähr 8 bis 9 cm lang und 7 bis 9 mm breit. Sie sind blaßgrün, mit dunkelbraunen Flecken und Warzen überzogen, bandförmig und nach unten gebogen, im apikalen Endbereich gedreht, spitz zulaufend.

Die deutlich sackförmige Lippe ist ungefähr 4,5 cm lang und 1,5 bis 1,8 cm breit. Sie ist blaß bis dunkelbraun und von dunkelbraunen Adern überzogen. Die nach innen gefalteten Seitenlappen sind weiß berandet.

Das knieförmige Staminodium erinnert an das von *Paphiopedilum rothschildianum*. Es hat eine gelbe Platte, die ungefähr 8 mm breit und 8 bis 9 mm hoch ist und dunkelbraun behaarte Ränder aufweist.

Paphiopedilum supardii wurde entlang des Kapuas-Flusses in einer Höhenlage von 600 bis 900 m ü. NN als Humusepiphyt in Felsspalten an steilen Hängen gefunden. Die Pflanzen wachsen an halbschattigen Stellen die nur morgens unter direkter Sonneneinstrahlung stehen.

Diese Art gedeiht am besten in einem durchlässigen, gut drainierten Pflanzsubstrat im temperierten oder im warmen Gewächshaus.

Paphiopedilum supardii braucht ständig hohe Luftfeuchtigkeit und reichliche Wassergaben.

Die Blütezeit ist von Mai bis Juli. Gut kultivierte Pflanzen bringen bis zu 12 Blüten hervor.

with wart-like structures, ribbon-shaped and drooping, twisted in the apical part, acuminate.

The lip is clearly saccate, about 4.5 cm long and 1.5 to 1.8 cm wide, pale to dark brown, reticulated darker brown. The inwardly folded lateral lobes with white margins.

The knee-shaped staminode is somewhat like that of *Paphiopedilum rothschildianum*, with a yellow shield, about 8 mm wide and 8 to 9 mm high, the margins of the platform covered with dark brown hairs.

Paphiopedilum supardii was found along the Kapuas-river growing as an epiphyte in leaf-mould in rock crevices at an altitude between 600 and 900 m above sea-level. The plants grow in half-shaded places that are exposed to direct sunlight only during the morning hours.

This species grows best in an open, well-drained compost in the intermediate or warm greenhouse.

Paphiopedilum supardii needs constant high humidity and should be watered abundantly.

The plants generally flower from May to June. Well established plants bear up to 12 flowers.

Paphiopedilum supardii Braem & Löb

Sektion POLYANTHA

Diese nur aus zwei Arten bestehende Sektion unterscheidet sich von den verwandten Sektionen durch die Ausprägung von seitlichen Aurikeln an der Lippe und durch schmale, löffelförmige, laterale Petalen, deren basaler bis Mittelteil unregelmäßig purpurn gefleckt ist.

Section POLYANTHA

This section, consisting of two species, is well distinguished from related sections by the presence of lateral ears on the main lobe of the lip, as well as by the narrow, spoon-shaped petals of which the basal to middle portion is roughly blotched with deep purple spots.

Schlüssel zu den Arten der Sektion POLYANTHA

1. Unterer Teil des dorsalen Sepalums (Fahne) mit großen braunen Flecken, Staminodiumplatte oval, unten tief eingeschnitten . . . . *Paphiopedilum haynaldianum*

1a. Dorsales Sepalum (Fahne) ohne braune Flekken, fast gänzlich einfarbig, Staminodiumplatte umgekehrt herzförmig . . . . *Paphiopedilum lowii*

Key to the species belonging to section POLYANTHA

1. Basal portion of the dorsal sepal with large brown blotches; staminodal shield ovate, deeply notched at the apex . . . . *Paphiopedilum haynaldianum*

1a. Dorsal (upper) sepal without brownish blotches, almost uniformly coloured throughout; staminodal shield inverted cordate . . . . *Paphiopedilum lowii*

Paphiopedilum haynaldianum (Rchb. f.) Stein

Stein's Orchideenbuch: 470 (1892)

Basionym: *Cypripedium haynaldianum* Rchb. f.
Xenia Orchidacea 2: 222 (1874)

Synonym: *Cordula haynaldiana* (Rchb. f.) Rolfe
Orchid Review 20 (1): 2 (1912)

Paphiopedilum haynaldianum (Rchb. f.) Stein

Diese Art wurde durch Gustav WALLIS im Jahre 1873 in San Isidro bei Manila auf der philippinischen Hauptinsel Luzon entdeckt und durch die Firma VEITCH nach England importiert.

REICHENBACH erhielt Pflanzenmaterial von VEITCH und beschrieb die Art im zweiten Band seiner **Xenia Orchidacea** zu Ehren des Kardinals HAYNALD, Erzbischof von Kaloesa (Ungarn).

Paphiopedilum haynaldianum ist sehr eng mit *Paphiopedilum lowii* verwandt, hat aber ein längeres, abweichend geformtes und abweichend gefärbtes Staminodium, breitere Sepalen, die deutlich gefleckt sind, andersgefärbte basale Teile der Petalen; größere, nach innen gefaltete Lippenseitenlappen und ledrigere Laubblätter. Auch die Stellung der Lippe in bezug auf die Blütenachse ist verschieden. Während die Lippe bei *Paphiopedilum lowii* nach vorne zeigt, ist sie bei *Paphiopedilum haynaldianum* eindeutig nach unten gerichtet.

Paphiopedilum haynaldianum wächst im allgemeinen als Humusepiphyt, oft auch in den Astgabeln größerer Bäume.

This species was discovered in 1873 by Gustav WALLIS in San Isidro near Manila on the main Philippine Island of Luzon and introduced to Europe by VEITCH.

Plant material was sent to REICHENBACH, who described the species in the second volume of his **Xenia Orchidacea** in honour of Cardinal HAYNALD, the Archbishop of Kaloesa (Hungary).

Paphiopedilum haynaldianum is very closely related to *Paphiopedilum lowii*, but differs in a more elongated, differently-shaped staminodal shield; broader, spotted sepals; the different colouration of the basal parts of the petals; larger, infolded lateral lobes of the lip and more leathery leaves.

There also is a difference in position of the lip to the axis of the inflorescence. Whereas the lip of *Paphiopedilum lowii* is clearly pointed forwards, the pouch of *Paphiopedilum haynaldianum* is distinctly directed downwards.

Paphiopedilum haynaldianum generally grows as an epiphyte in leaf-mould on the ground, or in crotches of larger trees.

Es hat bis zu 30 cm lange und ca. 4,5 cm breite, riemenförmige bis schmal elliptische, matt dunkelgrüne, mehr oder weniger aufrecht stehende, stumpf zweilappig endende Laubblätter. Die Ränder sind scharf, hart und rauh. Sie werden von der Basis bis zur Spitze hin immer deutlicher gesägt. Die Unterseite ist gekielt.

Die bis zu 75 cm hohe, braune Infloreszenz ist lang behaart, 3- bis 12blütig.

Die Brakteen sind halb so lang wie der 4 bis 4,5 cm lange Fruchtknoten und ebenso wie dieser flaumig behaart.

Die Blüten werden bis 16 cm breit.

Die Fahne ist ungefähr 5,5 cm lang und 4 cm breit. Sie ist eiförmig, spitz, an den Rändern in den unteren zwei Dritteln zurückgebogen, gelbgrün mit großen braunen Flecken entlang der Beaderung.

Das Synsepalum ist breit eiförmig, weißlich, mit grünen Adern, doppelt gekielt.

Die bis 9 cm langen und ca. 2 cm breiten linealisch spatelförmigen Petalen stehen waagerecht ab und neigen sich am apikalen Ende etwas nach unten. Sie sind mehr oder weniger gedreht, gelbgrün, an

Paphiopedilum haynaldianum (Rchb. f.) Stein

This herbaceous plant has up to 30 cm long and about 4.5 cm broad leaves. These are strap-shaped to narrow-elliptic, dull green, more or less erect, at the apex blunt bilobate, edges sharp, hard and rough, exhibiting an increasing serration towards the apex. The under side is sharply keeled.

The inflorescence is up to 75 cm high, brown, covered with long hairs; bearing 2 to 12 flowers with a natural spread of up to about 16 cm.

The floral bracts are half as long as the ovary which is 4 to 4.5 cm long. The bracts and the ovaries are hirsute.

The dorsal sepal is elongate oval, about 5.5 cm long and more or less 4 cm wide, acuminate, marginally recurved for about two-thirds of its height, yellowish-green with large brown spots along the veins.

The synsepalum is broadly oval, whitish, with green longitudinal veins, double keeled.

The petals are broadly linear-, lanceolate, up to about 9 cm long and more or less 2 cm wide, spreading sideways in a downward curve, the distal part sweeping up again, the lower margin twisted forward; yellowish-green, marked with up to 12 large dark brown bars, spots or dots in the basal half; the distal half broadened, purple.

The lip is invertedly helm-shaped, up to about 5 cm long and 2.7 to 3.9 cm wide, with a large mouth; greenish, suffused purple-brown. The sidelobes are infolded, glossy ivory white. The lip is pointed downwards.

The staminodal shield is elongate oval with an hairy basal tooth, the apical part deeply notched, green, marbled with darker green around the notch.

Paphiopedilum haynaldianum has its native habitat on the island Luzon in the Northern Phlippines. The plants occur from about sea-level to mountain areas of over 1000 meters elevation. They usually grow on moss-covered limestone hills and in rock cracks filled with moss and leafy debris.

In this native habitat, a cool, northeasterly winter monsoon blows during the months of December and January carrying with it heavy rain. In Februáry, it becomes somewhat drier and in March and April, the days are summerly clear.

April is the hottest month. Convectional rains occur during May and June. Beginning at the end of June and through July and August the southwestern monsoon brings havier rains lasting until the end of September. In October and November again, convectional rains bring showers and the temperatures drop.

der Basis mit bis zu 12 großen, dunkelbraunen Flecken, im verbreiterten apikalen Endbereich purpurn.

Die verkehrt helmförmige Lippe ist bis 5 cm lang und 2,7 bis 3,9 cm breit. Die Öffnung ist breit. Die Lippe ist grünlich, purpurbraun überlaufen. Die nach innen gebogenen Seitenlappen sind elfenbeinweiß bis gelblich braun. Das Labellum ist nach unten gerichtet.

Das Staminodium ist lang eiförmig, mit einem behaarten basalen Zahn, unten eingebuchtet, grün, dunkelgrün marmoriert an der unteren Einbuchtung.

Paphiopedilum haynaldianum ist auf der Insel Luzon (Philippinen) beheimatet und wurde von fast Meereshöhe bis in die Gebirgsregionen auf einer Höhe von über 1000 m ü. NN auf moosbewachsenen Kalksteinhügeln sowie in Moospolstern oder Humusansammlungen auf Felsen gefunden.

In diesem Gebiet werden die Monate Dezember und Januar vom nordöstlichen Wintermonsun geprägt, der sehr viel Regen mit sich führt. Ab Februar wird es etwas trockener, und im März und April gibt es meistens klare „Sommertage".

April ist der wärmste Monat. Im Mai und Juni gibt es wenig Regen; im Juli kommt eine deutliche Regenperiode, die bis September anhält, und in eine etwas kühlere Periode von Oktober bis November übergeht.

Die Tagestemperaturen schwanken im „Winter" um 20 °C, im Sommer um 30 °C. Die Nachttemperaturen erreichen in den kältesten Monaten 6 °C, im warmen April schwanken sie um 20 °C.

Die Blütezeit ist von Januar bis März.

Entsprechend ihrer Herkunft mögen die Pflanzen dieser Art warme Sommer und kühle Winter. Man sollte für hohe Luftfeuchtigkeit sorgen und während der Wachstumsperiode ausgiebig gießen. Schattig gehaltene Pflanzen bringen heller gefärbte Blüten hervor.

Die Pflanze blüht am besten, wenn sie im November und Dezember eine kühle Ruheperiode durchlebt.

Paphiopedilum haynaldianum var. *album*

The daily temperatures are from about 20 °C in "Winter" to about 30 °C in the "summer months". The nightly temperatues drop to 6 °C during the coldest months; in the warm month of April they are around 20 °C.

Paphiopedilum haynaldianum generally flowers from January through March.

In accordance to the climate in their natural habitat, the plants of *Paphiopedilum haynaldianum* like warm summers and cool winters. They should be grown under constant high humidity and watered abundantly during the growth period. Plants grown under high light intensity have more darkly coloured flowers as shaded ones.

Flowering is best induced by a cool resting period during November and December.

Varietäten:

Paphiopedilum haynaldianum var. *album* Hort. ex Asher
Orchid Digest 44 (6): 214 (1980)

Die Infloreszenz dieser Varietät ist hellgrün, die Blüten sind hellgrün, mit Ausnahme der weißen Petalenspitzen.

Varieties:

Paphiopedilum haynaldianum var. *album* Hort. ex Asher
Orchid Digest 44 (6): 214 (1980)

Inflorescence brightly green, flowers bright unicoloured green with the exception of the white apices of the lateral petals.

Paphiopedilum lowii (Lindley) Stein

Stein's Orchideenbuch: 476 (1892)

Basionym: *Cypripedium lowii* Lindley
Gardeners' Chronicle 7: 765 (1847)

Synonyma: *Cordula lowiana* (Lindley) Rolfe
Orchid Review 20 (1): 2 (1912)

Cypripedium cruciforme Zoll. & Mir. in DE VRIESE und PAHUD.: **Illustr. d'Orchid.,** t. 4 (1854)

Paphiopedilum lowii (Lindley) Stein

Die ersten in Europa bekannten Pflanzen dieser Art wurden durch Hugh LOW als Humusepiphyt „hoch oben in Astgabeln großer Bäume im dichten Urwald" gefunden, und zwar in Sarawak, im Norden der Insel Borneo. Die gesammelten Pflanzen erreichten England im Jahre 1846. Die erste Pflanze blühte in der Sammlung KENRICK und veranlaßte LINDLEY, die Art zu Ehren des Entdeckers zu beschreiben.

Paphiopedilum lowii ist eine meist als Humusepiphyt wachsende, krautige Pflanze mit schräg aufrechtstehenden, riemenförmigen, stumpfen, etwas ausgerandeten, am Grunde zusammengefalteten, oben gefurchten, unten gekielten Laubblättern. Diese sind bis ca. 40 cm lang und bis 5,5 cm breit, einfarbig grün und von einer ledrigen Struktur. Das erste und das letzte Blatt eines Triebes sind immer deutlich kleiner als die übrigen. Die Blattspitze ist

The first plants of this species were introduced to Europe in 1848 by Hugh LOW, who found them as epiphytes in leaf-mould "growing high up in the branches and forks of tall trees in thick jungle" in Sarawak (northern Borneo). The first specimen flowered in the KENRICK collection and caused LINDLEY to describe the species in honour of its discoverer.

Paphiopedilum lowii is an herbaceous, epiphytic plant, usually growing in leaf-mould. The leaves are suberect, strap-shaped, obtuse, somewhat serrated at the margins, folded at the base, the upper surface somewhat grooved, the under distinctly keeled; up to about 40 cm long and up to 5.5 cm broad, uniformly green, leathery. The first and last leaf of a growth are markedly smaller than the other ones. The leaf tip is unequally bilobed; with a blunt apical, variably shaped tooth.

ungleich zweilappig mit einem stumpfen, unterschiedlich geformten Zahn.

Die Infloreszenz ist aufrecht, bis 1 m hoch, hellgrün bis bräunlich purpurn, dicht behaart, locker vielblütig.

Die Brakteen bedecken ein Drittel bis die Hälfte des 6,5 bis 8 cm langen, gelbgrünen bis purpurnen, deutlich behaarten Fruchtknotens.

Die Blüten werden bis 14 cm breit und ungefähr 7 cm hoch.

Die Fahne ist etwa 4 cm hoch und 3,5 cm breit. Sie ist breit oval, spitz zulaufend. Die Ränder der unteren Hälfte sind nach hinten geschlagen, der obere Teil ist konkav und nach vorne gebogen. Die Grundfarbe ist gelb bis grüngelb, von der Basis aus nach oben hin hell- bis dunkelbraun überzogen, wobei braune Streifen bis in den oberen Teil durchlaufen.

Das Synsepalum ist so lang wie die Fahne, aber etwas schmaler; von der Lippe weggebogen, manchmal doppelspitzig, manchmal auch deutlich geteilt (so daß die lateralen Sepalen nicht verwachsen sind).

Die Petalen sind 6 bis 7,5 cm lang und bis 1,5 cm breit, spatelförmig, waagerecht abstehend, in der vorderen Hälfte nach unten gebogen, einmal gedreht. Im basalen Bereich grüngelb mit großen (teils augenförmigen) braunen Flecken, zur Spitze hin purpurn überlaufen.

Die Lippe steht im rechten Winkel zum Ovar. Sie ist deutlich pantoffelförmig, bis 4,5 cm lang und an der Öffnung ungefähr 3 cm breit, grünlich braun bis dunkelbraun, normalerweise mit einer deutlich sichtbaren dunkleren Beaderung. Die nach innen gefalteten Seitenlappen sind heller gefärbt und braun getüpfelt.

Die Staminodiumplatte ist gewölbt, umgekehrt herzförmig, von purpurvioletten Haaren berandet (in der Bucht mit einem stumpfen Zahn), cremefarben bis gelb, von unten bis zur Mitte braun überlaufen.

Paphiopedilum lowii ist inzwischen auf allen größeren Sunda-Inseln[1)] sowie in Malaysia gefunden worden. Die Pflanzen wachsen bis auf 1500 m ü. NN als Humusepiphyt in den Astgabeln größerer Bäume oder in Humusansammlungen auf moosbewachsenen Kalkfelsen. Die Pflanzen sind meist als Einzelpflanzen anzutreffen, mit einem deutlichen Abstand zum nächsten Exemplar. Kolonien sind selten.

[1)] Borneo (Sabah, Sarawak, W.-Borneo), Sumatra, Java, Celebes.

The flower scape is up to about 1 m high, erect, bright green to brownish-purple, densely hirsute, carrying up to six or more flowers.

The floral bracts cover one third to one half of the 6.5 to 8 cm long, yellowish green to purple, distinctly hirsute ovary.

The flowers have a natural spread of up to 14 cm and they are up to about 7 cm high.

The dorsal sepal is about 4 cm long and 3.5 cm broad, broadly ovate, shortly pointed; the edges of the basal half reflexed, the upper part concave and curved forward. The yellow to chartreuse green colour is suffused from the base upward by a more or less intense brown; radiating streaks of brown extending into the upper dorsal half.

The synsepalum is about as long as the dorsal sepal but somewhat narrower, concave, standing away from the lip, sometimes with a double apex, sometimes distinctly divided into two separate lateral sepals.

The petals are 6 to 7.5 cm long and up to about 1.5 cm wide, spathulate, extending sideways with a slight downward curve, upwards twisted in the distal half. The basal part yellowish-green, with brown (sometimes eye-shaped) spots; the broader end coloured violet to rose.

The lip, standing at a 90 degree angle to the ovary is distinctly saccate, up to about 4.5 cm long and about 3 cm broad at the mouth, greenish-brown to dark brown, usually with a clearly visible darker veining. The infolded sidelobes are lighter in colour and spotted brown.

The staminodal shield is inverted heart-shaped, convex, bordered with purple hairs; with a small erect hairy horn at the base, and a blunt tooth in the sinus of the apical edge; creme-coloured to yellow, from the lower areas to the center suffused brown.

Paphiopedilum lowii has been found on all of the larger Sunda Islands[1)] and on the Malay Peninsula. The plants generally grow epiphytically in forks of trees or in mould or leafy debris in the crevices of limestone rock in altitudes of up to 1500 meter above sea-level. Usually single plants are found far apart, groups are rare.

As noted above, this species is distributed over a vast area, making it very difficult to give generally valid information about the climate at each of its natural habitats. The plants generally prefer moist

[1)] Borneo (Sabah, Sarawak, W. Borneo), Sumatra, Java, Celebes.

Paphiopedilum lowii hat, wie oben angeführt, ein ausgedehntes Heimatareal. Klimatische Angaben sind daher sehr schwierig. Man kann im allgemeinen sagen, daß die Pflanzen sehr feuchte Standorte bevorzugen, vor allem dort, wo ein deutlicher Unterschied zwischen Regen- und Trockenzeit zu beobachten ist.

Die Tagestemperaturen in den Verbreitungsgebieten von *Paphiopedilum lowii* liegen zwischen 29 °C in der Regenzeit und 18 °C im tropischen „Winter". Die Nachttemperaturen variieren jahreszeitlich zwischen 18 und 12 °C.

Die Luftfeuchtigkeit ist über das ganze Jahr hinweg sehr hoch. Der jährliche Niederschlag beträgt zwischen 1900 und 2500 mm.

Paphiopedilum lowii blüht im allgemeinen von April bis Juni.

Die Pflanzen sollten in ein durchlässiges Substrat getopft und möglichst bei hoher Luftfeuchtigkeit im temperierten Haus kultiviert werden.

Während der sommerlichen Wachstumsperiode sollte viel Wasser gegeben werden, während des Winters weniger.

Eine kurze Ruheperiode nach Vollendung der Wachstumsperiode hilft bei der Blüteninduzierung.

places in areas where there is a distinct difference between the dry and rainy seasons.

The daily temperatures in the natural habitat are between 29 °C during the rainy season and 18 °C in the dry "winterly" parts of the year. At night, the temperatures are between 18 and 12 °C depending on the season.

The relative humidity remains very high throughout the entire year. Rainfall averages between 1900 and 2500 mm per year.

Paphiopedilum lowii usually flowers between April and June.

This species does well when potted in an open, well-drained substrate and grown under conditions of high humidity in the intermediate greenhouse.

During the summer months, abundant water should be given. During the winter the plants should be kept somewhat drier.

A brief rest after completion of the new growth will enhance flower induction.

Section MYSTROPETALUM
Pfitzer

Diese nur aus *Paphiopedilum parishii* bestehende Sektion ist sehr eng mit der Sektion **Polyantha** verwandt, was sich vor allem aus dem gesamten Blütenaufbau und aus der Karyologie ableiten läßt. Sie unterscheidet sich von letzterer Sektion vor allem durch die linealischen, stark gedrehten seitlichen Petalen.

Section MYSTROPETALUM
Pfitzer

This section, accomodating only *Paphiopedilum parishii* is very closely related to section **Polyantha,** as can easily be deduced from the similarities in the leaf morphology and from the karyological data. The difference to the latter section consists mainly in the leariform, strongly twisted lateral petals.

Paphiopedilum parishii (Rchb. f.) Stein

Stein's Orchideenbuch: 479 (1892)

Basionym: *Cypripedium parishii* Rchb. f.
Flora 52: 322 (1869)

Synonyma: *Cordula parishii* (Rchb. f.) Rolfe
Orchid Review 20 (1): 2 (1912)
Selenipedilum parishii E. André
Ill. Hort. 22: 122–123, t. 214 (1875)
Paphiopedilum dianthum Tang & Wang
Bull. Fan. Mem. Inst. Biol. Peiping, Bot. Ser. **10:** 23 (1940)

Paphiopedilum parishii (Rchb. f.) Stein

Paphiopedilum parishii wurde 1859 durch den britischen Missionar C. PARISH im Moulmein District als Epiphyt auf vermodernden Farnwurzeln gefunden. Erst 1868 aber wurden die ersten Pflanzen dieser Art durch die Fa. LOW nach England importiert.

Paphiopedilum parishii ist eine epiphytisch wachsende, krautige Pflanze mit bis zu 40 cm langen und ca. 6,5 cm breiten, länglich riemenförmigen, vorn eingeschnittenen, glänzend hellgrünen Laubblättern.

Die bis zu 60 cm hoch werdende, aufrecht stehende, robuste, hellgrüne, flaumig behaarte Infloreszenz trägt im allgemeinen 4 bis 8 Blüten.

Die Brakteen sind ungefähr 4 cm lang, aufgeblasen eiförmig, spitz. Sie bedecken den behaarten Fruchtknoten bis zu drei Viertel seiner Länge.

The merit for the discovery of this species is due to the Reverend C. PARISH who found the plants growing epiphytically on decaying fibrous roots of the fern *Drynaria quercifolia* in the Moulmein District in Burma back in 1859. The first plants reached England in 1868 in a consignment imported by the LOW company.

Paphiopedilum parishii is an epiphytic herbaceous plant. The leaves are oblong-ligulate, up to 40 cm long and up to 6.5 cm broad, twolobed or bifid at the apex, bright glossy green.

The inflorescence is erect, up to 60 cm high, stout, bright green, downy; generally bearing 4 to 8 flowers.

The bracts are about 4 cm long, ovate, acute, inflated, covering the hairy ovaries to about three-quarter of their length.

Die Blüten werden bis 17,5 cm hoch und bis zu 12 cm breit. Die Fahne ist 3 bis 5 cm lang und bis zu 2,5 cm breit, länglich elliptisch, spitz mit nach vorne gebogener oberer Hälfte. Die unteren Ränder sind etwas nach hinten gekrümmt.

Die Fahne ist hellgelb mit grünen Adern.

Das Synsepalum ist ähnlich, aber kleiner und hinten doppelt gekielt.

Die bis 12,5 cm langen und bis ungefähr 1,2 cm breiten Petalen sind linealisch, deutlich gedreht, an der Basis abstehend, später mehr oder weniger herabhängend. Sie haben gewellte Ränder und sind in der basalen Hälfte grün mit einigen schwarzen Flecken, zu der Spitze hin schwarzpurpurn mit bleichen Rändern.

Die Lippe ist bis zu 4,5 cm lang, umgekehrt helmförmig, mit weiter Öffnung, tiefgrün, purpurbraun überhaucht. Die schmalen Seitenlappen sind glatt, tiefgrün, oft braun purpurn gefleckt.

Die Staminodiumplatte ist länglich, umgekehrt eiförmig, an der oberen Seite mit einem deutlichen Zahn. Es ist unten deutlich eingebuchtet, hellgelb mit grüner Zeichnung.

The flowers are up to 17.5 cm high and up to 12 cm across. The dorsal sepal is 3 to 5 cm long and up to 2.5 cm wide, elliptic-oblong, acute, keeled at the back, the upper half bent forwards, the lateral margins revolute at the base; pale yellow with green veins.

The synsepalum is similar but smaller; double-keeled.

The lateral petals are linear, twisted, up to about 12.5 cm long and more or less 1.2 cm broad, spreading at the base, then quite pendulous; the basal half with undulate margins, green, with a few scattered blackish spots, the distal half blackish purple with a pale margin.

The lip is inverted helmet-shaped, up to 4.5 cm long, with a large mouth; deep green, suffused with brownish-purple. The narrow lateral lobes are smooth, infolded, deep green, often with brown-purple stains.

The staminodal shield is obovate-oblong, with a sinus in the broader front edge and a prominent tooth at the base; pale yellow mottled green.

Paphiopedilum parishii has by now been reported from Burma, Thailand and from the Chinese pro-

Paphiopedilum parishii (Rchb. f.) Stein

Paphiopedilum parishii ist inzwischen aus Birma, Thailand und aus der chinesischen Provinz Yunnan bekannt. Die Pflanzen sind auf Bäumen und auf Baumfarnen in ziemlich diffusem Licht, meist in Höhenlagen um 1400 m ü. NN, anzutreffen. Manchmal wächst *Paphiopedilum parishii* auch auf moosbewachsenen Felsen.

Das Klima in den Heimatarealen ist gekennzeichnet durch eine trockene „Winterperiode" von November bis April oder Mai. Der Dezember ist am kühlsten mit Tagestemperaturen bis 25 °C und Nachttemperaturen, die bis auf 4 °C absinken. Ab März wird es etwas wärmer, und der April ist der wärmste Monat mit durchschnittlichen Tagestemperaturen um 32 °C und Nachttemperaturen um 15 °C. Von Mai bis Oktober bringen die Sommermonsune viel Regen.

Die Luftfeuchtigkeit bleibt das ganze Jahr über sehr hoch und das Gebiet wird niemals ausgesprochen trocken.

Als Klimabeispiel zeigen die nachfolgenden Diagramme[1] die Verhältnisse im thailändisch-birmanischen Grenzgebiet, einem der Hauptverbreitungsareale von *Paphiopedilum parishii*.

vince of Yunnan. The plants grow on trees and tree-ferns in rather diffuse light, usually in elevations of about 1400 meters above sea-level. Sometimes, *Paphiopedilum parishii* grows on moss-covered rocks.

The climate in the distribution area is characterized by a dry "Winter" period from November to April or May. The coolest month is December with daily temperatures of about 25 °C and night time temperature dropping to about 4 °C. From March on it gets somewhat warmer and April is usually the hottest month with temperatures of about 32 °C during the day and about 15 °C at night. Between May and October the monsoon brings plenty of rain.

The relative humidity remains very high the whole year through and this area never becomes really dry.

The following diagram[1] shows the climatic conditions in one of the major habitats of *Paphiopedilum parishii*, in the Thailand-Burmese border region.

Wachstumszyklus von *Paphiopedilum parishii* / Growth Cycle of *Paphiopedilum parishii*

Ruhe Rest	Reproduktives Wachstum Reproductive growth	Blüte Flowering	Vegetatives Wachstum Vegetative growth	Ruhe Rest

Monat / Month

Januar Jan.	Februar Feb.	März Mar.	April Apr.	Mai May	Juni June	Juli July	August Aug.	Sept. Sept.	Oktober Oct.	Nov. Nov.	Dez. Dec.

Niederschlag / Rain Fall

Trockenzeit Dry Season	geringer Niederschlag Light Rain	viel Niederschlag (Regenzeit) Heavy Rain	geringer Niederschlag Light Rain

Max. Temperatur °C

26–30	32–35	26–28	25–26

Min. Temperatur °C

4–8	9–16	9–13	4–8

Wie man den obigen Angaben entnehmen kann, ist die Art relativ einfach zu kultivieren.

Paphiopedilum parishii benötigt sehr wenig Licht und blüht willig, auch nach einer nur kurzen Ruheperiode.

The above diagramme shows that the species is fairly easy to grow.

Paphiopedilum parishii needs very little light and flowers readily, even after only a short period of rest.

[1] Nach VORAURAI, **Orchid Digest 38:** 213 (1974).

[1] After VORAURAI, **Orchid Digest 38:** 213 (1974).

Man sollte die Luftfeuchtigkeit hoch halten, ein durchlässiges Substrat wählen und für eine gute Luftumwälzung und ausreichend Frischluft im Kulturraum sorgen.

Varietäten:

Paphiopedilum parishii var. *dianthum* (Tang & Wang) Karasawa & Saito
Bull. of the Hiroshima Botanical Garden Nr. 5: 1–69 (1982)

Diese Varietät wurde durch TANG und WANG als eigenständige Art beschrieben *(loc. cit.)*.

Pflanzen dieser „Abart" unterscheiden sich vom „typischen" *Paphiopedilum parishii* durch grünlichere, glänzendere Blütenteile, eine ausgesprochen konische Form der Lippe und einen unbehaarten Fruchtknoten.

Diese Pflanzen stellen eindeutig eine gesonderte, geographische Sippe der Art dar, aber sie sind keineswegs als eigenständige Art anzusehen.

The humidity should be kept high and the plants should be potted in a well-drained, open compost. Plenty of fresh air should be circulated throughout the greenhouse.

Varieties:

Paphiopedilum parishii var. *dianthum* (Tang & Wang) Karasawa & Saito
Bull. of the Hiroshima Botanical Garden Nr. 5: 1–69 (1982)

This variety was originally described as an autonomous species by TANG and WANG *(loc. cit.)*.

Plants of this "variety" differ from the nominal types of *Paphiopedilum parishii* by a deeper green, brightly shiny perianth with a distinct conically shaped lip as well as by glabrous ovaries.

It is quite clear that the plants of this taxon represent an autonomous geographic race, they are, however, by no means to be regarded as a separate species.

Untergattung PAPHIOPEDILUM

Diese Untergattung ist gekennzeichnet durch die Kombination von einblütigen Infloreszenzen, einfarbig grünen Laubblättern und gut entwickelten seitlichen Öhrchen (Aurikeln) an der Lippe.

KARASAWA und SAITO[1] vertreten die Meinung, daß die Untergattung **PAPHIOPEDILUM** aus der Untergattung **POLYANTHA** hervorging, was die einblütige Infloreszenz und die Lippenmorphologie (wohl vor allem das Vorhandensein der lateralen Öhrchen) beweisen sollen. Es fällt mir sehr schwer, diesen Gedanken nachzuvollziehen.

Was die geographische Verbreitung anbelangt, so ist diese Gruppe eher „kontinental". In der Tat sind die Arten der Untergattung **PAPHIOPEDILUM** nur auf dem indo-malaysischen Festland vertreten im Gegensatz zu den Arten der Untergattung **POLYANTHA**, die allesamt auf den philippinischen und indonesischen Inseln beheimatet sind.

Innerhalb dieser Untergattung gibt es eine Reihe von phylogenetischen Linien, die hier als Sektionen behandelt werden.

Subgenus PAPHIOPEDILUM

This subgenus is easily recognized and mainly characterized by the combination of uni-flowered racemes, uni-coloured green leaves and well-developed lateral auricles of the upper margin of the lip.

KARASAWA and SAITO[1] consider subgenus **PAPHIOPEDILUM** to have emerged from subgenus **POLYANTHA** as proven by the uni-flowered racemes and the presence of the lateral ears of the lip. I have great difficulty in following that hypothesis.

As far as the geographic distribution is concerned, this group is rather "continental". All of the species of the subgenus **POLYANTHA** are to be found in the Indo-Malayan Continent, India and adjacent regions whereas the species of the subgenus **POLYANTHA** have their natural distribution on the Philippine and Indonesian archipelagoes.

Within this subgenus, there are a series of phylogenetic lines, treated here as different sections.

Schlüssel zu den Sektionen der Untergattung PAPHIOPEDILUM

1. Seitliche Petalen von der Basis bis zur Spitze allmählich breiter werdend . . . . Sektion **PAPHIOPEDILUM**

1a. Apikaler Teil (Endteil) der seitlichen Petalen nicht breiter als der Mittelteil 2

2. Seitliche Petalen linealisch-lanzettförmig, etwas nach unten gebogen, an der Spitze eindeutig nach hinten verdreht; Fahne deutlich mehrfach gestreift, die Streifen sich an den Rändern berührend . . . . Sektion **CERATOPETALUM**

2a. Seitliche Petalen lanzett- bis spatelförmig. Spitze nicht nach hinten verdreht, Fahne nicht von mehreren Längsstreifen überzogen . . 3

3. Seitliche Petalen abstehend, im Endbereich mehr oder weniger gedreht. Seitliche Petalen und Fahne ohne deutlichen, dicken Mittelstrich . . . . Sektion **STICTOPETALUM**

3a. Seitliche Petalen eindeutig nach innen gebogen, die Spitze nicht gedreht. Seitliche Petalen und Fahne mit einer deutlichen, dicken Mittellinie . . . . Sektion **THIOPETALUM**

[1] **Bull. of the Hiroshima Botanical Garden** Nr. 5: 1–69, 1982.

Key to the sections of subgenus PAPHIOPEDILUM

1. Lateral petals gradually widening towards the apex . . . . . . section **PAPHIOPEDILUM**

1a. Apical portion of the lateral petals not wider than the middle portion 2

2. Lateral petals linear-lanceolate, somewhat drooping, the apex strongly distorted backwards; dorsal sepal with many distinct stripes, more of less connecting in the marginal areas . . . . section **CERATOPETALUM**

2a. Lateral petals lanceolate to spathulate, apex not distorted backwards; dorsal sepal not with several distinct stripes 3

3. Lateral petals spreading directly, more or less twisted; dorsal sepal not with a distinct middle line . . . . section **STICTOPETALUM**

3a. Lateral petals strongly incurved, not twisted in the apical portion; lateral petals and upper (dorsal) sepal with a distinct thick line at the middle . . . . section **THIOPETALUM**

[1] **Bull. of the Hiroshima Botanical Garden** No. 5: 1–69, 1982.

Sektion PAPHIOPEDILUM

Die Sektion **Paphiopedilum** zeichnet sich aus durch spatelförmige Petalen, eine relativ kurze Lippe, eine umgekehrt herzförmige Staminodiumplatte mit hornförmigen Anhängseln und eine Chromosomenzahl von 2n = 26.

Section PAPHIOPEDILUM

Section **Paphiopedilum** is well characterized by the lanceolate petals, the relatively short lip, the obcordate staminodal shield with a horn-shaped protuberance and a mitotic chromosome number of 2n = 26.

Schlüssel zu den Arten der Sektion PAPHIOPEDILUM

1. Fahne sehr breit, fast kreisrund, ungefleckt, einheitlich rötlich-purpurn überzogen, Färbung am dunkelsten an der Basis . . . . *Paphiopedilum charlesworthii*

1a. Fahne breit eiförmig, im Randbereich weiß, im Mittelbereich grün, von der Mitte bis zur Basis unregelmäßig dunkelbraun gefleckt . . . 2

2. Seitliche Petalen von einer schmalen Basis aus allmählich breiter werdend, deutlich asymmetrisch . . . . . . *Paphiopedilum villosum*

2a. Seitliche Petalen anders geformt, nicht so eindeutig verbreitert, mehr oder weniger symmetrisch . . . . *Paphiopedilum insigne*

Key to the species of section PAPHIOPEDILUM

1. Dorsal (upper) sepal very broad, nearly round, not blotched but uniformly tinged with purple, colouration darkest near the base . . . . *Paphiopedilum charlesworthii*

1a. Dorsal sepal broad ovate with white marginal portions, green in the middle portion, roughly blotched or tinged with dark brown at middle to basal portion 2

2. Lateral petals distinctly widened towards the apex from a narrower base, clearly asymmetric . . . . *Paphiopedilum villosum*

2a. Lateral petals shaped otherwise, not so strongly widened, more or less symmetric . . . . *Paphiopedilum insigne*

Paphiopedilum charlesworthii (Rolfe) Pfitzer

in **Engler's Bot. Jahrbücher 19:** 40 (1894)

Basionym: *Cypripedium charlesworthii* Rolfe
Orchid Review 1: 303 (1893)

Synonym: *Cordula charlesworthii* (Rolfe) Rolfe
Orchid Review 20 (1): 2 (1912)

Paphiopedilum charlesworthii (Rolfe) Pfitzer

Paphiopedilum charlesworthii (Rolfe) Pfitzer

Paphiopedilum charlesworthii wurde 45 km südwestlich des Inle Sees in den Shan Staaten[1] auf einem Kalksteinhügel in ungefähr 1700 m ü. NN von dem Engländer R. MOORE im Jahre 1893 entdeckt.

Die ersten Pflanzen wurden durch die Firma CHARLESWORTH nach England importiert und durch ROLFE erstmals beschrieben.

Paphiopedilum charlesworthii ist eine krautige Pflanze die als Humusepiphyt in Moos auf Kalksteinfelsen wächst.

Die Laubblätter sind riemenförmig, bis zu 25 cm lang und ungefähr 2,5 bis 3 cm breit. Sie sind glän-

[1] Jetzt Birma.

Paphiopedilum charlesworthii was discovered in 1893 by an Englishman named R. MOORE about 45 km to the south-west of Lake Inle in the Shan States[1] growing on limestone rock at an altitude of more than 1700 meters above sea-level.

The species was introduced to Europe by the English orchid dealer CHARLESWORTH, SHUTTLEWORTH and Co. who managed to flower the first specimens in the same year. The species was described by R. A. ROLFE *(loc. cit.)*.

Paphiopedilum charlesworthii is an herbaceous plant, growing epiphytically in leaf-mould and vegetable debris on limestone rock.

[1] Now Burma.

zend grün, an der Basis der Oberseite und auf der ganzen Unterseite purpur gefleckt.

Die Infloreszenz ist so lang wie die Laubblätter, rot gefleckt und flaumig behaart, im allgemeinen einblütig.

Die Braktee ist etwa 2,5 cm, der Fruchtknoten 3,8 cm lang.

Die Blüte ist vor allem gekennzeichnet durch eine enorme Fahne, die bis zu 5 cm lang und 5,5 cm breit wird. Sie ist breit eiförmig, weiß, rosa-purpur überzogen, mit mehr oder weniger intensiver karmesinroter Aderung.

Das Synsepalum ist ca. 3,5 cm lang und deutlich schmaler als die Fahne.

Die Petalen können 5 cm lang und 1,8 cm breit werden. Sie sind schmal spatelförmig, kastanienbraun, an den Rändern bewimpert.

Die umgekehrt helmförmige Lippe ist bis zu 3,5 cm lang und an der Öffnung etwas über 3 cm breit. Sie ist kastanienbraun.

Die Staminodiumplatte ist mehr oder weniger eiförmig, am Rande etwas konkav, elfenbeinweiß bis porzellanweiß, in der Mitte mit einem kleinen glänzenden, gelblich weißen, hornförmigen Zahn.

Paphiopedilum charlesworthii ist inzwischen von Birma und Ostassam bekannt, wo die Pflanzen in isolierten Kolonien auf kalkhaltigem Boden in Höhenlagen bis zu 2000 m ü. NN zu finden sind.

Das Klima in den Heimatregionen dieser Art ist geprägt von Wintermonsunen, die von September bis April viel Nieselregen und Nebel bringen. In dieser Periode ist der Himmel fast immer bedeckt. Ab April kommen die Sommermonsune mit tagelang andauernden Stürmen. Die Gebiete werden oft von Taifunen heimgesucht.

Die Temperaturen schwanken zwischen 5 und 17 °C im Winter und 16 und 31 °C im Sommer.

Entsprechend den klimatischen Bedingungen des Heimatareals kultiviert man *Paphiopedilum charlesworthii* am besten an einem hellen Platz im temperierten Gewächshaus. Im Winter sollte man für extreme Nachtabsenkung sorgen.

Die Pflanzen blühen im Herbst, und die Blüten sind sehr dauerhaft an einem schattigen Standort.

The leaves are strap-shaped, up to about 25 cm long and about 2.5 to 3 cm wide, the upper surface bright shiny green, the basal portion of the upper surface and the entire under surface with irregular purple-brown blotches and streaks.

The inflorescence is about as long as the leaves, spotted reddish, downy, generally bearing a single flower.

The floral bract is about 2.5 cm long, the ovary about 3.8 cm long.

The flower is mainly characterized by an enormous dorsal sepal, which is up to about 5 cm long and up to about 5.5 cm wide, broadly ovate to nearly round, white, suffused with light soft rose-purple, with more or less distinct, crimson red streaks.

The synsepalum is about 3.5 cm long and distinctly narrower than the dorsal sepal.

The petals are up to about 5 cm long and more or less 1.8 cm broad, narrowly spathulate, maroon-brown, the margins ciliate.

The lip is inverted helm-shaped, up to about 3.5 cm long and at the mouth about 3 cm across, maroon-brown.

The staminodal shield is more or less ovate, somewhat concave at the margins, ivory-white to nearly porcelain-white, with a curved, light yellow horn-like tooth near the middle.

Paphiopedilum charlesworthii is known to occur in Burma as well as in East-Assam. The plants grow in isolated colonies on limestone at altitudes of up to 2000 m above sea-level.

The climate in the natural habitats of this species is characterized by wintery monsoons bringing drizzling rain and fog as well as an almost continual cloud cover from September to April. At the end of April the south-east monsoon comes with heavy rainstorms often lasting for days. Typhoons are not unknown in these areas.

The temperatures are between 5 and 17 °C in winter and between 16 and 31 °C in the summer.

Paphiopedilum charlesworthii is an easy to grow species. The plants do best in a bright part of the intermediate greenhouse. They should be kept cooler during the winter nights.

Paphiopedilum charlesworthii generally flowers in the autumn. The blooms last very long when shaded.

A

B

C

D

Paphiopedilum insigne (Wallich ex Lindley) Pfitzer

Morphologische Studien über die Orchideenblüthe: 11 (1886)

Basionym: *Cypripedium insigne* Wall. ex Lindley
Collectanea Botanica, t. 32 (1821)

Synonyma: *Cordula exul* (Ridley) Rolfe
Orchid Review 20 (1): 2 (1912)
Cordula insignis Rafinesque
Flora Telluriana 4: 46 (1838)
Cypripedium exul (Ridley) O'Brien
Gardeners' Chronicle, 3. Ser., **11:** 522 (1892)
Cypripedium insigne var. *exul* Ridley
Gardener's Chronicle, 3. Ser., **10:** 94 (1891)
Paphiopedilum barbigerum Tang & Wang
Bull. Fan. Mem. Inst. Biol. Peiping, Bot. Ser. **16:** 24 (1940)
Paphiopedilum exul (Ridley) Kerchove
Le livre des Orchidees, Gent: 478 (1894)

Paphiopedilum insigne wurde durch WALLICH um das Jahr 1819 in der nordindischen Provinz Sylhet entdeckt. Einige Pflanzen wurden nach England gesandt, wo die ersten im Herbst 1820 im Botanischen Garten von Liverpool zur Blüte kamen.

John LINDLEY beschrieb die Art anhand der Angaben WALLICHs in seiner **Collectanea Botanica** *(loc. cit.).*

Paphiopedilum insigne ist eine äußerst variable Art, was Form und Farbe der Blüte anbelangt.

Im allgemeinen haben diese terrestrisch, meist als Humusepiphyt wachsenden, krautigen Pflanzen mehrblättrige Triebe.

Die Blätter sind 15 bis 40 cm lang, riemenförmig, stumpf endend mit einer mehr oder weniger deutlichen Einkerbung. Die Laubblätter sind einfarbig grün, an der Basis bräunlich violett überhaucht. Die jungen Blätter stehen aufrecht, die älteren hängen leicht über.

Paphiopedilum insigne was discovered by WALLICH in 1812 in the North-Indian province of Sylhet. Some of the plants were sent to England where the first ones came to flower in late 1820 in the Botanical Gardens of Liverpool.

The description by LINDLEY in his **Collectanea Botanica** was based on notes forwarded to him by WALLICH.

Paphiopedilum insigne is a very variable species, especially in respect to the colouration of the flowers.

Paphiopedilum insigne is an herbaceous plant, growing epiphytically in leaf-mould. The robust plants usually generate multifoliate growths.

The leaves are linear ligulate, obtuse, more or less distinctly notched at the apex, 15 to 40 cm long, uniformly pallid green, suffused with a brownish-violet near the base, suberect in the juvenile state, later somewhat arching.

Tafel Seite 110 / Plate page 110

A *Paphiopedilum insigne* var. *sanderae*
B *Paphiopedilum insigne* „var. *albo-marginata*"
C *Paphiopedilum insigne* var. *barbigerum*
D *Paphiopedilum insigne* – Nahaufnahme / close-up

Die aufrecht stehende Infloreszenz wird bis ungefähr 30 cm hoch. Sie ist dunkelgrün, vollständig mit kleinen rotbraunen Härchen überzogen und trägt meist eine einzelne, manchmal auch zwei Blüten.

Die zusammengepreßte Braktee ist mit einer Länge von bis zu 5,2 cm etwas länger als der Fruchtknoten.

Die Blüten sind 8 bis 12 cm im Durchmesser, glänzend.

Die Fahne ist breit oval, mit deutlich gewellten Rändern. Im oberen Teil leicht nach vorne gebeugt. 4,5 bis 5,5 cm hoch und bis ca. 3 cm breit, breit eiförmig. Die Grundfärbung variiert von gelben über gelbgrüne bis zu grünlichen Tönen, mit einem mehr oder weniger weißen Rand. Von der abgeflachten Spitze bis zur Basis verlaufen mehrere Längslinien. Die Fahne ist mehr oder weniger intensiv mit verschiedenen großen braunen Flecken überzogen, auf der Rückseite deutlich gekielt und flaumig behaart, innen glänzend.

Das Synsepalum ist kleiner, eiförmig, spitz, weniger befleckt, zur Spitze hin eingerollt, heller gefärbt.

Die spatelförmigen, abstehenden Petalen zeigen deutlich gewellte Ränder. Sie sind bis zu 6 cm lang und 1,5 bis 2 cm breit. Sie sind gelblich grün, kastanienbraun geadert.

Die Lippe ist verkehrt helmförmig, ca. 4,5 cm lang und 3 cm breit. Sie ist gelblich grün mit kastanienbraunem Überzug. Die nach innen gefalteten Seitenlappen sind ledergelb mit helleren Rändern.

Die mattgelbe Staminodiumplatte ist umgekehrt herzförmig, behaart, mit einer kleinen gelben Warze in der Mitte und einem kleinen gerundeten Anhängsel am unteren Ende.

Inzwischen hat sich herausgestellt, daß *Paphiopedilum insigne* ein großes Verbreitungsgebiet besiedelt. Die Art wurde bisher sowohl im Khasiagebirge Assams, im Sylhetgebiet des heutigen Bangladesch, im Staate Meghalaya als auch im südlichen Teil Chinas (Kweichow) gefunden.[1] *Paphiopedilum insigne* wächst meist auf Kalkfelsen fossilen Ursprungs, manchmal auch in der prallen Sonne, aber somit auch dem Monsunregen ausgesetzt. Das große Verbreitungsgebiet erlaubt es nicht, im Detail auf die klimatischen Bedingungen einzugehen.

Paphiopedilum insigne sollte im temperierten oder warmen Haus kultiviert werden. Für hohe Lichtintensität muß gesorgt werden. Von Januar bis März sind die Pflanzen etwas kühler und trockener zu halten, sonst sind ausreichende Wassergaben

[1] var. *barbigerum*.

The inflorescence is erect, up to about 30 cm high, deep green, completely covered with reddish-brown bristles, generally bearing one single, sometimes two flower(s).

The laterally compressed bract is about 5.2 cm long and sowith somewhat longer than the ovary.

The flowers are 8 to 12 cm across, with a glossy varnished surface.

The upper sepal is broadly ovate, the margins slightly revolute, apically bent forward. The ground colour varies from yellow and yellowish green to a distinct apple-green with more or less white marginal regions; the central and basal area with numerous brownish purple spots more or less regularly arranged along the longitudinal green veins; externally keeled, internally glossy.

The synsepalum is smaller, ovate, acute, less spotted, curved inwards towards the apex, coloured brighter than the upper sepal.

The petals are spreading, linear-oblong to spathulate, with clearly undulate margins, pale yellowish green with brownish purple longitudinal veins, up to about 6 cm long and 1.5 to 2 cm wide.

The lip is inverted helmet shaped, about 4.5 cm long and up to 3 cm broad, yellowish green, shaded with brown; the infolded lateral lobes deep tawny yellow with paler margins.

The staminodal shield is obcordate, dull yellow with a bright orange-yellow tubercle about one third from the top and a small blunt protuberance at the lower end.

Paphiopedilum insigne has been reported from a vast area. The species has been found in the Khasia Hills of Assam, the Sylhet-area, now belonging to Bangladesh, the Indian State of Meghalaya as well as in the southern part of China (Kweichow).[1] The plants usually grow on dolomite rocks, sometimes in full sun. Due to the large natural distribution, it is not possible to note generally valid climatic conditions.

Paphiopedilum insigne should be kept in a bright part of the intermediate or warm greenhouse. The plants need abundant water with the exception of a short resting period between January and March, during which the plants are to be kept cooler and drier. Relative humidity should be high over the entire year, and the plants should be potted in an open, well-drained compost, preferably mixed with moss.

Paphiopedilum insigne has the peculiar habit of interrupting the growth period after the appear-

[1] var. *barbigerum*

gebracht. Die relative Luftfeuchtigkeit sollte das ganze Jahr über sehr hoch sein. Die Pflanzen benötigen ein gut durchlässiges Substrat, möglichst mit etwas Sphagnum als Beimischung.

Paphiopedilum insigne hat die Eigenart, daß es, nachdem der neue Trieb im Frühjahr angefangen wurde, eine Ruhezeit einlegt, und das neue Wachstum erst im Spätherbst mit der Blüte abschließt.

Die Blütezeit fällt zwischen Oktober und Dezember.

Varietäten:

Wie bereits erwähnt, ist diese Art sehr variabel. Es ist daher kaum verwunderlich, daß eine ganze Reihe von Varietäten beschrieben worden sind. Die meisten basieren auf äußerst geringfügigen Unterschieden, die vor allem die Blütenfärbung betreffen. Die meisten dieser „Varietäten" verdienen aber kaum diesen taxonomischen Status. Ich werde hier nur die drei wichtigsten dieser „Abarten" behandeln. Für die anderen verweise ich den interessierten Leser auf die entsprechende Literatur.[1]

Paphiopedilum insigne var. *barbigerum* (Tang & Wang) Braem, *comb. et stat. nov.*

Basionym: *Paphiopedilum barbigerum* Tang & Wang
Bull. Fan. Mem. Inst. Biol. Peiping, Bot. Ser. **10:** 24 (1940)

Diese Varietät wurde bisher als eigenständige Art betrachtet. Eingehende Untersuchungen an blühendem Material lassen aber keinen Zweifel daran, daß hier eine geographische Variante von *Paphiopedilum insigne*[2] vorliegt, welche sich von den übrigen Formen der Art durch einen kleineren Wuchs und etwas kleinere, anders gefärbte Blüten unterscheidet.

Die Laubblätter sind bis 18 cm lang und mehr oder weniger 1,8 cm breit.

Die Blüte ist ungefähr 6 cm hoch und bis ca. 7,5 cm breit.

[1] DESBOIS, F., **Cypripedium, Selenipedium & Uropedium** – Monographie contenant la description de toutes les Espèces, varietées et hybrides existant jusqu'a ce jour., Gent (Gand) (1898).
LINDEN, J. J., **Lindenia**
SANDER, H. F. C., **Reichenbachia**
PUCCI, A., **Les Cypripedium et genres affines**
WARNER & WILLIAMS, **Orchid Album**

[2] Obwohl *Paphiopedilum barbigerum* durch ASHER in **Orchid Digest 44** (6): 218 (1980) irrtümlicherweise als Teil der Sektion **Stictopetalum** betrachtet wurde.

ance of new growth. The growth cycle therefore is not completed before autumn.

Paphiopedilum insigne usually flowers between October and December.

Varieties:

Paphiopedilum insigne is, as I have already stated, a very variable species. It is, therefore, not surprising that quite a number of varieties have been described. The great majority of these "varieties" are based on very minor differences, generally in respect to the colouration of the flower. Most of them cannot be upheld as autonomous taxonomic entities. I will only discuss the three most important ones, two of which have been considered to be separate species. For the others, I refer the interested reader to the pertinent literature.[1]

Paphiopedilum insigne var. *barbigerum* (Tang & Wang) Braem, *comb. et stat. nov.*

Basionym: *Paphiopedilum barbigerum* Tang & Wang
Bull. Fan. Mem. Inst. Biol. Peiping, Bot. Ser., **10:** 24 (1940)

This has, up till now, been considered to be an autonomous species. Detailed study of flowering plants however prove that this is a geographic variant of *Paphiopedilum insigne*[2] differing from the other forms of the species by a smaller habit and somewhat smaller, differently coloured flowers.

The leaves are up to about 18 cm long and about 1.8 cm wide.

The flower is about 6 cm high and up to about 7.5 cm across.

[1] DESBOIS, F., **Cypripedium, Selenipedium & Uropedium** – Monographie contenant la description de toutes les espèces, varietées et hybrides existant jusqu'a ce jour, Gent (1898).
LINDEN, J. J., **Lindenia**
SANDER, H. F. C., **Reichenbachia**
PUCCI, A., **Les Cypripedium et genres affines**
WARNER & WILLIAMS, **Orchid Album**

[2] Although erroneously considered part of the section Stictopetalum by ASHER in **Orchid Digest 44** (6): 218 (1980)

Die Fahne ist weiß, im basalen Teil grün überlaufen, hinten deutlich gekielt, die apikalen Ränder nach hinten gebogen, ungefähr 3,2 cm breit und bis ungefähr 3,6 cm hoch, im ausgebreitetem Zustand fast rund.

Das Synsepalum ist ungefähr 3 cm lang und bis zu 1,9 cm breit, eiförmig, grünlich weiß mit mehr oder weniger deutlichen Längsadern.

Die seitlichen Petalen sind bandförmig, gerundet, fast viereckig, bis ungefähr 4,4 cm lang und ungefähr 1,2 cm breit, grünlich braun mit einer regelmäßigen Beaderung und schmalen grünen Rändern, über die ganze Länge gewellt.

Die Lippe ist tief sackförmig, ungefähr 4 cm lang und bis 2,4 cm breit, braun mit dunkleren Adern. Die Lippe steht in einem Winkel von 90 Grad zu der Blütenstandsachse.

Die Staminodiumplatte ist undeutlich dreieckig bis fast verkehrt herzförmig, etwa 1 cm hoch und am unteren Rand ca. 1 cm breit, leicht gelb mit einem schwefelgelben Zahn im unteren Drittel.

Diese Varietät ist möglicherweise endemisch für das Kweichow-Gebiet im südlichen China.

The dorsal sepal is white with a green basal area, externally prominently keeled, the apical margins folded backwards; about 3.2 cm wide and up to about 3.6 cm high; suborbicular when spreaded.

The synsepalum is about 3 cm long and up to 1.9 cm wide, ovate, whitish-green with more or less distinct longitudinal veins.

The petals are ligulate, blunt, nearly quadrangular, up to about 4.4 cm long and more or less 1.2 cm wide, greenish-brown with a regular venation and narrow green margins, undulate over the whole length.

The deeply saccate lip is about 4 cm long and up to about 2.4 cm wide, brown with darker venation, standing in an angle of about 90 degrees to the axis of the inflorescence.

The staminodal shield is obtusely triangular to obcordate, about 1 cm high and at the lower margin about 1 cm wide, light yellow with a sulphureous tooth in the lower part.

This variety is said to be endemic in the Kweichow-Area in southern China.

Paphiopedilum insigne var. *exul* Ridley
Gardeners' Chronicle, 3. Ser., **10:** 94 (1891)

RIDLEY beschrieb diese Varietät aufgrund von Pflanzenmaterial, welches er aus Thailand erhalten hatte. Obwohl diese Varietät zum Artstatus erhoben wurde *(loc. cit.)*, besteht kein Zweifel daran, daß es sich hier um eine geographische Variante von *Paphiopedilum insigne* handelt; und als solche wurde sie auch beschrieben.

Paphiopedilum insigne var. *exul* unterscheidet sich von den „normalen" *Paphiopedilum insigne*-Formen lediglich durch etwas stärker niederliegende Laubblätter und eine etwas kräftiger getüpfelte Fahne.

Die Blütezeit dieser Variante ist im allgemeinen von April bis Juni.

Paphiopedilum insigne var. *exul* Ridley
Gardeners' Chronicle, 3. Ser., **10:** 94 (1891)

RIDLEY described this variety on the basis of some plants which he had received from Thailand. Although this variety was redescribed as an autonomous species by KERCHOVE *(loc. cit.)* these plants belong, without any doubt, to a geographic race of *Paphiopedilum insigne.*

Paphiopedilum insigne var. *exul* differs from the other forms of the species merely by leaves standing more horizontally and a flower with a more intensively spotted dorsal sepal.

This variant flowers, in general, between April and June.

Paphiopedilum insigne var. *sanderae* (Rchb. f.) Pfitzer
in ENGLER, A.: **Das Pflanzenr. 4** (50) Heft 12: 71 (1903)

Blütenblätter, außer der Fahne, reingelb. Fahne gelb mit weißer Spitze und einigen wenigen dunklen, bräunlichen Flecken.

Aufgrund des Vorhandenseins dieser braunen Flecken ist diese Varietät keineswegs als Albinoform zu betrachten. Die Angaben im **Orchideenatlas** von BECHTEL, CRIBB und LAUNERT treffen somit nicht zu.

Paphiopedilum insigne var. *sanderae* (Rchb. f.) Pfitzer
in ENGLER, A.: **Das Pflanzenr. 4** (50) Heft 12: 71 (1903)

Synsepalum, lateral petals and lip immaculate bright yellow; dorsal sepal bright yellow with some indistinct darker brownish spots, the tip white.

Because of these dark spots, this variety cannot be considered as an albino as is claimed in the **Orchid Atlas** by BECHTEL, CRIBB and LAUNERT.

Paphiopedilum villosum (Lindley) Stein

Stein's Orchideenbuch: 490 (1892)

Basionym: *Cypripedium villosum* Lindley
Gardeners' Chronicle 14: 135 (1854)

Synonyma: *Cordula boxallii* (Rchb. f.) Rolfe
Orchid Review 20 (1): 2 (1912)

Cordula gratrixiana (Masters) Rolfe
Orchid Review 20 (1): 2 (1912)

Cordula villosa (Lindley) Rolfe
Orchid Review 20 (1): 2 (1912)

Cypripedium boxallii Rchb. f.
Gardeners' Chronicle, 2. Ser., **7:** 367 (1877)

Cypripedium gratrixianum Masters
Gardeners' Chronicle, 3. Ser., **33:** 76–77 (1905)
non Warner & Williams, **Orchid Album 11:** t. 524 (1897)[1)]

Paphiopedilum affine De Willdeman
Tribune Horticole 1: 57 (1906)

Paphiopedilum boxallii (Rchb. f.) Pfitzer in ENGLER & PRANTL: **Natürliche Pflanzenfam. 2:** 6: 83 (1889)

Paphiopedilum gratrixianum (Masters) Guillaumin
Bull. Soc. Bot. France 71: 556 (1924)

Paphiopedilum villosum wurde durch Thomas LOBB in der Nähe von Moulmein in einer Höhe von ca. 1600 m ü. NN gefunden. 1853 erfolgte, und zwar durch die Firma VEITCH, der erste Import nach England. Deren Kontaktmann, der Anglikanerpfarrer PARISH, fand kurz darauf noch andere Standorte.

1877 importierte die Firma LOW aus dem selben Gebiet *Paphiopedilum*-Pflanzen, die REICHENBACH zu Ehren des Sammlers der Firma LOW als *Paphiopedilum boxallii* beschrieb *(loc. cit.)*. Diese und die durch MASTERS und DE WILDEMAN beschriebenen *Paphiopedilum gratrixianum* bzw. *Paphiopedilum affine* sind ohne jeglichen Zweifel als geographische Sippen von *Paphiopedilum villosum* zu betrachten und weichen von der Nomi-

Paphiopedilum villosum was discovered by Thomas LOBB on the mountains in the Moulmein district (Burma) at about 1600 meters above sea-level. The first plants of this species were imported to England by VEITCH and it has been recorded that their correspondent, the reverent C. PARISH found more populations in the same region.

In 1877, LOW imported *Paphiopedilum*-plants from that same district. These plants were described by REICHENBACH as *Paphiopedilum boxallii (loc. cit.)* in honour of LOWs plant collector. This concept, and the concepts of *Paphiopedilum gratrixianum* from MASTERS as well as DE WILDEMANs *Paphiopedilum affine* are without the slightest doubt to be considered geographic variants of *Paphiopedilum villosum.* They differ from "normal"

[1)] Die als *Cypripedium gratrixianum* beschriebene Pflanze im **Orchid Album 11,** t. 524 stellt eine gärtnerische Hybride zwischen *Paphiopedilum bellatulum* und *Paphiopedilum* Enfieldense (= *Paphiopedilum hookerae* × *Paphiopedilum lawrenceanum*) dar.

[1)] The plant described as *Cypripedium gratrixianum* in the **Orchid Album 11,** t. 524, is an artificial hybrid between *Paphiopedilum bellatulum* and *Paphiopedilum* Endfieldense (= *Paphiopedilum hookerae* × *Paphiopedilum lawrenceanum*).

A

B

C

D

nalform nur geringfügig (vor allem in der Blütenfärbung) ab.

Paphiopedilum villosum ist eine epiphytisch wachsende Pflanze, die ungefähr 30 cm hoch wird.

Die Laubblätter sind schmal lanzettlich, 25 bis 40 cm lang und bis 4 cm breit, vorne spitz zweilappig, halb aufrecht, dunkelgelbgrün, auf der Unterseite der Basis purpurn getüpfelt.

Die Infloreszenz ist bis ca. 20 cm hoch, dunkelgrün, lang purpurrot behaart, meist einblütig.

Die Braktee ist mit 5 cm Länge fast so lang wie der Fruchtknoten.

Die Blüten sind auffallend groß, bis 15 cm breit, von glänzendem, lackiertem Aussehen.

Die Fahne ist ca. 5,5 cm hoch und bis 3,5 cm breit. Sie ist breit eiförmig, bewimpert und unten an den Rändern zurückgebogen. Die Grundfärbung ist ein leuchtendes Grün, in den unteren zwei Dritteln intensiv purpurn überzogen. Die Ränder sind weißlich.

Das Synsepalum ist bis 4,5 cm lang und schmaler als die Fahne, eiförmig, hell gelblich grün, zurückgeschlagen.

Die Petalen sind bis zu 6,5 cm lang und 3 cm breit, spatelförmig, gerundet mit bewimperten Rändern, onduliert, an der Basis braun behaart, etwas nach vorne gebogen, mit einer breiten braunpurpurnen Mittelader, die obere Hälfte gelblich braun, die untere Hälfte heller.

Die umgekehrt helmförmige Lippe ist bis 5,5 cm lang und 3 cm breit, spitz, mit weiter Öffnung, in der Grundfärbung hell Goldgelb, mit dunklerer Beaderung und einem ledergelben Rand um die Öffnung.

Die Staminodiumplatte ist verkehrt länglich herzförmig, ledergelb, mit einer Warze unterhalb der Mitte und einem kurzen Spitzchen am unteren Ende.

Paphiopedilum villosum hat seine Verbreitung in Assam, Birma und Thailand in Höhenlagen um 1600 m ü. NN, immer in Gebieten mit sehr hohen Niederschlägen.

Diese Art, die wohl bekannteste unter den Paphiopedilen, ist relativ einfach zu kultivieren. *Paphiopedilum villosum* stellt keine besonderen Ansprüche an Substrat oder Licht.

Die Pflanzen gedeihen am besten in einem durchlässigen Substrat mit viel Moos als Beimischung.

plants of the species only by minor variations in the colour of the flower.

Paphiopedilum villosum is an herbaceous epiphyte, usually growing in leafy debris and vegetable mould.

The plants attain an overall height of about 30 cm.

The leaves are narrow-lanceolate to linear-ligulate, 25 to 40 cm long and up to 4 cm wide, at the end acutely bilobed, semi-erect, deep yellow-green, the base of the under side speckled purple.

The inflorescence is up to about 20 cm high, dark green, covered by long reddish-purple bristles, generally carrying a single flower.

The bract is about 5 cm long and nearly as long as the ovary.

The flowers are remarkably large, up to 15 cm across, with a glossy varnished surface.

The dorsal sepal is about 5.5 cm high and up to 3.5 cm wide, broad oval, ciliate with reflexed margins in the basal region. The ground colour is a bright green, suffused by an intense purple in the lower two-thirds and a narrow white marginal band.

The synsepalum is up to 4.5 cm long and narrower than the dorsal sepal, ovate, bright yellowish-green, reflexed.

The lateral petals are up to 6.5 cm long and about 3 cm wide, spathulate, rounded, the margins ciliate, undulate, at the base with brown hairs, somewhat bent forward, with a broad brown-purple mid-vein; the superior half yellow brown, the inferior half paler.

The lip is inverted helmet-shaped, about 5.5 cm long and about 3 cm wide, more or less pointed, with a wide mouth; bright brownish yellow with darker venation, and a tawny yellow rim around the aperture.

The staminodal shield is inverted oblong-cardiform, tawny yellow, with a small yellow or green hornshaped tooth below the center and a short protuberance at the lower margin.

Paphiopedilum villosum has its natural distribution in Assam, Burma, and Thailand. The plants grow at elevations of up to about 1600 meters above sea-level. The habitat areas are characterized by a large amount of precipitation.

This species, maybe the best known of the genus, is very easy to grow. *Paphiopedilum villosum* has

Tafel Seite 116 / Plate page 116

A *Paphiopedilum villosum*

B *Paphiopedilum villosum* var. *boxallii*

C *Paphiopedilum villosum* var. *aureum*

D *Paphiopedilum villosum* var. *gratrixianum*

Es sollte vor allem während der Wachstumsperiode ausgiebig gegossen werden. Für gute Luftzirkulation ist zu sorgen.

Gut kultivierte Pflanzen bringen mehrere Neutriebe in einem Jahr. Bei einer Kultur unter intensivem Licht sollte regelmäßig gedüngt werden.

Die Blütezeit ist normalerweise von Januar bis März, manchmal auch bereits ab Oktober.

no special requirements in respect to compost and light conditions.

The plants do best in the intermediate house, potted in a well-drained open substrate with lots of moss. Water should be given abundantly, especially in the growth period. Good air-circulation in the greenhouse is indicated.

Well cultivated specimens will flower more than once a year. When cultivating under intense light, fertilization should be regular.

Paphiopedilum villosum usually flowers between January and March. Depending on the state of the plants, other blooming times are possible.

Varietäten:

Auch von *Paphiopedilum villosum* sind eine Reihe von Varietäten beschrieben worden. Alle weichen nur geringfügig von der Nominalform ab. Der interessierte Leser sei an die zutreffende Fachliteratur verwiesen. Ich behandle hier lediglich drei Varietäten, die im allgemeinen irrtümlicherweise als autonome Arten betrachtet werden.

Paphiopedilum villosum var. *boxallii* (Rchb. f.) Veitch
A manual of Orchidaceous Plants 4: 54 (1889)

Synonym: *Cypripedium delectum* Rchb. f.
Gardeners' Chronicle, 3. Ser., **3:** 330 (1888)

Paphiopedilum delectum (Rchb. f.) Pfitzer
in ENGLER, A.: **Das Pflanzenr. IV** (50): 71 (1903)

Diese als *Cypripedium boxallii* beschriebene (siehe oben) Abart unterscheidet sich von der „normalen" Form durch:

1. eine schmalere Fahne, mit mehr nach hinten gedrehten Rändern in der unteren Hälfte, eine deutliche Befleckung, mit zum Teil ineinander verlaufenden schwarzbraunen Flecken und einen deutlichen weißen Rand,
2. deutlichere Marmorierung der Petalen,
3. ein spitzeres, oft zweiteilig endendes Synsepalum.

Varieties:

A number of variants of *Paphiopedilum villosum* have been described. All of them differ only slightly from the "normal" forms. The interested reader is again referred to the pertinent literature. I wish to discuss here three varieties having been considered erroneously as autonomous species in the past.

Paphiopedilum villosum var. *boxallii* (Rchb. f.) Veitch
A Manual of Orchidaceous Plants 4: 54 (1889)

Synonym: *Cypripedium delectum* Rchb. f.
Gardeners' Chronicle, 3. Ser., **3:** 330 (1888)

Paphiopedilum delectum (Rchb. f.) Pfitzer
in ENGLER, A.: **Das Pflanzenr. IV,** (50): 71 (1903)

This variant was originally described as *Cypripedium boxallii (loc. cit.)* and differs from the other forms of the species by:

1. a narrower dorsal sepal, with more reflexed margins in the lower half, more distinct blackish-brown spots partly merging in the upper region and more distinct white margins,
2. a more distinct tesselation of the lateral petals,
3. a more acute synsepalum, often showing a double apex.

Paphiopedilum villosum var. *gratrixianum* (Masters) Braem *comb. nov.*

Basionym: *Cypripedium gratrixianum* Masters[1)]
Gardeners' Chronicle, 3. Ser., **33:** 76–77 (1905)

Synonym: *Paphiopedilum gratrixianum* (Masters) Guillaumin
Bull. Soc. Bot. France 71: 556 (1924)

Dies ist die geographische Variante aus dem Pak Song-Gebiet in Laos. Diese Varietät ist vor allem an der hell gefärbten Blüte und der stark purpurrot gefleckten Unterseite der Laubblattbasis zu erkennen. Die Blütezeit dieser Variante ist von März bis Ende April.

Laut KARASAWA & SAITO[2)] könnte dieses Taxon eine Naturhybride sein.

Paphiopedilum villosum var. *affine* (De Wildeman) Braem *comb. nov.*

Basionym: *Paphiopedilum affine* De Wildeman
Tribune Horticole 1: 57 (1906)

Eine geographische Variante aus dem laotisch-chinesischen Grenzgebiet. Die Blüte ist gleichbleibend glänzend maronenbraun, mit Ausnahme des breiten, weißen Fahnenrandes.

Die Blütezeit ist Februar bis März.

[1)] non WARNER & WILLIAMS, siehe Fußnote Seite 115

[2)] **Bull. of the Hiroshima Botanical Garden** Nr. 5: 1–69 (1982)

Paphiopedilum villosum var. *gratrixianum* (Masters) Braem *comb. nov.*

Basionym: *Cypripedium gratrixianum* Masters[1)]
Gardeners' Chronicle, 3. Ser., **33:** 76-77 (1905)

Synonym: *Paphiopedilum gratrixianum* (Masters) Guillaumin
Bull. Soc. Bot. France 71: 556 (1924)

This is the geographical race from the Pak Song Area in Laos. This variety differs from the other forms mainly by its brightly-coloured flower and the underside of the leaves being strongly-spotted purple-red at the base. This variant flowers from March through April.

According to KARASAWA & SAITO[2)] this taxon could represent a natural hybrid.

Paphiopedilum villosum var. *affine* (De Wildeman) Braem *comb. nov.*

Basionym: *Paphiopedilum affine* De Wildeman
Tribune Horticole 1: 57 (1906)

This is a geographic race from the Laos-China border region. The flower colour is fairly uniform maroon with the exception of the wide, white marginal area of the dorsal sepal.

This variety flowers from February through March.

[1)] non WARNER & WILLIAMS, see footnote page 115

[2)] **Bulletin of the Hiroshima Botanical Garden,** No. 5: 1–69 (1982)

Sektion CERATOPETALUM
Hallier

Diese monospezifische Sektion, die lediglich *Paphiopedilum fairrianum* umfaßt, unterscheidet sich deutlich durch die fast vertikal herabhängenden seitlichen Petalen mit deutlich nach hinten gebogener Spitze.

BRIEGER (in SCHLECHTER, **Die Orchideen,** 3. Ausgabe) setzt *Paphiopedilum fairrieanum,* wahrscheinlich wegen der bewimperten Petalenränder, in die Untergattung BARBATA. Die Art gehört aber aufgrund der einfarbigen Laubblätter und vor allem der Karyologie eindeutig in die Untergattung PAPHIOPEDILUM.

Section CERATOPETALUM
Hallier

This monospecific section includes only *Paphiopedilum fairrieanum.* It is readily distinguished from all other sections by almost vertically pendulous lateral petals with strongly recurved apical portions.

BRIEGER (in SCHLECHTER, **Die Orchideen,** 3. Edition) includes *Paphiopedilum fairrieanum* in the subgenus BARBATA, probably because of the ciliate margins of the lateral petals. However, the uni-coloured green leaves and especially the karyotype are strong support for including the species in the subgenus PAPHIOPEDILUM.

Paphiopedilum fairrieanum (Lindley) Stein

Stein's Orchideenbuch: 467 (1892)

Basionym: *Cypripedium fairrieanum* Lindley
Gardeners' Chronicle 17: 740 (1857)

Synonym: *Cordula fairrieana* (Lindley) Rolfe
Orchid Review 20 (1): 2 (1912)

Paphiopedilum fairrieanum (Lindley) Stein

Paphiopedilum fairrieanum erreichte Europa im Jahre 1857. Einige angeblich aus Assam stammende Pflanzen wurden in London verkauft und kamen kurz darauf in mehreren Sammlungen zur Blüte. Eine dieser Pflanzen wurde durch einen Mr. FAIRRIE im Oktober desselben Jahres anläßlich einer Versammlung der Londoner Gartenbaugesellschaft ausgestellt. Dies veranlaßte John LINDLEY die Art zu Ehren des Ausstellers zu beschreiben.

Im selben Jahr schrieb der belgische Gartenbauunternehmer VAN HOUTTE,[1] daß es sich um Pflanzen handele, die seine Firma aus Bhutan erhalten und an einige seiner „Korrespondenten" veräußert habe.

[1] **Flore des Serres 12:** 120, sub t. 124 (1857).

Paphiopedilum fairrieanum first reached England in 1857 and several plants, allegedly originating from Assam, were auctioned in London and flowered in several collections. One of these plants was exhibited in October of that same year at the London Horticultural Society by Mr. FAIRRIE, to whom LINDLEY dedicated the species in his description *(loc. cit.).*

In the same year, the Belgian horticulturalist VAN HOUTTE wrote[1] that the plants originated from Bhutan and were sold without a name to some of his "correspondents".

In any case, the origin remained uncleared for many years, and by the end of the century few or no plants of the species remained in culture. This

[1] **Flore des Serres 12:** 120, sub t. 124 (1857).

Jedenfalls war der genaue Standort der Art für viele Jahre unbekannt. Durch die Firma SANDER wurde 1904 eine Belohnung von 1000 £ für eine Sendung Pflanzen von *Paphiopedilum fairrieanum* und Angabe des Standortes ausgesetzt, da inzwischen kaum noch eine Pflanze der Art die damals übliche falsche Kultur in den englischen „Hothouses" bei extremer Hitze und wenig Luftzufuhr überlebt hatte.

Die für damalige Verhältnisse enorme Summe machte eine ganze Reihe von „Orchideenjägern" hellhörig, und bereits 3 Monate später wurde über entsprechende Funde berichtet. SANDER erhielt die Pflanzen und schaffte es damit, seine kurz vor dem Ruin stehende Firma zu retten. Die Belohnung wurde unter einem ziemlich obskuren Vorwand nur teilweise ausbezahlt.

Neben diesem unschönen Verhalten SANDERs ging aber auch die Gier der anderen Orchideenhändler damals sehr weit. Ein in Kalkutta residierender Pflanzenhändler erfuhr die Standorte von *Paphiopedilum fairrieanum* und befahl seinem Sammler LOONGOOK, so viele Pflanzen wie möglich zu holen. Die Pflanzen, die dann noch am Standort waren und nicht mitgenommen werden konnten, sollten aus dem Erdreich gezogen und in den Fluß geworfen werden.

Paphiopedilum fairrieanum ist eine kompakte, zwergwüchsige, als Humusepiphyt wachsende, krautige Pflanze.

Die Laubblätter sind länglich bandförmig, spitz, 10 bis 15 cm lang und ungefähr 2,5 cm breit. Sie sind einfarbig mattgrün, auf der Unterseite deutlich gekielt.

Die meist einblütige Infloreszenz ist 10 bis 20 cm hoch, dunkelgrün, purpurviolett behaart.

Die Blüten werden 6 bis 7,5 cm hoch und ca. 4 cm breit.

Die Fahne ist ungefähr 3,5 cm lang und 2,5 cm breit, eiförmig, an der Spitze zweilappig, an den Rändern gewellt und bewimpert, grünlichweiß mit einem gelblichgrünen Fleck an der Basis und purpurnen Längsadern.

Das Synsepalum ist etwas kleiner, ca. 2,5 cm lang, hellgrün mit purpurnen Streifen.

Die ca. 4 cm langen und 1 cm breiten Petalen sind lanzettförmig, herabgebogen und an der Spitze rückwärts gekrümmt, gewellt und bewimpert, grünlich weiß, purpurn beadert und purpurn berandet. Die becherförmige Lippe ist bis 4 cm hoch und hat eine relativ weite Öffnung. Sie ist bräunlich grün und netzartig mit purpurnen Adern überzogen. Die

was mainly due to the maltreatment of the plants in typical English "hothouses" by extreme heat and stale air conditions. In 1904, the English orchid firm SANDER set out a price of 1000 £ for a shipment of *Paphiopedilum fairrieanum* and the exclusive knowledge of its origin.

This small fortune caused a number of "orchid hunters" to become extremely active and a scant 3 months after the offer, rediscovery of the species was announced.

SANDER received the shipment and managed to save his firm from financial ruin, but for some very dubious reason, he only paid half of the reward.

An Indian nurseryman named CHATTERJEE got the information about the rediscovery and travelled to Darjeeling. There he hired a collector to go to the Am-mochu Valley where *Paphiopedilum fairrieanum* was to be found. He ordered his collector LOONGOOK to bring as many plants as he could carry and to uproot the rest and throw them into the river!

Paphiopedilum fairrieanum is a compact-growing herbaceous plant of dwarf habit, growing as an epiphyte in humus.

The leaves are oblong-ligulate, acute, 10 to 15 cm long and about 2.5 cm wide, uni-coloured dull green, the underside distinctly keeled.

The inflorescence is 10 to 20 cm high, dark green, covered with purple bristles, usually bearing one, seldom two flower(s).

The flower is 6 to 7.5 cm high and about 4 cm across.

The dorsal sepal is about 3.5 cm long and 2.5 cm wide, ovate, the apex bilobed, ciliate, undulate at the margin, reflexed at the apex, with a hairy keel behind, greenish-white with a pale yellow-green stain at the base and purple longitudinal veins.

The synsepalum is somewhat smaller, about 2.5 cm long, brightly green with purple veins.

The petals are oblong-ligulate to lanceolate, up to about 4 cm long and 1 cm wide, deflexed with a recurved apex, fringed at the undulate margins with minute blackish hairs, greenish white to yellowish white with longitudinal streaks and marginal bands of purple.

The lip is calceiform, about 4 cm high, with a relativelly broad aperture, brownish-green with purplish reticulation; the infolded lateral lobes creamy white, often spotted purple.

The staminodal shield is orbiculate-lunate, at the bottom with a tooth between the horns of the nach innen gefalteten Seitenlappen sind cremeweiß, oft purpurn gefleckt.

Die Staminodiumplatte ist halbmondförmig, unten mit zwei langgezogenen Lappen und einem Zahn, elfenbeinweiß mit grüner Zeichnung, am unteren Rand eine purpurne Linie.

Paphiopedilum fairrieanum wächst in Bhutan, Assam und Sikkim auf Dolomitfelsen an steilen Hängen, in Höhenlagen von 1400 bis 2000 m ü. NN, meist in Humusansammlungen in Felsspalten.

Die Heimatgebiete werden sowohl vom nassen Sommermonsun als auch von trockenen Winden im Winter heimgesucht. Von September bis Oktober gibt es starke Regenfälle und Temperaturen von bis zu 32 °C am Tage und über 18 °C in der Nacht. Ab Ende Oktober wird es kühler und etwas trockener, ohne daß die Regenfälle vollständig aufhören. Dezember und Januar sind die kühlsten Monate mit Tagestemperaturen um 20 °C und einer Nachtabkühlung bis auf 4 °C. Ab März wird es wieder wärmer und niederschlagreicher.

Die Luftfeuchtigkeit ist das ganze Jahr über sehr hoch.

Paphiopedilum fairrieanum ist sehr einfach zu kultivieren. Man sollte die Pflanzen während der Wachstumsperiode ausreichend gießen und für eine gute Luftumwälzung und viel Licht sorgen. Von Dezember bis Mitte Januar sollte man *Paphiopedilum fairrieanum* kühler und trockener halten, um eine gute Blüteninduktion zu erreichen.

Am besten stellt man die Pflanzen in den kühlsten Teil des temperierten Gewächshauses.

Paphiopedilum fairrieanum blüht im allgemeinen von Januar bis März.

crescent, ivory-white, mottled with green, and with a purple band along the front.

Paphiopedilum fairrieanum grows on steep dolomite cliffs in humus in the cracks and crevices at an elevation of 1400 to 2000 m above sea-level in Bhutan, Assam and Sikkim.

This distribution area is influenced by a wet summer monsoon, blowing from the southwest as well as by a dry northeastern monsoon. From June to October heavy rains fall, often lasting for days. During that time the temperatures are up to about 32 °C during the day and 18 °C at night.

In late October, the northeastern monsoon brings cooler and dryer weather, but rainfall never ceases completely. December and January are the coolest months with daytime temperatures averaging 20 °C and night time temperatures dropping to 4 °C. From March on, it becomes warmer and wetter again.

The relative humidity is very high over the whole year.

Paphiopedilum fairrieanum is very easy to grow. The plants should be watered abundantly during the growth period and be placed in a bright spot with good air circulation.

Between December and January, the plants should be given a rest, keeping them cooler and drier. This results in good flower induction.

Paphiopedilum fairrieanum does best in the coolest part of the intermediate house.

Varietäten:

Von allen „Varietäten", die von *Paphiopedilum fairrieanum* beschrieben worden sind, ist nur eine so verschieden, daß sie als autonomes Taxon anerkannt werden kann.

Varieties:

Of all the "varieties" of *Paphiopedilum fairrieanum* published, only one deserves to be mentioned here:

Paphiopedilum fairrieanum var. *bohlmannianum* Matho
Die Orchidee 33: 17 (1942)

Dies ist die gültige Benennung der Albinoformen von *Paphiopedilum fairrieanum*. Die Blüte ist gelb bis gelblich grün, ohne jegliche roten Pigmente.

Paphiopedilum fairrieanum var. *bohlmannianum* Matho
Die Orchidee 33: 17 (1942)

This is the valid designation for the albinoform of *Paphiopedilum fairrieanum*. The flower is uni-coloured yellow to yellow-green, devoid of any red pigmentation.

Sektion STICTOPETALUM
Hallier

Noch eine Sektion mit nur einer Art – *Paphiopedilum hirsutissimum* –, die durch die lange, etwas eingeschnürte Lippe, die breitlanzettförmigen, horizontal abstehenden, etwas gebuchteten lateralen Petalen, die deutlich gewellte obere Ränder aufweisen, und eine dichte Behaarung der ganzen Blüte charakterisiert ist.

Section STICTOPETALUM
Hallier

This is another monospecific section, containing *Paphiopedilum hirsutissimum* only. This section is characterized by the long, slightly constricted lip; the broad lanceolate, vertically spreading and more or less sinuating lateral petals with distinct undulations at the basal to middle portion of the upper margin and by the dense hairiness of the entire surface of the flower.

Paphiopedilum hirsutissimum (Lindley ex Hooker) Stein

Stein's Orchideenbuch: 470 (1892)

Basionym: *Cypripedium hirsutissimum* Lindley ex Hooker
Curtis's Botanical Magazine 83, t. 4990 (1857)

Synonyma: *Cordula esquirolei* (Schlechter) Hu
Rhodora 27: 105 (1925)

Cordula hirsutissima (Lindley ex Hooker) Rolfe
Orchid Review 20 (1): 2 (1912)

Paphiopedilum chiwuanum Tang & Wang
Acta Phytotaxonomica Sinica 1 (1): 56 (1951)

Paphiopedilum esquirolei Schlechter
FEDDE: **Repertorium,** Beiheft **4:** 39, 86 (1919)

Paphiopedilum hirsutissimum (Lindley ex Hooker) Stein

Paphiopedilum hirsutissimum wurde um 1857 durch den Pflanzensammler SIMONS ohne Angabe des Fundortes nach England gesandt. Der Fundort blieb zunächst unbekannt, bis John DAY einige Pflanzen dieser Art von seinem Neffen, Hauptmann WILLIAMSON, erhielt, der diese auf den Khasia-Hügeln Assams gefunden hatte.

Im Jahre 1919 schuf Rudolf SCHLECHTER ein neues Taxon als *Paphiopedilum esquirolei,* aufgrund einer Pflanze, die er aus Südchina erhalten hatte. Obwohl der Typus von *Paphiopedilum esquirolei* im letzten Weltkrieg zerstört wurde, ist sicher, daß es sich dabei um eine Variante von *Paphiopedilum hirsutissimum* handelt.[1)]

Die chinesischen Botaniker TANG und WANG beschrieben im Jahre 1951 ein *Paphiopedilum chiwuanum (loc. cit.).* Der Beleg, der sich im Herbar in Beijing (Peking) befindet, stammt von einer Auf-

[1)] Siehe unter Varietäten Seite 127.

Plants of this species were first sent to England by a plant collector named SIMONS. The exact origin remained unclear until John DAY received some plants of the same species from his nephew, Captain WILLIAMSON, collected in the Khasia Hills in the Indian State of Assam.

In 1919, Rudolf SCHLECHTER published a new concept of a *Paphiopedilum*-species as *Paphiopedilum esquirolei,* based on a plant which he had obtained from southern China. Although the type specimen of *Paphiopedilum esquirolei* was destroyed during the World War II, it is certain that *Paphiopedilum esquirolei* is to be considered as a variant of *Paphiopedilum hirsutissimum.*[1)]

In 1951, the Chinese botanists TANG and WANG described a *Paphiopedilum chiwuanum (loc. cit.).* The type specimen, being kept in the herbarium of Beijing (Peking), originates from a collection from

[1)] See under described varieties page 127.

Paphiopedilum hirsutissimum (Lindley ex Hooker) Stein

sammlung WANGs aus dem südlichen Teil Yunnans und besteht aus einer Pflanze in Knospe. Sowohl S. C. CHEN[1] als auch C. Z. TANG und P. J. CRIBB[2] haben den Originaltypus untersucht und betrachten *Paphiopedilum chiwuanum* als Synonym zu *Paphiopedilum hirsutissimum*. CRIBB[3] betrachtet aber neuerdings das Taxon als eigenständige Varietät.

Paphiopedilum hirsutissimum ist ein krautiger Humusepiphyt.

Die Laubblätter sind 16 bis 40 cm lang, ca. 1,5 bis 3 cm breit. Sie sind einfarbig grün, linealisch länglich, vorne spitz.

Die Infloreszenz ist bis zu 30 cm hoch, grün, mit dunkelvioletter Behaarung, meistens einblütig.

Die Braktee ist mit 2,6 cm Länge ungefähr ein Drittel so lang wie der Fruchtknoten. Braktee und Fruchtknoten sind dunkel purpurn behaart.

Die Blüte ist 10 bis 15 cm hoch.

Die Sepalen (Fahne und Synsepalum) sind, wie Fruchtknoten und Braktee, außen purpurn behaart und an den Rändern bewimpert.

Die Fahne ist bis ca. 4 cm lang und 3 cm breit, hinten gekielt, breit herzförmig, an der Basis und in der Mitte dicht dunkelpurpurn getupft; in der breiten Randregion heller oder dunkler grün.

Das Synsepalum ist etwas kleiner, hellgrün mit purpurner Beaderung.

Die ungefähr 7,5 cm langen und 2 cm breiten, breit spatelförmigen, waagerecht abstehenden oder

[1] **Iconographia Cormophytorum Sinicorum 2:** 609–610 (1976).

[2] **The Plantsman 4:** 165–170 (1982).

[3] **The Genus Paphiopedilum** (1987).

the southern part of Yunnan, and consists of a plant in bud. S. C. CHEN,[1] as well as C. Z. TANG and P. J. CRIBB[2] have viewed the type and concluded that *Paphiopedilum chiwuanum* is synonymous to *Paphiopedilum hirsutissimum*. CRIBB, in his latest publication[3], however, considers the taxon to be an autonomous variety.

Paphiopedilum hirsutissimum is an herbaceous plant, generally growing epiphytically in humus.

The leaves are 16 to 40 cm long, about 1.5 to 3 cm wide, uniformly green, linear-oblong, acute; the basal parts of the under surface suffused with purple.

The inflorescence is up to about 30 cm high, usually green and clothed with dark purple hairs, generally bearing a single large, 10 to 15 cm high flower, having all the segments ciliated.

The bract is more or less 2.6 cm long, about a third the length of the ovary. Bract and ovary are, just like the scape and the entire outer side of the flower, clothed with blackish purple hairs.

The dorsal sepal is broadly cordate, up to about 4 cm long and 3 cm across, keeled behind, the central and basal area densely spotted with blackish purple, in the broad marginal regions deeper or paler green.

The synsepalum is somewhat smaller, ovate, pale green with some purplish marks along the veins.

The lateral petals are about 7.5 cm long and 2 cm broad; they are broadly spathulate, spreading hori-

[1] **Iconographia Cormophytorum Sinicorum 2:** 609–610 (1976).

[2] **The Plantsman 4:** 165–170 (1982).

[3] **The Genus Paphiopedilum** (1987)

etwas nach unten gerichteten Petalen sind leicht gedreht, an den Rändern bewimpert und in der unteren Hälfte gewellt und gekräuselt, grün mit tiefpurpurnen Flecken und von vielen schwarzen Haaren besetzt, im apikalen Teil hell purpurn.

Die Lippe ist bis 5 cm lang und ungefähr 3 cm breit, umgekehrt helmförmig, mattgrün, braunpurpurn überhaucht und mit winzigen, schwärzlichen Warzen überzogen.

Die fast quadratische Staminodiumplatte ist grün mit zwei weißen „Augen" im oberen Bereich.

Paphiopedilum hirsutissimum hat eine Verbreitung von Nordostindien bis Nordbirma, Nordthailand und Südchina. Die Pflanzen wachsen im allgemeinen als Humusepiphyten am Waldboden oder auf moosbedeckten Felsen in Höhenlagen von 1000 bis 1400 m ü. NN. Sie bevorzugen schattige Standorte.

Entsprechend den klimatischen Bedingungen im Heimatareal sollte man dieser Art von Juni bis Oktober viel Wasser geben, die Temperaturen am Tage bis auf 30 °C steigen lassen und für eine Nachtabsenkung bis auf 14 °C sorgen. Ab Mitte Oktober kann man die Pflanzen kühler stellen und weniger gießen. Im Dezember hält man die Temperaturen tagsüber bei maximal 17 °C und nachts bei 5 °C. Dabei ist das Gießen fast ganz einzustellen.

Ab Januar wird wieder mehr gegossen und die Temperaturen werden allmählich gesteigert.

Paphiopedilum hirsutissimum blüht von März bis Mai aus zwei bis drei Jahre alten Trieben. Die Varietät *esquirolei* kann entsprechend ihrer Herkunft etwas wärmer gehalten werden.

Varietäten:

Von *Paphiopedilum hirsutissimum* sind eine Reihe von „Varietäten" beschrieben worden. Die meisten basieren lediglich auf einer „besonders schönen" Pflanze und können kaum den Varietätsstatus für sich in Anspruch nehmen. Von diesen „Pseudovarianten" weicht nur folgendes Taxon insoweit ab, daß es vom Normaltyp deutlich zu unterscheiden ist.

Paphiopedilum hirsutissimum var. *esquirolei* (Schltr.) Karasawa & Saito
Bull. of the Hiroshima Botanical Garden Nr. 5: 1–69, (1982)

Diese Varietät unterscheidet sich von der Normalform durch spärliche und kurze Behaarung, schmalere, hell gefärbte Laubblätter ohne purpurnen Überzug auf dem basalen Teil der Blattunterseite, und eine dunkelpurpurne, grün umrandete Fahne.

zontally, slightly twisted, the margins crisp and undulate along the basal half, the narrower basal part green, blotched and spotted with deep purple and studded with numerous blackish hairs, the dilated apical part bright violet-purple.

The lip is inverted helm-shaped, up to 5 cm long and about 3 cm broad, dull green stained with brownish purple and dotted with minute blackish warts.

The staminodal shield is nearly square, green with two white spots ("eyes") near the basal edge.

Paphiopedilum hirsutissimum is distributed from the north-eastern part of India to the northern parts of Burma and Thailand and to the south of China.

The plants usually grow epiphytically in leafy debris and humus-like topsoil in woods as well as on moss covered rocks, in altitudes of 1000 to 1400 meters above sea-level, generally in shady habitats.

According to the climatic conditions in the area of natural distribution, the plants should be grown in the intermediate house.

From July to October, *Paphiopedilum hirsutissimum* requires plenty of water and the temperatures can be as high as 30 °C during the day and as low as 14 °C during the night. From the middle of October, the plants should be kept somewhat cooler and drier. In December the temperatures should be maximal 17 °C during the day and a minimum 5 °C at night. During this cold period, waterings should be reduced to a minimum.

From January on, water can be given more abundantly and the temperatures can be raised gradually. The variety *esquirolei* should be kept somewhat warmer.

Paphiopedilum hirsutissimum usually flowers from March through May from 2 to 3 year old growths.

Varieties:

Although a number of pseudovariations, mainly based on more or less noticeable differences of the flower colour have been described, only the var. *esquirolei* can be considered a distinct geographic race and therefore deserves to be treated here.

Paphiopedilum hirsutissimum var. *esquirolei* (Schltr.) Karasawa & Saito
Bull. of the Hiroshima Botanical Garden No. 5: 1–69 (1982)

This geographical race differs from the normal form in being more sparsely covered with shorter hairs; narrower and brighter coloured leaves, without any purple in the basal regions of the under surface; and a deep purple dorsal sepal with a green margin.

Sektion THIOPETALUM Hallier

Diese aus zwei Arten bestehende Sektion ist ähnlich der Sektion **Paphiopedilum.** Sie unterscheidet sich von dieser durch eine einzelne, deutliche Mittellinie auf der Fahne und auf den seitlichen Petalen. Die Chromosomenzahl ist 2n = 30.

Schlüssel zu den Arten der Sektion THIOPETALUM

1. Fahne gänzlich grün bis gelb-grün mit einer einzigen dicken vertikalen, braungefärbten Mittellinie; obere Ränder der Petalen kaum gewellt; Schuh gelb; Staminodiumplatte herzförmig, gelb überhaucht *Paphiopedilum druryi*

1a. Die obere Hälfte der Fahne weiß, die untere Hälfte grün, mit einem dünnen vertikalen Mittelstrich; obere Ränder der Petalen eindeutig gewellt; Schuh dunkelbraun überhaucht; Staminodiumplatte rautenförmig, die oberen Teile stark nach innen gebogen, mit einer charakteristischen kreuzartigen purpurnen Zeichnung . . . . *Paphiopedilum spicerianum*

Section THIOPETALUM Hallier

This section comprises two species and is very similar to section **Paphiopedilum.** It differs however by a distinct longitudinal middle stripe on the dorsal sepal and lateral petals and a somatic chromosome count of 2n = 30.

Key to the species belonging to section THIOPETALUM

1. Dorsal sepal uni-coloured green to greenish-yellow with the exception of the thick longitudinal brown middle-stripe; upper margins of the lateral petals not or only faintly undulate; lip yellow; staminodal shield obcordate, suffused yellow . . . . *Paphiopedilum druryi*

1a. The upper half of the dorsal sepal white, the lower half green, with a thin longitudinal stripe in the middle; upper margins of the lateral petals distinctly undulate; lip suffused dark brown; staminodal shield rhombic, the upper part distinctly turned inwards, with a characteristic cross-shaped purple marking in the middle . . . . *Paphiopedilum spicerianum*

Paphiopedilum druryi (Beddome) Stein

Stein's Orchideenbuch: 466 (1892)

Basionym: *Cypripedium druryi* Beddome
Icon. Plant. Ind. Orien. 23, t. 112 (1874)

Synonym: *Cordula druryi* (Beddome) Rolfe
Orchid Review 20 (1): 2 (1912)

Paphiopedilum druryi (Beddome) Stein

Paphiopedilum druryi wurde um 1865 in den Travancore-Hügeln im Süden Indiens durch den englischen Armee-Oberst H. DRURY entdeckt. Ihm zu Ehren wurde sie durch R. H. BEDDOME beschrieben.

Bis heute ist diese Art eine Seltenheit in den Gewächshäusern geblieben. Das wird leider auch in Zukunft so sein, da das Verbreitungsgebiet dieser Art sehr begrenzt und abgelegen ist. Nur durch Samenvermehrung der sich in Kultur befindlichen Exemplare ließe sich hier Abhilfe schaffen.[1)]

[1)] Der Orchideenhobbyist sollte beim Kauf von *Paphiopedilum druryi* Jungpflanzen vorsichtig sein. Es werden Pflanzen angeboten, die angeblich aus einer Selbstung einer *Paphiopedilum druryi* stammen sollen, aber in Wirklichkeit das Resultat einer Kreuzung mit *Paphiopedilum villosum* sind. Diese Hybriden erkennt man eindeutig an den im Endbereich überhängenden Laubblättern.

Paphiopedilum druryi was discovered about the year 1865, on the Travancore Hills, in the extreme south of India by the British Army Colonel DRURY and described in his honour by R. H. BEDDOME.

To date, *Paphiopedilum druryi* is still a rarity in collections. The reason for this being the extreme isolation of the species in nature. It is therefore necessary that the artificial propagation of this species be intensified.[1)]

Paphiopedilum druryi is an herbaceous plant, growing as a leaf humus epiphyte.

[1)] The interested orchid amateur should be aware of the fact that plants have been offered for sale allegedly originating from a selfing of *Paphiopedilum druryi,* but in fact being the progeny of a cross between *Paphiopedilum druryi* and *Paphiopedilum villosum.* These hybrids can readily be recognized by the leaves arcing in the apical part.

Paphiopedilum druryi (Beddome) Stein

Paphiopedilum druryi ist eine bodenbewohnende, als Humusepiphyt wachsende, krautige Pflanze mit einem relativ kräftigen, kriechenden Rhizom.

Die kurzen Stämme tragen 4 bis 5 länglich riemenförmige bis 25 cm lange und 3,5 cm breite, lederartige, halbaufrecht bis abstehende, vorn abgerundete, einfarbig hellgrüne Laubblätter.

Die einblütige Infloreszenz wird bis zu 30 cm hoch. Der Infloreszenzschaft ist purpurn, stielrund, am Grunde mit einer kurzen, zusammengedrückten grünen Scheide, dicht flaumig und lang behaart.

Die hellgrüne Braktee ist bis 2 cm lang.

Die Blüte ist bis 7,5 cm hoch und 5,5 cm breit. Sie ist etwas geneigt, außen flaumig behaart, hell mattgelb in der Grundfärbung, mit braunen oder purpurvioletten Linien und Tupfen.

Die Fahne ist ungefähr 3,5 cm lang und 3,3 cm breit, breit eiförmig, stumpf und vorwärts gekrümmt, bleich grünlich gelb mit einem breiten, tiefbraunen Längsstreifen in der Mitte, hinten deutlich gekielt und mit vielen schwärzlichen Haaren überzogen.

Das Synsepalum ist ca. 3 cm lang und bis ca. 2,3 cm breit, elliptisch, spitz zulaufend, sehr bleich

The plants are readily recognized by their stoutish, creeping and ascending rhizome.

The short stems carry up to 5, ligulate, acute leaves. These are up to about 25 cm long and 3.5 cm wide, leathery, suberect to spreading, rounded at the apex, uniformly bright green.

The inflorescence is up to 30 cm high, terete, purple, at the base with a short compressed green sheath, densely downy, usually carrying one single flower.

The bract is small and sheathing, up to about 2 cm long.

The flower is up to 7.5 cm long and about 5.5 cm across, somewhat inclined, the outer surface downy, of a dull yellow ground colour, striped and spotted brown to purple-violet.

The dorsal sepal is broadly ovate, obtuse and inflexed above, up to about 3.5 cm long and 3.3 cm broad, ciliolate at the margin, greenish yellow with a broad black median band, keeled behind and studded with numerous blackish hairs.

The synsepalum is about 3 cm long and up to about 2.3 cm wide, elliptoid, acuminate, paler in in der Grundfärbung, oft mit zwei dunkelpurpurnen Längsstreifen.

Die umgekehrt lanzettförmigen, länglichen Petalen sind etwa 4,2 cm lang und 2 cm breit, gleichzeitig nach vorne und nach unten gebogen, an den Rändern bewimpert und in der basalen Hälfte gewellt, an der Spitze 3zähnig, gelb, mit einem mittleren, braunen Längsstreifen, purpurn getupft. Sie sind an der Basis mit einigen braunen Härchen versehen.

Die Lippe ist sackförmig, ungefähr 3,3 cm lang und 2 cm breit, 3lappig. Die Seitenlappen sind nach innen gebogen. Der Mittellappen ist umgekehrt helmförmig, außen glänzend gelb, innen purpurn getupft.

Die Staminodiumplatte ist umgekehrt herzförmig, oben 2lappig, 3zähnig, dunkelgelb mit einem dottergelben Mittelfleck, ca. 1 bis 1,2 cm lang und ebenso breit.

Paphiopedilum druryi wächst in Höhenlagen um 1500 m ü. NN in hartem, lehmigem Boden auf Kalkgestein. Das Verbreitungsgebiet sind die Kalikaden- und Travancore-Hügel im Süden Indiens.

Die Regenperiode in diesem Gebiet ist von Mai bis Ende Oktober. Die Temperaturen schwanken in dieser Periode zwischen 30 °C am Tage und 14 °C in der Nacht.

Von November bis April herrschen trockene Winde. Während dieser Zeit fallen die Temperaturen nachts auf rund 6 °C und steigen am Tage bis auf 18 °C an.

Die Pflanzen wachsen zwischen Gräsern, die sie in der warmen Jahreszeit vor der intensiven Sonneneinstrahlung schützen.

Paphiopedilum druryi gedeiht am besten im temperierten Gewächshaus mit einem etwas kühleren Standort im Dezember. Die Pflanzen brauchen eine hohe Lichtintensität und bevorzugen einen Pflanzstoff mit Moosbeimischung.

Paphiopedilum druryi blüht im allgemeinen in den Monaten Januar bis April.

colour than the dorsal sepal and frequently with two blackish-purple longitudinal lines.

The lateral petals are elongated lanceolate to ligulate, about 4.2 cm long and more or less 2 cm wide, inflexed, undulate and somewhat drooping, the margins ciliate, the apex 3-dentate, ochreous, almost golden yellow, with a broad blackish median line, spotted purple; at the base with some brown hairs.

The lip is deeply saccate, about 3.3 cm long and about 2 cm wide, 3-lobate, the lateral lobes infolded. The median lobe is inverted helm-shaped, the outside bright yellow, the inside spotted reddish to purple.

The staminodal shield is obcordate, bilobate at the top, 3-toothed, deep yellow with a yolk-coloured medium part, 1 to 1.2 cm long and wide.

Paphiopedilum druryi grows in altitudes of up to 1500 m on rather hard loamy soil on a limestone underground. The natural distribution is probably restricted to the Kalikad Hills and the Travancore Highlands in the outmost southern part of India.

The rainy period in this area is between May and the end of October. During this time the temperature ranges between 30 °C during the day and 14 °C at night.

From November to April dry winds prevail. During this season, the temperature drop to about 6 °C at night and raise to 18 °C during the daytime.

The plants grow between grasses, shading them during the warm season.

Paphiopedilum druryi does best when cultivated in the intermediate house and moved to a colder spot during December. This species needs a high amount of light and prefers an open substrate mixed with moss.

The flowering period of *Paphiopedilum druryi* is between January and April.

Paphiopedilum spicerianum (Rchb. f.) Stein

Stein's Orchideenbuch: 485 (1892)

Basionym: *Cypripedium spicerianum* Rchb. f. **Gardeners' Chronicle,** 2. Ser., **13:** 363 (1880)

Synonym: *Cordula spiceriana* (Rchb. f.) Rolfe **Orchid Review 20** (1): 2 (1912)

Paphiopedilum spicerianum (Rchb. f.) Stein

Paphiopedilum spicerianum wurde durch einen Plantagen-Angestellten in Bhutan entdeckt. Eine Pflanze wurde durch Herrn SPICER, den Besitzer der Plantage, an seine Eltern in England verschickt. Die Pflanze blühte im Jahre 1878. Ein Teilstück wurde durch die Firma VEITCH aufgekauft und durch REICHENBACH als *Cypripedium spicerianum* benamt,[1] aber nicht sofort veröffentlicht. Am 10. Januar 1880 erschien im **Gardeners' Chronicle**[2] eine kurze Beschreibung und eine Strichzeichnung. Die ausführliche Beschreibung REICHENBACHs erschien erst am 20. März desselben Jahres wiederum im **Gardeners' Chronicle**.

[1] Laut REICHENBACH am 1. Dezember 1878, *fide* **Gardeners' Chronicle,** 2. Ser., **13:** 363 (1880).

[2] **Gardeners' Chronicle,** 2. Ser., **13:** 40, 41.

Paphiopedilum spicerianum was discovered by a plantation worker in Bhutan. The owner of the plantation, Mr. SPICER sent a plant to his parents in England. The specimen flowered in 1878, and a division was purchased by the horticulturalist VEITCH, who sent some material to REICHENBACH. REICHENBACH named the plant as *Cypripedium spicerianum*[1] but did not publish his concept until two years later. A short description, accompanied by a line drawing appeared on January 10th., 1880[2] and a detailed one was published on March 20th of the same year, again in the **Gardeners' Chronicle.**

[1] **According to REICHENBACH on the 1. December 1878, *fide* Gardeners' Chronicle,** 2. Ser., **13:** 363 (1880).

[2] **Gardeners' Chronicle,** 2 Ser., **13:** 40, 41.

Die genaue Herkunft der im VEITCHschen Besitz befindlichen Pflanze war nicht bekannt, und die Firma begnügte sich mit einer vegetativen Vermehrung, um ab und zu einen zahlungskräftigen Kunden zu beglücken.

Auch die Konkurrenzfirma SANDER lockte der zu erwartende Gewinn, und so wurde versucht, der Lady SPICER die Herkunft der Pflanze zu entlokken. SANDER gelang dies schließlich bei einem Teebesuch. Sofort sandte er seinen Sammler FOERSTERMANN nach Indien. Es gelang ihm, den Fundort ausfindig zu machen und 40000 Pflanzen nach England zu verfrachten. Es ist klar, daß der Markt solche Menge nicht verkraften konnte; der Preis für ein *Paphiopedilum spicerianum* sackte rapide ab, so daß die Reise FOERSTERMANNs trotz des Fundes ein wirtschaftliches Fiasko wurde, welches die Firma SANDER fast in den Ruin trieb.

Paphiopedilum spicerianum hat eng stehende Sprossen, die meist 4 bis 5 Blätter tragen.

Die länglichen Laubblätter werden bis ungefähr 22 cm lang und bis 5,5 cm breit. Sie sind dunkelgrün, oberseitig von feinen, helleren Längsadern durchzogen und deutlich gefurcht mit einem meist dunkelpurpurnen Mittelnerv.

Die deutlich gekielte Unterseite ist etwas heller grün mit intensiver, dunkelpurpurroter Befleckung an der Basis. Die Blätter sind oft von der Basis an aufwärts am Rande leicht gewellt. Die Blattspitzen sind unregelmäßig gezahnt.

Die purpurne Infloreszenz wird 20 bis 25 cm hoch und trägt meist eine, selten auch zwei, bis ungefähr 8 cm hohe und 7 cm breite Blüte(n).

Die Fahne ist ca. 4 cm hoch und 5,6 cm breit, herzförmig. Der obere Rand ist nach vorne, der untere nach hinten gebogen, so daß die grüne Basis unten deutlich keilförmig verschmälert erscheint; weiß, mit Ausnahme der Mittelfalte, die durch ein rotpurpurnes Band gekennzeichnet ist, manchmal an der Spitze rosa überhaucht. Die Fahnenhälften sind längs der deutlich purpurrot gefärbten Mittelader leicht nach hinten gefaltet.

Das Synsepalum ist breit eiförmig, hellgrün mit weißem Rand, spitz zulaufend.

Die Petalen sind leicht nach vorne gebogen, nach unten hängend, ca. 4 cm lang und ungefähr 0,8 cm breit, schmal länglich, vorne abgerundet, am oberen Rande gewellt. Sie sind gelbgrün, kräftig purpurn marmoriert und mit purpurnem Mittelstreifen.

Die Lippe ist bis 4,2 cm lang und ungefähr 3 cm breit, sackförmig, glänzend bronzegrün, mit gel-

The detailed provenance of the plant was not known and for quite some time, VEITCH was quite satisfied in duplicating the plant by vegetative division to satisfy the needs of some of his more wealthy customers.

However the competition did not sleep, and SANDER, envious of a prosperous affair, managed to be invited for tea at the SPICERs. A remark from Mrs. SPICER about her son's plantation was enough reason to send his best collector – FOERSTERMANN – to India. FOERSTERMANN managed to seek out the habitat and send a shipment of 40000 plants back home. Of coarse, the market could not cope with such an enormous amount of plants and they were sold at extremely low prices (whereas the divisions from VEITCH were sold only a few months earlier for a small fortune). The result was that FOERSTERMANNs trip, in spite of the discovery, nearly ruined SANDER.

Paphiopedilum spicerianum is an herbaceous orchid with a tufted habit, the growths usually bearing 4 to 5 leaves.

The leaves are linear-oblong, up to about 22 cm long and up to 5.5 cm wide, more or less undulate at the margins, dark green above, distinctly grooved; the underside spotted with purple towards the base, keeled, lighter green; the apices irregularly dentated.

The inflorescence is erect, up to about 20 to 25 cm high, purple, usually bearing one, rarely two to about 8 cm high and 7 cm wide flower(s).

The dorsal sepal is broadly obcordate, about 4 cm high and up to 5.6 cm across, folded at the middle, with the lateral margins much reflexed at the base, and with the apical margin bent forwards; white except for the fold where there is a crimson purple band, and at the base where there is a large green blotch speckled with dull red; sometimes with the apex suffused rose.

The synsepalum is broadly ovate, bright green with a white margin, acuminate.

The lateral petals are ligulate, deflexed and curved forwards, undulate at the margins, yellowish-green, spotted with dull red and with a crimson mid-line; about 4 cm long and more or less 0.8 cm wide.

The lip is up to about 4.2 cm long and up to 3 cm broad, saccate, shiny bronze-green tinged with crimson, the infolded lateral lobes yellow spotted purple.

The staminodal shield is suborbicular, at the upper margin with a v-shaped notch, the center rose to ben Rändern. Die Seitenlappen sind gelb mit purpurnen Flecken.

Die Staminodiumplatte ist eiförmig, oft fast rund, oben V-förmig eingeschnitten. Die rosa bis violettfarbene Mittelfläche trägt ein unregelmäßiges grünliches Sternmuster. Der Rand ist weiß.

Paphiopedilum spicerianum wächst in Indien und Assam an steilen Abhängen und in engen Flußschluchten, auf Kalkfelsen zwischen Gräsern und Farnen bei ziemlich diffusen Lichtverhältnissen in Höhenlagen von 600 bis 1300 m ü. NN. Mit ihrem kräftigen Wurzelwerk durchdringen die Pflanzen humusgefüllte Felsspalten und verankern sich am Kalkgestein.

Im Heimatareal herrschen von September bis Ende Februar die nordöstlichen Monsune, die Nebel, Sprühregen und Kühle mit sich bringen. Im Dezember sinken die Temperaturen auf 4 °C in der Nacht und steigen nur bis 18 °C am Tage.

Ab März wird das Klima etwas milder, und im April herrschen Temperaturen bis zu 32 °C am Tage und 18 °C in der Nacht.

Im Mai beginnt eine ausgesprochene Regenperiode, wobei es während der Monate Juni, Juli und August oft tagelang ohne Pause regnet. Die Luftfeuchtigkeit ist also über das ganze Jahr hinweg sehr hoch.

In Kultur benötigen diese Pflanzen ein mooshaltiges Substrat, ständig hohe Luftfeuchtigkeit und ausreichende Luftumwälzung. In unserem Frühjahr und im Frühsommer sollten die Temperaturen nicht über 27 °C am Tage ansteigen und nicht unter 13 °C in der Nacht absinken.

Zur Anregung der Blütenbildung ist in den Monaten Juni bis Juli eine ausgeprägte Nachtabsenkung erforderlich.

Paphiopedilum spicerianum sollte ständig feucht, aber nie naß gehalten werden.

Die Blütezeit von *Paphiopedilum spicerianum* fällt im allgemeinen in die Zeit von Oktober bis Dezember.

Varietäten:

Von *Paphiopedilum spicerianum* sind eine ganze Reihe von „Varietäten" beschrieben worden (meist als *Cypripedium spicerianum* var.). Keine davon unterscheidet sich wesentlich vom „Normaltyp".

violet with a irregular greenish cross-shaped pattern, the margins white.

Paphiopedilum spicerianum has its natural distribution in India and Assam. The plants grow on steep cliffs and in narrow ravines, on limestone rock between grasses and ferns in relatively diffuse light. The habitats are usual at elevations between 600 and 1300 meters above sea-level. The stout roots penetrate the humus in the crevices and keep the plants tightly attached to the rocky underground.

In the areas were *Paphiopedilum spicerianum* is to be found, a northeasterly monsoon blows between September and the end of February, bringing fog, drizzles and cold. In December, the temperatures are as low as 4 °C at night and climb to about 18 °C during the day.

Starting at the beginning of March, the weather becomes somewhat milder and in April the temperatures become as high as 32 °C during the day and about 18 °C at night.

May brings a distinct rainy season with storms in June, July and August, sometimes lasting for days at a time. The relative humidity is high at all times.

For successful cultivation, the plants should be potted in an open, well drained, but moisture retaining mossy substrate. *Paphiopedilum spicerianum* does best in the intermediate house with high humidity and a high rate of air exchange. During the spring months and in early summer, the daytime temperature should not exceed 27 °C and at night, it should not be cooler than 13 °C.

The plants should be kept moist all over the year without leaving them wet.

Paphiopedilum spicerianum will only flower well after a resting period with lower temperatures between May and July. The normal flowering time is between October and December.

Varieties:

A whole series of varieties of *Paphiopedilum spicerianum* has been described (most of them as *Cypripedium spicerianum* var.). None of these differ enough from the "normal" types as to warrant being upheld as autonomous taxa.

Untergattung SIGMATOPETALUM Hallier

Dies ist die artenreichste Untergattung von *Paphiopedilum* mit insgesamt 6 Sektionen, 20 Arten und 1 Naturhybride.

Die Arten dieser Untergattung sind gekennzeichnet durch gefleckte Laubblätter, einblütige Infloreszenzen, schmale Petalen, gut entwickelte seitliche Lippenöhrchen (Aurikeln), ein Synsepalum, welches kleiner ist als die Fahne und eine mehr oder weniger deutlich halbmondförmige Staminodiumplatte.

Schlüssel zu den Sektionen der Untergattung SIGMATOPETALUM

1. Blattunterseite vollständig tief purpurn befleckt, Blattoberseite fahl graugrün und tief purpurgrün marmoriert . . . . Sektion SIGMATOPETALUM

1a. Blattunterseite nicht oder nur an der Basis purpurn gefleckt, Blattoberseite fahl grau oder fast weiß, marmoriert mit Sattgrün 2

2. Laterale Petalen spatelförmig, mit einer mehr oder weniger verbreiterten Endregion und einem leicht gewellten Mittelteil des Randes; Fahne grünlich mit undeutlichen dunkelgrünen Längsstreifen und einem stark nach innen gedrehten oberen Randteil Sektion SPATHOPETALUM

2a. Laterale Petalen anders geformt, ohne gewellten Rand; Fahne weiß oder weißlila mit deutlichen dunkelgrünen oder dunkelpurpurnen Längsstreifen und einem fast geraden oder etwas zurückgebogenen oberen Rand . 3

3. Laterale Petalen mit fast parallel verlaufenden Rändern, in der oberen (apikalen) Hälfte meistens dunkler gefärbt als an der Basis, nicht oder nur sehr undeutlich gefleckt . . . . Sektion BLEPHAROPETALUM

3a. Laterale Petalen fast einfarbig, mit großen, mehr oder weniger verdickten, tief purpurnen Flecken 4

4. Laterale Petalen kaum bewimpert, fast gänzlich ganzrandig, schmal spatelförmig, zur Spitze hin allmählich breiter werdend; Blüte fahl braun . . . . Sektion PUNCTATUM

4a. Laterale Petalen mehr oder weniger regelmäßig bewimpert, schmal länglich bis linealisch, Blüte nicht fahl braun 5

Subgenus SIGMATOPETALUM Hallier

This is the largest subgenus of *Paphiopedilum,* comprising 6 sections with a total of 20 species and 1 natural hybrid.

The species of this subgenus are characterized by tesselate leaves, uni-flowered racemes, narrow petals, well-developed lateral ears of the lip, a synsepalum distinctly smaller than the dorsal petal as well as by a more or less semi-lunate staminodal shield.

Key to the sections of the subgenus SIGMATOPETALUM

1. Under surface of the leaves tinged dark purple throughout, leaf upper surface tesselate with pale grayish green parts and deep green parts . . . . section SIGMATOPETALUM

1a. Under surface of leaves pale green throughout, or slightly tinged deep purple at the base, leaf upper surface tesselated with pale grayish or almost white parts and green parts . . . 2

2. Lateral petals spathulate, more or less widened in the spical region, margin slightly undulate at middle; dorsal sepal greenish, margin of upper half strongly incurved, longitudinal stripes indistinct, deep green in colour . . . . section SPATHOPETALUM

2a. Lateral petals shaped otherwise, margin not undulate; dorsal sepal white or purplish, margin of upper half not incurved, almost straight or slightly incurved, longitudinal striped distinct, deep green or deep purple 3

3. Lateral petals without large, more or less swollen distinct spots, almost same width from base to tip, upper halves usually deeper in colour than the lower halves . . . . section BLEPHAROPETALUM

3a. Lateral petals almost the same colour throughout, with large, more or less swollen deep purple spots 4

4. Lateral petals hardly ciliate, the margins almost entire throughout, narrow spathulate, gradually widened towards the apex; flower pale brown in colour . . . . section PUNCTATUM

4a. Lateral petals more or less regularly ciliate, narrow oblong or linear oblong; flower colour not pale brown 5

5. Laterale Petalen schmal länglich, an der Basis fast ebenso breit wie an der Spitze, nicht gedreht, horizontal abstehend, Staminodiumplatte mehr oder weniger halbmondförmig . . . . Sektion PLANIPETALUM

5a. Laterale Petalen linealisch, zur Spitze hin allmählich schmaler werdend, leicht herunterhängend; Staminodiumplatte halbkreisförmig oder oval . . . . Sektion BARBATA

5. Lateral petals narrow oblong, almost same width from the base to the apex, not twisted, horizontally spreading; staminodal shield more or less semi-lunate . . . . section PLANIPETALUM

5a. Lateral petals linear oblong, gradually narrowed toward the apex, slightly declivitous; staminodal shield subcircular to oblong . . . . section BARBATA

Sektion SIGMATOPETALUM

Diese nur *Paphiopedilum venustum* beherbergende Sektion unterscheidet sich von allen anderen durch die gänzlich purpurn überzogene Blattunterseite. Sobald die Art in Blüte ist, fällt die hervorstechende Beaderung der gelb braunen Lippe auf.

Was die verwandtschaftlichen Beziehungen anbelangt, so ist anzunehmen, daß diese Sektion am engsten mit der Sektion **Barbata** verwandt ist.

Die Sektion ist außerdem die nördlichste der Untergattung und hat ihre Verbreitung in Nepal, Bangladesh, Assam und Ostindien.

Durch diese Anpassung an temperierte Zonen gilt sie als die am weitesten entwickelte innerhalb der Untergattung.

Section SIGMATOPETALUM

This section includes *Paphiopedilum venustum* only. The species differs from all other by the deep purplish green, tesselate leaves of which the underside is entirely suffused with purple; as well as by the grotesquely-veined, yellowish brown lip.

As far as the phylogenetic relationships are concerned, section **Sigmatopetalum** seems to be closely related to section **Barbata.**

Another noteworthy character of *Paphiopedilum venustum* is its distribution area. It is the species which grows in the northernmost area within the subgenus, being found in Nepal, Bangladesh, Assam and East India.

Because of this adaptation to temperate regions, this species is considered the most specialized within the group.

Paphiopedilum venustum (Wallich ex Sims) Pfitzer

in PRINGSHEIM: **Jahrbuch für wissenschaftliche Botanik** (1888)

Basionym: *Cypripedium venustum* Wallich ex Sims
Botanical Magazine 47, t. 2129 (1820)

Synonyma: *Cordula venusta* (Wallich ex Sims) Rolfe
Orchid Review 20 (1): 2 (1912)
Cypripedium pardinum Rchb. f.
Gardener's Chronicle: 554 (1869)
Paphiopedilum pardinum (Rchb. f.) Pfitzer
in ENGLER, A.: **Bot. Jahrb. 19:** 40 (1894)
Stimegas venustum (Wallich ex Sims) Rafinesque
Flora Telluriana 4: 45 (1838)

Paphiopedilum venustum wurde durch WALLICH im Jahre 1816 bei Sylhet[1] entdeckt.

Die erste in Europa bekannte Pflanze dieser Art wurde vom Botanischen Garten Calcuttas an die englische Firma WHITLEY, BRAMES and MILNE gesandt, wo die Pflanze kurz nach der Ankunft blühte. Das Material wurde zwecks Beschreibung an SIMS weitergeleitet, der den durch WALLICH vorgeschlagenen Namen aufgriff.

Pflanzen von *Paphiopedilum venustum* mit Blattfächern von bis zu 45 cm von Blattspitze zu Blattspitze sind keine Seltenheit. Die einzelnen Blätter erreichen eine Länge von bis zu 22 cm und eine Breite von 4 bis 5 cm. Bei den etwas älteren Blättern sind die Blattspitzen leicht gewellt und nach unten gebogen. Die Blattoberfläche ist matt silbergrün mit unregelmäßiger dunkelgrüner Marmorierung. Die Blattunterseite ist meistens intensiv bis flächendeckend mit purpurnen Flecken überzogen. Der deutliche Blattkiel und die Blattränder sind fast über die gesamte Länge glatt, nur im Endbereich sind sie leicht gezahnt. Die Triebe stehen dicht beisammen und geben der Pflanze somit eine gedrängte Wuchsform.

Die 15 bis 20 cm hohe Infloreszenz ist dunkelpurpurn gefärbt und fein behaart. Sie trägt meistens eine oder zwei 6 bis 7,5 cm große Blüte(n).

Das auffallendste Merkmal der *Paphiopedilum venustum*-Blüte ist die dunkelgrüne Äderung, die fast die gesamte Blüte überzieht und nur auf dem Endteil der Petalen durch orange oder rote Färbung maskiert wird.

[1] Damals Nord-Indien, jetzt Bangladesh.

Paphiopedilum venustum was discovered in 1816 by WALLICH in Sylhet.[1]

The plants were introduced to Europe from the Botanic Garden at Calcutta by the British Company of WHITLEY, BRAMES and MILLNE. The first plants flowered shortly after reaching England and material was forwarded to Dr. SIMS who described the species *(loc. cit.)* giving it the name suggested by WALLICH.

Although usually smaller, it is not incommon to find plants of *Paphiopedilum venustum* with a span of 45 cm. The single leaves are up to 22 cm long and 4 to 5 cm broad. The somewhat older leaves are slightly undulate and curved downwards in the apical region. The surface colour is a dull, silvery green with a random dark green blotchy pattern; the underside is usually richly spotted purple, the denser areas corresponding with the darker spots on the upper surface. The sharp keel and the leaf edges are smooth for about their complete length, slightly serrated only at the end.

The growths are close to one another, giving the plant a heavily-clustered habit.

The inflorescence is 15 to 20 cm high, terete, dark purple, shortly hirsute and usually carries one or two flowers with a natural spread of up to 7.5 cm.

The most striking feature of the *Paphiopedilum venustum*-flower is the deep green venation covering almost the entire flower and masked only in the apical regions of the petals by an orange to red colouration.

[1] At the time North India, now Bangladesh.

A

B

A–D Varianten von / variants of *Paphiopedilum venustum* (Wallich ex Sims) Pfitzer [B–C var. *measuresianum*]

C

D

E–G Blattvarianten von / leaf variants of *Paphiopedilum venustum* (Wallich ex Sims) Pfitzer

E

F

G

Die Fahne ist verkehrt herzförmig, oft mit einer breiten Basis, 2,5 bis 3 cm hoch und 2 bis 3,5 cm breit. Sie endet spitz und ist leicht schüsselförmig. Die Ränder sind manchmal leicht gewellt. Die Fahne ist weiß und wird von 12 bis 20 dunkelgrünen Adern durchzogen.

Das Synsepalum ist durch die etwas vorstehenden Ränder deutlich konkav, 2,5 bis 3 cm lang und 1,2 bis 1,5 cm breit, weißlich, mit 5 bis 10 dunkelgrünen Adern.

Die im allgemeinen ziemlich horizontal stehenden Petalen sind 4,5 bis 5,5 cm lang und 1,2 bis 1,7 cm breit. Der obere Rand ist im Endbereich meist gewellt. Die Spitzen sind oft deutlich nach hinten geschlagen, die Ränder sind unregelmäßig behaart.

Die Petalen sind über die gesamte Fläche von kleinen, unregelmäßigen, dunkelpurpurnen, warzenartigen Flecken in wechselnder Häufigkeit überzogen. Die hellgrüne bis weißliche Grundfärbung wird außerdem von einer dunkelgrünen Äderung durchzogen. In den Endregionen dominiert eine orange bis rote Färbung.

Die umgekehrt helmförmige Lippe ist 4 bis 5 cm hoch und 2,2 bis 3 cm breit. Sie steht in einem 45°-Winkel zur Infloreszenz. Die Ränder sind glatt. Die nach innen gefalteten Seitenlappen sind gerundet und leicht bräunlich gefärbt. Die Lippengrundfärbung ist ein leichtes bräunlich Creme, manchmal von rötlichen Tönen überzogen und immer von einer dunkelgrünen, netzförmigen Aderung durchzogen, so daß eine marmorierte Oberflächenstruktur entsteht.

Die Staminodiumplatte ist hufeisenförmig, ungefähr 13 mm breit und 6 mm hoch. Die nach unten weisenden Enden sind deutlich gerundet. Die Staminodiumplatte ist leicht konkav, mit einer hervorstehenden Mittelregion und einem abgerundeten, nach innen zeigenden Zahn am unteren Rand. Die glänzend helle, grüngelbe Grundfärbung ist in der Mitte durch eine dunkelgrüne, netzartige Zeichnung überzogen. Die Rückseite der Staminodiumplatte ist purpurbraun.

Paphiopedilum venustum ist an den südöstlichen Hängen des Himalaya in einer Höhe von 1000 bis 1600 m ü. NN beheimatet. Die Verbreitung ist im Vergleich zu den anderen *Paphiopedilum*-Arten ziemlich groß (siehe Sektionsbesprechung).

Die Pflanzen wachsen in Humus und Moos am Boden dichter Wälder oder in schattigen Lagen an Steilhängen, oft zwischen Bambus. Diese Art blüht im allgemeinen von Januar bis März.

Die Heimatareale von *Paphiopedilum venustum* werden von Juni bis Oktober von schweren, oft tagelang andauernden Regenfällen heimgesucht.

The dorsal sepal is obcordate, basally often quite broad, 2.5 to 3 cm high and 2 to 3.5 cm wide, pointed, slightly cupped, the margins sometimes slightly waved; white with 12 to 20 longitudinal deep green veins.

The synsepalum is distinctly concave through the forward rolling margins, 2.5 to 3 cm long and 1.2 to 1.5 cm wide, whitish with 5 to 10 deep-green longitudinal veins.

The petals are generally more or less horizontally spreading, 4.5 to 5.5 cm long and 1.2 to 1.7 cm wide, the upper margins usually undulate distally. The apical region of the petals are more or less recurved.

The petal rim is irregularly ciliate and interrupted by sparse, randomly-placed warts. Some brown spots are scattered irregularly over the light green basal area from which the dark green veins stretch out and fade into the orange to red apical region of the petal. There is a brown median line with a higher concentration of spots alongside of it, separating the more intensely coloured upper half from the lower one.

The lip is inverted helmet-shaped, 4 to 5 cm high and 2.2 to 3 cm broad, standing in a 45 degree angle to the scape. The margins are smooth. The infolded sidelobes are rounded, light brown.

The ground colour of the lip is a light browny-cream, sometimes suffused with reddish tones and always covered by a dark green, web-like veining which gives this flowerpart a distinct marbled surface appearance.

The staminodal shield has the shape of a moon crescent with strongly rounded tips. It is about 13 mm wide and more or less 8 mm high, slightly concave with a raised center and a rounded, inward pointing tooth at the lower rim. The glossy, light green colour is, around the central area, marked with a dark green, web-like pattern. The rear of the staminodal disc is purple-brown.

Paphiopedilum venustum has its natural distribution on the southeastern slopes of the Himalayas in altitudes between 1000 to 1600 meters above sea-level. The distribution area is very extensive in comparison to most other *Paphiopedilum*-species.

The plants grow epiphytically in humus and moss in dense forests or in shaded spots on steep cliffs, often between Bamboo. The flowering time is generally from January through March.

In the areas were *Paphiopedilum venustum* is found, heavy rains, often lasting for days at the time, dominate the climate between June and October. In this period, the temperatures go up to

In dieser Zeit erreichen die Tagestemperaturen 32 °C, wobei nachts noch Temperaturen um 18 °C herrschen. Im späten Oktober erreichen die nordöstlichen trockeneren Monsune das Gebiet und bringen Kühlung. Bis November gibt es noch unregelmäßige Regenschauer. Von November bis Januar schwanken die Tagestemperaturen um 20 °C, und in der Nacht sinkt das Thermometer teilweise bis auf 5 °C ab. Ab Januar fällt wenig Regen, aber es gibt andauernde Nebel, so daß die relative Luftfeuchtigkeit immer hoch bleibt. Ab März gehen die Temperaturen wieder in die Höhe, und die Niederschläge werden wieder häufiger.

Paphiopedilum venustum ist einfach zu kultivieren. Als Pflanzstoff empfiehlt sich ein gut durchlässiges Substrat mit guter Drainage, so daß die Wurzeln genügend Luftkontakt haben und Staunässe vermieden wird. Man sollte *Paphiopedilum venustum* nie vollkommen austrocknen lassen und möglichst an einem schattigen Standort belassen.

Eine Ruheperiode während der Monate November und Dezember, in der die Pflanzen etwas kühler gestellt werden und weniger Wasser erhalten, fördert die Blüteninduktion.

Varietäten:

Paphiopedilum venustum ist eine ausgesprochen variable Orchideenart was die Blütenfärbung und die Blattmarmorierung anbelangt (siehe Tafel Seite 139: E, F, G). Man wird also kaum zwei identische Wildpflanzen finden. Leider verleitet solche Variabilität immer wieder dazu, „Varietäten" zu beschreiben, die im Grunde genommen innerhalb der natürlichen Bandbreite der Art anzusiedeln sind.

Von den 9 beschriebenen Varietäten von *Paphiopedilum venustum* ist nur eine als solche anzuerkennen und auch hier etwas ausführlicher besprochen.

Paphiopedilum venustum var. ***measuresianum*** Hort.
Amer. Gard. 865 (1900)

Dies ist die **Albinoform** der Art. Die Blüten sind nicht weiß und die Pflanzen sollten daher niemals als *Paphiopedilum venustum „album"* bezeichnet werden (obwohl dies leider allgemein üblich ist).

Bei dieser Mutationsform von *Paphiopedilum venustum* ist die Synthese der roten Farbpigmente blockiert, so daß die Blüte nur Weiß, Gelb und Grün aufweist.[1]

[1] Oft nimmt man es, aus welchen Gründen auch immer, nicht so genau mit den Kriterien für eine Albinoform. So werden des öfteren Pflanzen als Albinos angeboten, die bei genauem Hinsehen noch schimmerhafte bräunliche Flecken aufweisen.

about 32 °C during the day and do not drop below 18 °C at night. In late October, dryer, northeasterly monsoon winds reach the area bringing somewhat lower temperatures. Until November there is intermittent rainfall. From November to January, the temperatures are between 20 °C during the day and 5 °C at night. January and February are the driest months but there is a constant mist assuring high humidity. From March on, the temperatures rise again and rainfall becomes more abundant.

Paphiopedilum venustum is an easy to grow species. The plants do well when potted in a well-drained, open compost assuring enough fresh air around the roots. The plants should never be allowed to become completely dry and should be kept in a well-shaded spot.

A resting period from November through December, during which the plants are kept somewhat cooler and are given less water will be rewarded by good flowering.

Varieties:

Paphiopedilum venustum is an extremely variable orchid species, especially in respect to the colouration of the flower and the tesselation of the leaves (see plate page 139: E, F, G). It is not possible to find two identical clones in the wild. Unfortunately, such high degree in variation does tempt various authors to describe "varieties", based on plants, which are in fact differing only slightly.

Of the 9 described "varieties" of *Paphiopedilum venustum*, only one deserves that status and is included here.

Paphiopedilum venustum var. ***measuresianum*** Hort.
Amer. Gard.: 865 (1900)

This is the correct designation of the **albino** of the species. The flowers are greenish-yellow but not white. Therefore, the plants should never be referred to as *Paphiopedilum venustum* var. *album* (although that is the name you will find in most literature).

In this mutant the synthesis of the reddish pigments are blocked genetically. The flower, therefore, exhibits only white, yellow and green.[1]

[1] The orchid enthusiast should be very careful when buying a plant of an albino-form. The plants offered are not always true albinos and sometimes show, although faintly, the remnants of reddish or brown pigmentation.

Sektion SPATHOPETALUM
Pfitzer

Dieser Pflanzenkomplex besiedelt Thailand, die malaysische Halbinsel, Borneo, Sulawesi und vielleicht noch einige andere Inseln des Malaiischen Archipels.

Die wesentlichen Unterschiede der Sektion zu den anderen Gruppen innerhalb der Untergattung sind:

1. eine undeutliche Streifung der Fahne,
2. in der oberen Hälfte stark nach innen gedrehte Fahnenränder,
3. ganzrandige, spatelförmige, im basalen Bereich manchmal bewimperte seitliche Petalen, die im apikalen Bereich, kurz vor der Spitze am breitesten sind und
4. eine fast runde bis länglich rautenförmige Staminodiumplatte.

Zu dieser ziemlich natürlichen phylogenetischen Gruppe wurden bisher eine ganze Reihe von Taxa gestellt, die zum Teil sehr schwer auseinanderzuhalten sind, und bei denen es sich teilweise um wiederholte Beschreibungen derselben Arten handelt.

1. *Paphiopedilum amabile* Hallier
2. *Paphiopedilum appletonianum* (Gower) Rolfe
3. *Paphiopedilum bullenianum* (Rchb. f.) Pfitzer
4. *Paphiopedilum celebesensis* Fowlie & Birk
5. *Paphiopedilum hainanensis* Fowlie
6. *Paphiopedilum hookerae* (Rchb. f.) Stein
7. *Paphiopedilum johorensis* Fowlie & Yap
8. *Paphiopedilum linii* Schoser
9. *Paphiopedilum robinsonii* (Ridley) Ridley
10. *Paphiopedilum tortipetalum* Fowlie
11. *Paphiopedilum volonteanum* (Sander) Stein
12. *Paphiopedilum wolterianum* (Krzl.) Pfitzer

Diese Pflanzengruppe kann in zwei Komplexe aufgeteilt werden. KARASAWA & SAITO haben dementsprechend zwei Untersektionen aufgestellt.[1)]

[1)] **Bull. of the Hiroshima Botanical Garden** Nr. 5: 1–69 (1982).

Section SPATHOPETALUM
Pfitzer

This complex group of orchids is distributed over Thailand, the Malayan peninsula, Borneo, Sulawesi and maybe some other islands of the Malayan Archipelago.

The characteristic differences to the other sections of the subgenus are mainly:

1. indistinct longitudinal stripes on the dorsal sepal,
2. margins of the dorsal sepal strongly incurved in the upper half,
3. lateral petals spathulate, much widened just behind the apex, margins almost entire, the base rarely ciliated,
4. suborbicular or oblong rhomboid staminodal shield.

Quite a number of taxa have been included in this group. Part of them are very difficult to differentiate and some are repeated descriptions of one and the same species. The following concepts have been described within this section:

1. *Paphiopedilum amabile* Hallier
2. *Paphiopedilum appletonianum* (Gower) Rolfe
3. *Paphiopedilum bullenianum* (Rchb. f.) Pfitzer
4. *Paphiopedilum celebesensis* Fowlie & Birk
5. *Paphiopedilum hainanensis* Fowlie
6. *Paphiopedilum hookerae* (Rchb. f.) Stein
7. *Paphiopedilum johorensis* Fowlie & Yap
8. *Paphiopedilum linii* Schoser
9. *Paphiopedilum robinsonii* (Ridley) Ridley
10. *Paphiopedilum tortipetalum* Fowlie
11. *Paphiopedilum volonteanum* (Sander) Stein
12. *Paphiopedilum wolterianum* (Krzl.) Pfitzer

This group of plants can be divided into two complexes and KARASAWA & SAITO[1)] divided the section into two subsections accordingly.

[1)] **Bull. of the Hiroshima Botanical Garden** No. 5: 1–69 (1982)

Schlüssel zu den Untersektionen der Sektion SPATHOPETALUM

1. Staminodiumplatte fast kreisrund, sehr groß im Vergleich zu den anderen Blütenteilen . . . . Untersektion MACRONODIUM

1a. Staminodiumplatte rautenförmig oder herzförmig, nicht kreisrund und nicht sehr groß im Vergleich zu den anderen Blütenteilen . . . . Untersektion SPATHOPETALUM

Key to the subsections of section SPATHOPETALUM

1. Staminodal shield suborbicular, enormous compared to the size of the other flower parts . . . . subsection MACRONODIUM

1a. Staminodal shield rhomboid or cordate, not suborbicular, not enormous compared to the size of the other flower parts . . . . subsection SPATHOPETALUM

Untersektion MACRONODIUM Karasawa & Saito

Diese Untersektion ist an der übergroßen Staminodiumplatte, welche manchmal oben, manchmal unten und manchmal oben und unten eingekerbt ist, einfach zu erkennen.

Zwei beschriebene Taxa – *Paphiopedilum hookerae* und *Paphiopedilum volonteanum* – gehören in diese Untersektion. Diese zwei Konzepte sind in Wirklichkeit Beschreibungen von zwei Varianten derselben Art, so daß den Regeln der Priorität entsprechend, *Paphiopedilum hookerae* anzuerkennen ist.

Subsection MACRONODIUM Karasawa & Saito

This subsection is easily recognized by the enormous, suborbicular staminodal shield, sometimes notched above, below, or above and below.

Two described taxa – *Paphiopedilum hookerae* and *Paphiopedilum volonteanum* – belong to this subsection. The concepts are in fact descriptions of two variants of one species and therefore, according to the rules of priority, *Paphiopedilum hookerae* is accepted.

Paphiopedilum hookerae (Rchb. f. ex Hooker f.) Stein

Stein's Orchideenbuch: 470 (1892)

Basionym: *Cypripedium hookerae* Rchb. f. ex Hooker f.
Curtis's Botanical Magazine 89, t. 5362 (1863)

Synonyma: *Cordula hookerae* (Rchb. f. ex Hooker f.) Rolfe
Orchid Review 20 (1): 2 (1912)

Cypripedium volonteanum Sander
Gardeners' Chronicle, 3. Ser., **8:** 7 (1890)

Cypripedium hookerae var. *volonteanum* (Sander) Rolfe
Gardeners' Chronicle, 3. Ser., **8:** 66 (1890)

Paphiopedilum volonteanum (Sander) Stein
Stein's Orchideenbuch: 491 (1892)

Paphiopedilum hookerae (Rchb. f. ex Hooker f.) Stein

Paphiopedilum hookerae wurde erstmals durch Sir Hugh LOW im Norden Borneos gesammelt und nach England gebracht.

Auch die Sammler der Firma VEITCH wurden auf die Suche nach Pflanzen dieser Art gesandt und fanden sie zusammen mit *Paphiopedilum stonei* auf Kalkfelsen in Sarawak.

Die Art wurde dann durch REICHENBACH, sicherlich auf Wunsch des Herrn LOW, zu Ehren der Frau des damaligen Direktors am Königlichen Botanischen Garten in Kew, Lady HOOKER, benamt.

Paphiopedilum hookerae hat länglich elliptische, tiefgrüne Laubblätter mit deutlicher graugrüner bis gelblich grüner Marmorierung, unterseitig oft rot überlaufen. Sie sind 10 bis 15 cm lang und 3,8 bis 5,2 cm breit, ledrig fleischig, oft mit zurückgebogenen Rändern.

Die meist einblütige Infloreszenz wird 20 bis 40 cm hoch. Sie ist bräunlich purpurn, und genau wie der Fruchtknoten mit grauen Härchen überzogen.

Die Braktee bedeckt nur etwa die Hälfte des Fruchtknotens.

Die Blüte ist bis 10 cm breit und 6,5 bis 7,5 cm hoch.

This species was discovered by Sir Hugh LOW in North Borneo in 1862, and introduced shortly after by Messrs. LOW & Co.

It was subsequently also gathered on limestone cliffs growing together with *Paphiopedilum stonei* in Sarawak by one of the collectors of VEITCH.

The taxon was described by REICHENBACH who most certainly followed LOWs request in dedicating the species to Lady HOOKER, the wife of the director of the Royal Botanic Gardens at Kew.

The leaves of *Paphiopedilum hookerae* are elliptic-oblong, deep green, distinctly tesselate with grayish-green to yellowish-green, the under surface often suffused with red; up to 15 cm long and 3.8 to 5.2 cm wide, leathery, fleshy, often with recurved margins.

The inflorescence is 20 to 40 cm high, brownish-purple, just like the ovary covered with stiff grey hairs; usually bearing one single flower.

The bract covers only half of the ovary.

The flower is up to 10 cm wide and 6.5 to 7.5 cm high.

The dorsal sepal is cordate, acute, slightly striped or unstriped, yellowish-white with greenish center, distinctly keeled and hairy behind.

Paphiopedilum hookerae (Rchb. f. ex Hooker f.) Stein

Die herzförmige Fahne ist spitz endend, nur leicht oder gar nicht gestreift, gelblich weiß mit grüner Mitte, deutlich gekielt und auf der Rückseite behaart.

Das Synsepalum ist kleiner, eiförmig, blaß gelbgrün.

Die Petalen sind spatelförmig, mehr oder weniger nach unten gerichtet. Der schmale Basisteil hat gewellte Ränder. Er ist grün mit purpurnem Rand und vollständig von dunklen Flecken überzogen. Das breite Endteil ist rosapurpurn.

Die umgekehrt herzförmige Lippe ist gelblich grün, im unteren Bereich braunpurpurn und im oberen Teil dunkel olivpurpurn überlaufen. Die nach innen geschlagenen Seitenlappen sind gelblich braun mit rotpurpurnen Flecken.

Beim Vergleich mit den anderen Arten der Gattung fällt auf, daß die Staminodiumplatte in bezug auf die allgemeinen Ausmaße der Blüte sehr groß ist. Die Staminodiumplatte ist fast kreisrund, meistens mit mehr oder weniger ausgeprägten gerundeten Aussparungen am oberen und/oder unteren Rand. Es ist bräunlich purpurn mit blaßgelber Mitte und Randregion.

Paphiopedilum hookerae wächst vor allem in Nordborneo und Sarawak als Humusepiphyt auf Kalkfelsen in Höhenlagen von 150 bis über 1500 m ü. NN, meistens an halbschattigen Stellen in der Nähe eines fließenden Gewässers.

Was die Kultur anbelangt, gehört *Paphiopedilum hookerae* zu den einfachen Pflanzen. Diese Art wünscht einen hellen bis halbschattigen Standort

The synsepalum is smaller, ovate, pale yellow-green.

The lateral petals are spathulate, more or less depressed, the narrower basal part undulate; green with blackish spots and purple margin, the dilated distal part rose-purple.

The saccate lip is inverted helmet-shaped, yellowish-green, the lower parts suffused brown-purple, the upper part suffused deep olive. The infolded lateral lobes are yellowish-brown spotted reddish-purple.

Comparing this species to the others within the genus, the staminodal shield is found to be enormous compared to the size of the other floral parts or in respect to the overall size of the flower.

The staminodal shield is suborbicular, usually with more or less distinct notches at the upper, the lower or both margins; brown-purple with a pale yellow center and pale yellow margins.

Paphiopedilum hookerae grows in North Borneo and Sarawak, epiphytically in leaf mould on limestone rock in altitudes of 150 to 1500 meters above sea-level, usually in semi-shaded spots near running water.

As far as the cultivation of this species is concerned, it can be said that *Paphiopedilum hookerae* belongs to the more easy to grow plants. This species does best in a bright to semi-shaded spot without full sun. The relative humidity should be kept above 60% at all times, and air circulation should be abundant.

ohne direkte Sonneneinwirkung und eine relative Luftfeuchtigkeit von 60 bis 80 %. Für ausreichende Luftumwälzung sollte gesorgt werden.

Die Art ist ziemlich temperaturunempfindlich und kann mit Erfolg bei Tagestemperaturen zwischen 22 und 38 °C und einer deutlichen Nachtabsenkung auf 18 bis 12 °C kultiviert werden (also im temperierten oder warmen Kulturbereich).

Das Substrat sollte mäßig naß gehalten werden und niemals austrocknen.

Paphiopedilum hookerae blüht im allgemeinen zwischen März und Juni.

Varietäten:

Von *Paphiopedilum hookerae*[1] sind 11 Varietäten beschrieben worden. Vier davon gehören zu *Paphiopedilum bullenianum* und werden dort abgehandelt. Von den sieben weiteren verdient einzig und allein ***Paphiopedilum hookerae*** var. ***volonteanum*** den Varietätsstatus. Die anderen sind lediglich Farbvarianten, die innerhalb der natürlichen Bandbreite der Art anzusiedeln sind.

Paphiopedilum hookerae var. ***volonteanum*** (Sander) Braem, *comb. nov.*[2]

Basionym: *Cypripedium volonteanum* Sander
Gardeners' Chronicle, 3. Ser., **8:** 7 (1890)
(Inserat der Firma SANDER)

Synonym: *Cypripedium hookerae* var. *volonteanum* (Sander) Rolfe
Gardeners' Chronicle, 3. Ser., **8:** 66 (1890)

Obwohl die var. ***volonteanum*** seit 1890 immer wieder als eigenständige Art angegeben wird, kann kein Zweifel daran bestehen, daß es sich hier höchstens um eine geographische Variante von *Paphiopedilum hookerae* handelt.

ROLFE beschrieb die Varietät *(loc. cit.)* wie folgt:

„Die Blätter sind verhältnismäßig schmaler als bei den typischen Formen, die Petalen sind breiter und gerundeter an den Enden, die Lippe zeigt eine Einschnürung unter der Öffnung, und das Staminodium ist ziemlich rund, ohne Einkerbungen. In anderer Beziehung ist es wie das typische *Cypripedium hookerae.*"

[1] Alle als *Cypripedium hookerae* var.

[2] Dieses Taxon wird oft irrtümlicherweise als *Paphiopedilum hookerae* var. *volonteanum* (Sander) Rolfe zitiert. ROLFE kombinierte aber nur innerhalb der Gattung *Cypripedium* um.

This species is rather temperature tolerant and can be grown in the warm or intermediate greenhouses at daytime temperatures between 22 and 38 °C with a nightly drop to 12 respectively 18 °C.

The well drained substrate should be kept moderately moist and the plants should never be allowed to dry out completely.

Paphiopedilum hookerae generally flowers from March through June.

Varieties:

Eleven "varieties" of *Paphiopedilum hookerae*[1] have been described. Four of these belong to the *Paphiopedilum bullenianum*-complex and will be discussed there. Of the remaining seven, only the concept of *Paphiopedilum hookerae* var. *volonteanum* could deserve the varietal status. All others are based on single plants differing from the nominal species only in extremely minor changes in flower colour.

Paphiopedilum hookerae var. ***volonteanum*** (Sander) Braem, *comb. nov.*[2]

Basionym: *Cypripedium volonteanum* Sander
Gardeners' Chronicle, 3. Ser., **8:** 7 (1890)

Synonym: *Cypripedium hookerae* var. *volonteanum* (Sander) Rolfe
Gardeners' Chronicle, 3. Ser., **8:** 66 (1890)

Although the var. ***volonteanum*** has repeatedly been considered an autonomous species, there is no doubt that this taxon is to be regarded as a geographical variety of *Paphiopedilum hookerae.*

ROLFE described the variety as follows *(loc. cit.):*

"The leaves are proportionally narrower than in the typical form, the petals broader and more obtuse, the lip a little constricted below the horizontal mouth, and the staminode quite orbicular, without notches. In other respects, it is quite like the typical *Cypripedium hookerae.*"

[1] All as *Cypripedium hookerae* var.

[2] This taxon is usually but erroneously cited as *Paphiopedilum hookerae* var. *volonteanum* (Sander) Rolfe. ROLFE, however, transferred the taxon to the varietal status within the genus *Cypripedium* only.

Untersektion SPATHOPETALUM

Diese Untersektion wird durch eine rauten- oder herzförmige Staminodiumplatte gekennzeichnet.

Die Taxa, welche in dieser Untersektion beschrieben worden sind, können nach kritischer Analyse der Beschreibungen, eingehender Untersuchung der Typusbelege und Beobachtungen an ausreichendem lebendem Pflanzenmaterial, auf der Basis der Eigenarten der Staminodiumplatten in drei Gruppen aufgeteilt werden.

A. Gruppe mit breiter Einbuchtung am unteren Rand der Staminodiumplatte:

Paphiopedilum appletonianum
Paphiopedilum hainanensis
Paphiopedilum johorensis
Paphiopedilum tortipetalum
Paphiopedilum wolterianum

B. Gruppe mit schmaler Einbuchtung am unteren Rand der Staminodiumplatte:

Paphiopedilum amabile
Paphiopedilum bullenianum
Paphiopedilum celebesensis
Paphiopedilum linii

In beiden Gruppen ist unschwer festzustellen, daß es sich jeweils um unterschiedliche Benennungen von Blütenfarbvariationen desselben Taxons handelt, die sich höchstens noch in der Intensität der Petalendrehung unterscheiden, so daß nach der Prioritätsregel *Paphiopedilum appletonianum* und *Paphiopedilum bullenianum* anzuerkennen sind. Die übrigen Konzepte sind als Synonyma einzustufen.

C. Staminodiumplatte deutlich herzförmig, ohne Ausbuchtung am unteren Rand, aber mit einem Mittelzahn:

Paphiopedilum robinsonii

Subsection SPATHOPETALUM

This subsection is characterized by a rhomboid to cordate staminodal shield.

The taxa described in this subsection can, after critical review of the descriptions and detailed examination of the type specimens as well as after examination of abundant life material, be separated into three groups on the basis of the morphological peculiarities of the staminodal shield.

A. Group of plants with wide notch at the under margin of the staminodal shield:

Paphiopedilum appletonianum
Paphiopedilum hainanensis
Paphiopedilum johorensis
Paphiopedilum tortipetalum
Paphiopedilum wolterianum

B. Group of plants with narrow notch at the under margin of the staminodal shield:

Paphiopedilum amabile
Paphiopedilum bullenianum
Paphiopedilum celebesensis
Paphiopedilum linii

For both groups, it is readily recognized that the concepts are, in fact, descriptions of variants of one single species, differing merely in some aspects of flower colouration and in the intensity with which the lateral petals are twisted. According to the rules of priority, *Paphiopedilum appletonianum* and *Paphiopedilum bullenianum* are to be recognized, and the other taxa are to be reduced to synonymy or at the outmost varietal status.

C. Staminodal shield distinctly cordate, the under margin not notched but with a tooth at the middle.

Paphiopedilum robinsonii

Paphiopedilum appletonianum (Gower) Rolfe

Orchid Review 4: 364, 1896

Basionym: *Cypripedium appletonianum* Gower
The Garden 1: 93 (1893)

Synonyma: *Cordula appletoniana* (Gower) Rolfe
Orchid Review 20 (1): 2 (1912)

Cypripedium bullenianum var. *appletonianum* (Gower) Rolfe
Orchid Review 1: 135 (1893)

Cypripedium poyntzianum O'Brien
Gardeners' Chronicle, 3. Ser., **15:** 36 (1894)
Orchid Review 2: 54 (1894)

Cypripedium waltersianum Krzl.
lapsus calamiti in DESBOIS:
Cypripedium, Selenipedium & Uropedium – Monographie: 477 (1898)

Cypripedium wolterianum Krzl.
Gardeners' Chronicle, 3. Ser., **17:** 166 (1895)

Paphiopedilum hainanensis Fowlie
Orchid Digest 51 (2): 69–70 (1987)

Paphiopedilum johorensis Fowlie & Yap
Orchid Digest 36 (2): 72 (1972)

Paphiopedilum tortipetalum Fowlie
Orchid Digest 49 (4): 153–156 (1985)

Paphiopedilum wolterianum (Krzl.) Pfitzer
in ENGLER, A.: **Das Pflanzenr. IV** (50) Heft 12: 79 (1903)

Paphiopedilum ceramensis Birk
Paph. Grower's Man.: 103 (1983)
nomen nudum

Paphiopedilum appletonianum (Gower) Rolfe

Die Entdeckungsgeschichte von *Paphiopedilum appletonianum* ist unklar. Einerseits wird gemutmaßt, daß die Pflanzen aus Borneo kamen, da sie zusammen mit *Paphiopedilum hookerae* importiert wurden. Es ist aber viel wahrscheinlicher, daß die Pflanzen aus Thailand stammten, wo´ auch *Paphiopedilum × siamense,* die Naturhybride zwi-

Little is known about the discovery of this species. It was said to have been imported with *Paphiopedilum hookerae*, suggesting its origin in Borneo. However, subsequent importations of this species together with *Paphiopedilum callosum* and the putative natural hybrid *Paphiopedilum × siamense* indicate that the plants originated from Thailand.

schen *Paphiopedilum appletonianum* und *Paphiopedilum callosum* vorkommt.

Die Art wurde nach W. M. APPLETON benamt, dem es als erstem gelang, eine Pflanze dieser Art in Kultur zur Blüte zu bringen.

Paphiopedilum appletonianum ist eine epiphytisch, am Waldboden und auf Felsen, in Humus wachsende krautige Pflanze mit kurzen, 3 bis 8 Blätter tragenden, nur 2 bis 3 cm hohen Sprossen auf einem verkürzten Rhizom. Diese Laubblätter werden bis 23 cm lang und 2,5 bis 5 cm breit. Sie sind zweireihig angeordnet, schmal elliptisch mit einem spitzen, dreizähnigen Ende. Die Blätter sind unbehaart, oben mittel- bis hellgrün mit undeutlicher, dunkelgrüner Aderung und Marmorierung; unten graugrün, an der Basis purpurn überlaufend, deutlich gekielt.

Die meist einblütige Infloreszenz wird 18 bis 60 cm hoch. Der Schaft ist braun, kurz behaart.

Die eiförmige Braktee ist 1,8 bis 2,5 cm lang, kurz behaart, gekielt, grün, mit purpurner Basis.

Der Fruchtknoten ist 3,5 bis 6,5 cm lang, kurz behaart, braunpurpurn bis braunrot.

Die Fahne und das Synsepalum sind leicht gekielt, am Rande bewimpert. Ihre Oberflächen sind leicht papillös bis kurz behaart.

Die Fahne ist eiförmig, spitz. Die Ränder im basalen Teil sind deutlich nach hinten gefaltet. Die ganze Struktur ist deutlich kapuzenförmig und nach vorne gebogen, 3,5 bis 4 cm hoch und ungefähr 2,6 cm breit, apfelgrün mit dunkleren Adern.

Das Synsepalum ist länglich eiförmig, spitz, 2,5 bis 3 cm lang und 1,5 cm breit, apfelgrün.

Die länglich spatelförmigen Petalen sind in der Endregion am breitesten, mehr oder weniger stumpf, meist leicht gedreht, 5 bis 6 cm lang und 1,4 bis 2 cm breit, unbehaart mit leicht gewellten Rändern im basalen Bereich, an den Rändern bewimpert, mehr oder weniger horizontal abstehend oder leicht nach unten gekurvt, hellgrün mit kleinen braunen Flecken im basalen Bereich, übergehend in einen rosapurpurnen Endbereich.

Die umgekehrt helmförmige Lippe ist ungefähr 4,5 cm lang und 1,5 bis 1,8 cm breit, dreilappig, mit nach innen gebogenen, von glänzenden Wärzchen überzogenen Seitenlappen.

Die Außenseite des sackförmigen Hauptlappens ist glatt, die Innenseite kurz behaart. Die Lippe ist purpurbraun, im unteren Bereich oft etwas blasser, häufig mit cremeweißer Spitze. Der Rand der Öffnung ist olivgrün.

The species was named in honour of W. M. APPLETON, who first flowered a plant of this species in cultivation.

Paphiopedilum appletonianum is a perennial herb, growing epiphytically in leaf mould on forest ground or on rocks. The stems are short, 2 to 3 cm high, bearing 3 to 8 leaves. The leaves are distichous, up to about 23 cm long and 2.5 to 5 cm wide, narrowly elliptic to ligulate elliptic, apex acute, tridentate; glabrous, the margins minutely papillose, upper surface medium to bright green with obscure darker green venation and tesselation, under surface grey-green, suffused purple at the base, distinctly keeled.

The inflorescence is 16 to 60 cm high, the scape brown, shortly pubescent; usually bearing one single flower.

The bract is 1.8 to 2.5 cm long, ovate, shortly pubescent, keeled, green, purple at its base.

The ovary is 3.5 to 6.5 cm long, shortly pubescent, purple-brown to reddish-brown.

The dorsal sepal as well as the synsepalum is slightly keeled, the margins ciliate, their upper surface minutely papillose to minutely pubescent, ciliate at the base, their lower surface minutely papillose and shortly pubescent.

The dorsal petal is ovate, acute, the basal margins strongly recurved, the whole structure distinctly cucullate and directed forwards, 3.5 to 4 cm long and about 2.6 cm wide, apple-green with darker veins.

The synsepalum is ovate-elliptic, acute, 2.5 to 3 cm long and about 1.5 cm wide, apple-green.

The petals are ligulate to spathulate, broadest and sometimes obscurely toothed towards the apex, more or less obtuse, gently twisted, 5 to 6 cm long and 1.4 to 2 cm wide, glabrous with slightly undulate margins proximally, the margins ciliate; spreading more or less horizontally or slightly drooping, bright green towards the base with small brown spots, shading to pinkish-purple distally.

The saccate lip is inverted helm-shaped, about 4.5 cm long and 1.5 to 1.8 cm wide, trilobate, the infolded lateral lobes acute, converging, with shiny warts, the mid-lobe bulbous, often bidentate, outer surface glabrous, inner surface shortly hirsute, purplish-brown, paler below, the apex often creamy-white, margin of the mouth olive-green.

The staminodal shield is oblong rhomboid, with a middle tooth in the ventral sinus, ventral lobes variable in length. The shield is green, darker at the center, with white margins, minutely hirsute, about 7 mm long and 7 to 8 mm wide.

Die Staminodiumplatte ist länglich, umgekehrt herzförmig, im vorderen Einschnitt mit einem Mittelzahn, die Seitenzähne unterschiedlich lang. Die Platte ist grün, weiß berandet, leicht behaart, ca. 0,7 cm lang und 0,7 bis 0,8 cm breit.

Paphiopedilum appletonianum ist in Thailand, Laos, Kambodscha, auf der malaysischen Halbinsel, und im südlichen China anzutreffen. Die Pflanzen wachsen an schattigen, feuchten Standorten in Höhenlagen von 300 bis 2000 m ü. NN als Humusbesiedler am Waldboden oder in Felsspalten.

Die wärmsten Monate in dieser Region sind März und April, mit Temperaturen von bis zu 32 °C am Tage und selten unter 17 °C in der Nacht.

Während der Monsunmonate von November bis Februar steigen die Temperaturen am Tage nur bis auf 18 °C und fallen in der Nacht auf ca. 8 °C ab.

Die Luftfeuchtigkeit ist das ganze Jahr über sehr hoch.

Paphiopedilum appletonianum läßt sich im temperierten oder warmen Gewächshaus, in einem durchlässigen Substrat bei hoher Luftfeuchtigkeit kultivieren. Von Mai bis Oktober sollte viel Wasser gegeben werden. In den Monaten November bis Januar sollten die Pflanzen kühler und trockener gehalten werden. Die Lichtverhältnisse spielen bei *Paphiopedilum appletonianum* keine besondere Rolle, jedoch sollen die Pflanzen nicht zu hell stehen.

Die Blütezeit ist zwar variabel, jedoch blühen die meisten Pflanzen zwischen Januar und Mai.

Varietäten:

Paphiopedilum appletonianum ist eine variable Art. *Paphiopedilum hainanensis, Paphiopedilum johorense, Paphiopedilum tortipetalum* und *Paphiopedilum wolterianum* sind geographische Varianten, die sich von der Nominalform eben nur in der Blütenfärbung und machmal in der Intensität der Petalendrehung unterscheiden. Diese Unterschiede sind aber keineswegs so bedeutend, daß man die Beschreibung als eigenständige Art vertreten könnte. Nur die Albinoform und die möglicherweise als Naturhybride einzustufende var. *poyntzianum* verdienen es, hier abgehandelt zu werden.

Paphiopedilum appletonianum var. *immaculata* Braem, *var. nov.*
Typus: Icones Seite 152.

The natural distribution area of *Paphiopedilum appletonianum* is Thailand, Laos, Cambodia, the Malayan Peninsula and the southern part of China. The plants grow in shaded, moist habitats in altitudes of between 300 and 2000 m above sea-level, in humus on forest ground or in cracks and crevices of rocks.

In this region, the warmest period is from March through April with temperatures of up to 32 °C during the day and rarely below 17 °C at night.

During the monsoon months from November to February, the temperatures do not exceed 18 °C during the day and drop to about 8 °C at night.

The relative humidity remains high over the entire year.

Paphiopedilum appletonianum does best potted in an open, well drained substrate in the intermediate or warm greenhouse. Humidity should be constantly high. From May through October waterings should be abundant. From November through January, the plants should be kept somewhat cooler and drier. Light conditions are not very important in respect to the cultivation of this species as far as the plants are not subject to very bright light.

The flowering periods can vary considerably; most plants, however, flower between January and May.

Varieties:

Paphiopedilum appletonianum is, like most other species of the genus, very variable. *Paphiopedilum hainanensis, Paphiopedilum johorense, Paphiopedilum tortipetalum* and *Paphiopedilum wolterianum* are geographic races, differing from the typical form only in the variation of the flower colour and sometimes in the intensity of the twisting of the lateral petals. These differences are by no means important enough to consider these taxa to be separate autonomous species. Only the albino and the var. *poyntzianum,* which maybe a natural hybrid deserve to be discussed in detail.

Paphiopedilum appletonianum var. *immaculata* Braem, *var. nov.*
Typus: Icones pagina 152.

Paphiopedilum appletonianum var. *immaculata* Braem

Similis *Paphiopedilum appletonianum* typicum sed flores virides uniformiter.

Eine typische Albinoform ohne jegliche Rotfärbung. Sepalen hellgrün mit dunkelgrüner Aderung. Petalen hellgrün mit weißlicher Endregion. Lippe hellgrün mit dunklerer Aderung und weißlicher Spitze.

Paphiopedilum appletonianum var. *poyntzianum* (O'Brien) Pfitzer
in ENGLER, A.: **Das Pflanzenr. IV** (50) Heft 12: 79 (1903)

Petalen blaßgrün mit lila Spitze, Labellum weißlich, purpurn überhaucht und mit rosa Basis. Petalen bis zu 12 cm lang.

Dieses Taxon wurde anhand einer bisher einmalig gefundenen großblütigen, hellgefärbten Variante beschrieben. Die Pflanze wurde in einem Import von *Paphiopedilum callosum* aus Thailand gefunden. Sie ist laut KERCHOVE[1] eine Naturhybride zwischen *Paphiopedilum callosum* und *Paphiopedilum hookerae.*

[1] **Le Livre des Orchidées:** 482 (1894)

Similis *Paphiopedilum appletonianum* typicum sed flores virides uniformiter.

A typical albino lacking any reddish pigmentation. The sepals are bright green with darker green venation; the petals are a clear immaculate apple-green with whitish apical regions; the saccate lip is apple-green with darker green venation and a whitish tip.

Paphiopedilum appletonianum var. *poyntzianum* (O'Brien) Pfitzer
in ENGLER, A.: **Das Pflanzenr. IV** (50) Heft 12: 79 (1903)

The petals are up to 12 cm long, pale green with purple apical region. Saccate lip whitish, suffused purple and with a reddish basal part.

This taxon was described on the basis of a single specimen with extremely large and pale coloured flowers. The plant was discovered in a shipment of *Paphiopedilum callosum* originating from Thailand. Only one specimen was ever found. KERCHOVE[1] suggested this plant to be a natural hybrid between *Paphiopedilum callosum* and *Paphiopedilum hookerae.*

[1] **Le livre des Orchidées:** 482, (1894)

Paphiopedilum bullenianum (Rchb. f.) Pfitzer

in ENGLER: **Bot. Jahrbücher 19:** 40 (1894)

Basionym: *Cypripedium bullenianum* Rchb. f.
Bot. Zeitung: 99 (1865)

Synonyma: *Cordula amabilis* (Hallier) Merrill
Journ. Str. Br. Roy. As. Soc. Spec. No. 135 (1921)

Cordula bulleniana (Rchb. f.) Rolfe
Orchid Review 20 (1): 2 (1912)

Cypripedium hookerae var. *amabile* Krzl.
Orchidacearum Genera et Species 1: 59 (1897)

Cypripedium hookerae var. *bullenianum* Veitch
A Manual of Orchidaceous Plants 4: 32 (1889)

Paphiopedilum amabile Hallier
Natuurk. Tijdschr. Ned. Ind. 54: 450–452 (1895)

Paphiopedilum bullenianum var. *celebesense* (Fowlie & Birk) Cribb
The Genus Paphiopedilum: 167 (1987)

Paphiopedilum celebesensis Fowlie & Birk
Orchid Digest 44 (1): 23–30 (1980)

Paphiopedilum hookerae var. *bullenianum* (Rchb. f.) Kerchove
Le livre des Orchidées: 454 (1894)

Paphiopedilum linii Schoser
Die Orchidee 17 (3): 177–181 (1966)

Dieser Pflanzenkomplex besteht aus vier Taxa, wovon aus Prioritätsgründen *Paphiopedilum bullenianum* als Name zu verwenden ist. Sie grenzt sich durch eine schmale untere Staminodiumplatten-Einbuchtung vom *Paphiopedilum appletonianum*-Komplex ab.

Paphiopedilum bullenianum wurde durch Hugh LOW in Saba, im Norden der Insel Borneo, gesammelt und durch H. G. REICHENBACH im Jahre 1865 zu Ehren des englischen Gartenbauunternehmers BULLEN beschrieben *(loc. cit.)*. Der Status dieser Art war von jeher umstritten, und VEITCH *(loc. cit.)* sowie eine Reihe von anderen Autoren, darunter z. B. auch STEIN, betrachteten die Pflanzen als Varietät von *Paphiopedilum hoo-*

This complex consists of four described taxa of which *Paphiopedilum bullenianum* is the valid designation in view of the priority-rule of the International Code of Botanical Nomenclature. This complex differs from *Paphiopedilum appletonianum* by the narrow sinus at the bottom edge of the staminodal shield.

Paphiopedilum bullenianum was discovered by Hugh LOW in Sabah (North Borneo) and described by REICHENBACH in 1865 *(loc. cit.)* in honour of the English horticulturalist BULLEN. The taxonomic status of this species has been a matter of discussion for many years and VEITCH *(loc. cit.)* as well as a number of other authors, including STEIN, consider the taxon as a variety of *Paphiopedilum kerae*, ein Standpunkt, der kaum noch haltbar ist. Eher wäre es zu vertreten, daß der Unterschied zwischen dem *Paphiopedilum appletonianum*-Komplex und dem *Paphiopedilum bullenianum*-Komplex zu gering ist, um zwei getrennte Arten aufrechtzuerhalten.

Dann aber wäre die Benennung als *Paphiopedilum bullenianum* aus Prioritätsgründen für den ganzen Komplex zu verwenden und *Paphiopedilum appletonianum* zum Synonym zu reduzieren, ein Standpunkt, der bereits durch ROLFE[1] vertreten wurde.

Paphiopedilum bullenianum ist eine als Humusepiphyt wachsende, krautige Pflanze mit bis zu 15 cm langen und bis zu 3,5 cm breiten, schmal elliptischen Blättern. Diese sind hell bläulich grün, oberseitig mit schwacher dunkelgrüner Marmorierung, manchmal hellgrün, dunkler marmoriert. Die Unterseite kann purpurn gefleckt sein. Die Blattspitzen sind dreizähnig.

Die ein- oder zweiblütige Infloreszenz wird bis zu 50 cm hoch. Der Schaft ist purpurfarben, grün gefleckt, dicht purpurn behaart.

Die Braktee ist bis zu 2 cm lang, an der Basis knotig, etwa ein Drittel so lang wie der Fruchtknoten, dicht flaumig behaart.

Die Blüten werden bis zu 7,5 cm groß.

Die Fahne ist ca. 3,5 cm hoch und ungefähr 2 cm breit, eiförmig, an den Rändern nach unten hin zurückgebogen, hellgrün mit dunkelgrünem Mittel-

[1] **Orchid Review 1:** 115 (1893)
The Garden 1: 93 (1893)

hookerae, a point of view hardly to be substantiated in modern terms.

More justifiable would be the opinion that the difference between the *Paphiopedilum appletonianum*-complex and the *Paphiopedilum bullenianum*-complex is not enough to substantiate the upholding of two autonomous species.

In such a case, the whole group should be united under the concept of *Paphiopedilum bullenianum,* this having been published before *Paphiopedilum appletonianum. Paphiopedilum appletonianum* would then be a synonym, as already suggested by ROLFE.[1]

Paphiopedilum bullenianum is a perennial herb. The leaves are up to 15 cm long and up to about 3.5 cm wide, oblong-elliptic, dull bluish-green to bright green, the upper surface indistinctly mottled with darker green, the under side sometimes spotted purple; the apices three-toothed.

The inflorescence is up to 50 cm high and bears one or two flowers. The scape is green, spotted purple, densely purple pubescent.

The bract is up to 2 cm long, noded at the base, about one third of the length of the densily vilose ovary.

The flowers are up to 7.5 cm across.

The dorsal sepal is about 3.5 cm high and about 2 cm wide, ovate, the margins in the basal section reflexed, bright green with a darker green longitudi-

[1] **Orchid Review 1:** 115 (1893)
The Garden 1: 93 (1893)

Paphiopedilum bullenianum
(Rchb. f.) Pfitzer

Paphiopedilum bullenianum (Rchb. f.) Pfitzer

band und mehr oder weniger deutlich hervorstehenden, dunkelgrünen Adern.

Das Synsepalum ist kürzer und schmaler, eiförmig, hellgrün wie die Fahne.

Die bis ca. 5,5 cm langen und 1,5 cm breiten, waagerecht abstehenden und mit den etwas verbreiterten Enden leicht abwärts geneigten Petalen sind lang spatelig, grünlich bis ledergelb in der Grundfärbung, mehr oder weniger intensiv purpurn getupft. Sie sind mit einigen kleinen schwarzen Wärzchen auf den mehr oder weniger gewellten Seitenrändern versehen und deutlich purpurn oder rosapurpurn in den Endregionen überhaucht. Die bewimperten Ränder sind hellgelb.

Die umgekehrt helmförmige, bis ca. 5 cm lange und 3,5 cm breite Lippe ist purpurrot mit hellerer Spitze. Sie ist vorne mehr oder weniger dunkel marmoriert, am Rand der Öffnung grünlich, dreilappig. Die nach innen gebogenen Seitenlappen sind mit großen Warzen besetzt.

Die Staminodiumplatte ist eiförmig, oben und unten eingekerbt, unten mit einem mehr oder weniger deutlich ausgebildeten Mittelzahn, hellgrün mit

nal middle stripe and more or less prominent green veins.

The synsepalum is shorter and narrower, narrowly ovate, bright green.

The lateral petals are more or less spreading horizontally, the apices somewhat drooping, up to about 5.5 cm long and about 1.5 cm wide, oblong-spathulate, greenish to tawny yellow in their ground colour, more or less intensively spotted purple, suffused purple to rose-purple in the distal part. The margins are somewhat undulate and interrupted by some small blackish warts, ciliate, bright yellow.

The saccate lip is inverted helmet-shaped, up to about 5 cm long and up to about 3.5 cm wide, purple-red with lighter coloured apex, the front more or less darkly mottled, the rim of the mouth greenish. The infolded sidelobes are covered with large warts.

The staminodal shield is ovate, up to about 9 mm long and more or less 6 mm wide, notched above and below, the under sinus with a more or less distinct tooth in the middle, green with darker green pattern.

dunkelgrüner Zeichnung, bis ungefähr 9 mm hoch und 6 mm breit.

Paphiopedilum bullenianum hat seine Verbreitung auf den Inseln Borneo und Sulawesi (Celebes). Die Pflanzen wachsen an hellen, leicht beschatteten Stellen im Blatthumus oder zwischen den Wurzeln der Mangrovengewächse an Flußufern bis in Höhenlagen um ca. 1000 m ü. NN.

Paphiopedilum bullenianum kann im temperierten oder im warmen Gewächshaus kultiviert werden. Bei schattigen Wachstumsbedingungen ergeben sich eine deutliche Blattmarmorierung und eine längere Infloreszenz mit intensiv gefärbter Blüte. Man sollte ein durchlässiges Substrat wählen und relativ häufig gießen.

Während der Wintermonate werden die Wassergaben vermindert und die Temperaturen gesenkt.

Die Blütezeit liegt je nach geographischer Varietät zwischen März und Oktober.

Varietäten:

Wie man in der Synonymliste sehen kann, sind die bekannten geographischen Varianten als autonome Arten beschrieben worden. Diese „Varietäten“ sind jedoch alle innerhalb der natürlichen Bandbreite einer Orchideenart anzusiedeln.

Paphiopedilum bullenianum is to be found on the Islands of Borneo and Sulawesi (Celebes).

The plants grow in bright, lightly-shaded spots in leaf-moulder, sometimes in between Mangroves, in altitudes up to about 1000 meters above sea-level.

Paphiopedilum bullenianum can be grown in the intermediate as well as in the warm greenhouse. Shaded plants will show beautifully mottled leaves and generate longer inflorescences with a more intensively coloured flower.

The plants should be potted in an open substrate and kept rather moist during the vegetative period.

During the winter months, the plants should be given a slight rest with less water and lower temperatures.

Paphiopedilum bullenianum flowers between March and October, depending on the geographical origin of the specimen.

Varieties:

As can be seen in the list of synonyms, most of the known geographical variants have been described as autonomous species. These "varieties" however, do not differ considerably from the type.

Paphiopedilum robinsonii (Ridley) Ridley

Flora of the Malayan Peninsula 4: 232 (1924)

Basionym: *Cypripedium robinsonii* Ridley
Journ. Fed. Mal. States Museum 6:
183 (1915)

Paphiopedilum robinsonii
(Ridley) Ridley

Paphiopedilum robinsonii wurde durch den englischen Botaniker und damaligen Leiter des botanischen Gartens in Singapur, RIDLEY, beschrieben. Die Pflanzen sind in dem Komplex um *Paphiopedilum appletonianum* und *Paphiopedilum bullenianum* anzusiedeln, aber sie unterscheiden sich von diesen zwei Arten durch die Form der Staminodiumplatte.

Paphiopedilum robinsonii ist ein am Waldboden oder auf Felsen wachsender krautiger Humusepiphyt mit verkürzten Sprossen, die je bis zu acht länglich ovale, bis 10 cm lange und bis 4 cm breite, am Ende dreizähnige Laubblätter tragen. Diese sind deutlich marmoriert, dunkelgrün in der Grundfärbung mit unregelmäßigen, silbrig und gräulich grünen Flecken. Laut HOLLTUM[1] bilden die Pflanzen oft lange Ausläufer, die teilweise kleine Laubblätter tragen.

Die aufrecht stehende Infloreszenz ist bis zu 30 cm hoch, rund. Sie ist über die ganze Länge kurz be-

[1] **Flora of Malaya 1:** 75 (1964)

Paphiopedilum robinsonii was discovered by the British botanist RIDLEY, at the time director of the Botanical Gardens at Singapore. The plants belong into the complex surrounding *Paphiopedilum appletonianum* and *Paphiopedilum bullenianum* but are readily distinguished by the rather different shape of the staminodal shield.

Paphiopedilum robinsonii is a perennial herb, growing epiphytically on the ground or on rocks in leaf-mould. The plants have short stems, bearing up to 8 leaves. These are oblong-ovate, up to 10 cm long and about 4 cm wide, the apices 3 toothed, distinctly tesselate, deep green with irregular silvery green or grayish green areas. Holltum[1] claims that the plants produce stolons, sometimes bearing many short leaves.

The inflorescence is terete, erect, up to about 30 cm high, usually bearing one single, up to 9 cm wide and 6.5 cm high flower.

The bract is only about 0.7 cm long, green, pubescent, curved at the apex.

[1] **Flora of Malaya 1:** 75 (1964)

haart und trägt normalerweise eine einzige, bis 9 cm breite und bis 6,5 cm hohe Blüte.

Die Braktee ist nur 0,7 cm lang, behaart, grün, am Ende gebogen.

Der Fruchtknoten ist ungefähr 4 cm lang, behaart, purpurbraun.

Die Fahne ist breit eiförmig, nach vorne gebogen, spitz, ungefähr 3,5 cm lang. Die Ränder im basalen Bereich sind nach hinten geschlagen. Die Fahne ist grün, in der unteren Hälfte mit einem leichten purpurnen Überzug. Die Mittellinie ist dunkelpurpurn, die Ränder sind weißlich.

Das Synsepalum ist fahl grün, eiförmig, kürzer als die Fahne.

Die Petalen sind bis ungefähr 5 cm lang und bis 1,8 cm breit, spatelförmig breiter werdend nach dem Ende zu, leicht gedreht. Das basale Drittel ist gelblich grün in der Grundfärbung mit 2 bis 3 deutlichen Falten und vielen schwärzlichen Flecken in der oberen Hälfte. Die apikalen zwei Drittel sind einheitlich rötlich purpurn überzogen.

Die Lippe ist umgekehrt helmförmig. 2,5 bis 3,5 cm hoch und ungefähr 1,8 cm breit, fahl grün bis honigbraun mit einem purpurnen oder purpurbraunen Schatten unter der Öffnung. Die Spitze ist heller gefärbt und zeigt ein Netz von grünlichen Adern.

Die Staminodiumplatte ist herzförmig mit einer Einbuchtung am oberen, aber ohne solche am unteren Rand. Unten aber mit einem deutlich gerundeten Zahn. Sie ist honigfarben mit einem elfenbeinfarbenen Mittelbereich, der von grünlich purpurnen, netzartigen Markierungen umringt ist.

Paphiopedilum robinsonii ist bisher nur von zwei Standorten bekannt. Beide wurden durch RIDLEY in seinen Veröffentlichungen angegeben. In seiner Originalbeschreibung schrieb RIDLEY: „Ziemlich verbreitet in den Wäldern in der Nähe der Flüsse, in tiefen Moospolstern an schattigen Stellen in einer Höhenlage von 1900 m auf dem Padang."

In der zweiten Veröffentlichung anläßlich des Transfers zur Gattung *Paphiopedilum* teilt RIDLEY mit, daß die Pflanzen auch bei Gunong Tahan in Höhenlagen von 1700 bis 2000 m ü. NN wachsen. *Paphiopedilum robinsonii* scheint selten in Kultur zu sein. Dem Autor liegen keine Kulturerfahrungen vor.

Entsprechend der Herkunft kann man annehmen, daß die Pflanzen halbschattig im temperierten Gewächshaus mit ausreichenden Wassergaben in der Wachstumsperiode und einer nicht unbedingt ausgeprägten, aber etwas kühleren und trockeneren Winterruhe zu kultivieren sind.

The ovary is about 4 cm long, pubescent, purple-brown.

The dorsal sepal is broadly ovate, about 3.5 cm long, acute, bent forwards; the margins of the lower part reflexed; green, in the basal half somewhat suffused purple and with a deep purple median stripe, the margins whitish.

The synsepalum is dull green, ovate, shorter than the dorsal sepal.

The lateral petals are up to about 5 cm long and up to 1.8 cm wide, spathulate, widening towards the apex, slightly twisted; the basal one third dull yellowish-green with two or three distinct folds and numerous blackish spots on the upper half; the distal two-thirds uniformly suffused reddish-purple.

The saccate lip is inverted helmet-shaped, 2.5 to 3.5 cm high and about 1.8 cm wide, dull green to honey-brown with a purple to purple-brown rim around the aperture, the tip of a paler colour; with a network of greenish veins.

The staminodal shield is cordate with a sinus at the upper rim and a protruding tooth at the lower margin, honey-coloured with an ivory middle section surrounded by a greenish weblike pattern.

Paphiopedilum robinsonii has been reported from two locations only. Both were indicated by RIDLEY in his publications. In the original description he wrote: "fairly widely distributed in the forest along the larger rivers, in deep moss patches in shaded places in an altitude of about 1900 meters above sea-level on the Padang".

In the second publication, when transferring the species to *Paphiopedilum,* RIDLEY reported also finding the plants near Gunong Tahan at altitudes between 1700 and 2000 meters above sea-level.

Paphiopedilum robinsonii appears seldom in cultivation. No cultural experience can be reported.

According to the supposed natural habitats, it would be reasonable to grow the plants semi-shaded in the intermediate house, giving them plenty of water during the vegetative period and a slight resting period with somewhat lower temperatures and less water during the winter months.

Sektion BLEPHAROPETALUM Pfitzer

Diese Sektion ist der Sektion **Spathopetalum** sehr ähnlich, unterscheidet sich aber durch länglich linealische Blätter, seitliche Petalen, die keine großen, verdickten oder deutlichen Flecken aufweisen und deren obere Hälften meist dunkler in der Färbung sind als die unteren Hälften, und einer weißen oder purpurnen Fahne, deren Ränder im apikalen Bereich nicht deutlich nach innen gefaltet sind.

Die Staminodiumplatte ist hufeisen-, halbmond- oder umgekehrt herzförmig.

Die Systematik dieser Sektion ist in den letzten Jahren etwas durcheinandergebracht worden. Eine geographische Sippe von *Paphiopedilum violascens* wurde unter dem Namen *Paphiopedilum papuanum* auf den Markt gebracht. Diese Pflanzen haben aber nichts mit dem Originalkonzept von *Cypripedium papuanum* Ridley ex Rendle, welches später durch L. O. WILLIAMS zu *Paphiopedilum* transferiert wurde, gemeinsam. Zwei andere geographische Varianten von *Paphiopedilum violascens* wurden als *Paphiopedilum bougainvilleanum* bzw. *Paphiopedilum wentworthianum* veröffentlicht, und eine „neue Art" aus Neu Guinea, welche unter dem Namen *Paphiopedilum zieckianum* beschrieben wurde, erweist sich als identisch mit dem echten *Paphiopedilum papuanum*.

Vor kurzem wurde eine neue Art auf Celebes entdeckt und als *Paphiopedilum sangii* beschrieben.

Section BLEPHAROPETALUM Pfitzer

This section is very similar to section **Spathopetalum** but differs in the linear oblong leaves, lateral petals not distinctly broadened towards the apex and without large, more or less swollen and distinct spots, the upper halves usually deeper in colour than the lower halves and the white or purple upper sepal, of which the upper part of the margin is not strongly incurved.

The staminodal shield is horseshoe-shaped, lunate or obcordate.

Over the last years, the taxonomy of this section has been rather confused. One geographical race of *Paphiopedilum violascens* originating from New Guinea is being sold as *Paphiopedilum papuanum,* but these plants have nothing in common with the original concept of *Cypripedium papuanum* Ridley ex Rendle, which was later transferred to *Paphiopedilum* by L. O. WILLIAMS. Two other geographical variants of *Paphiopedilum violascens* were published as *Paphiopedilum bougainvilleanum* and *Paphiopedilum wentworthianum* and a "new" species from New Guinea was described as *Paphiopedilum zieckianum* and turns out to be the real *Paphiopedilum papuanum.*

Recently, a new species from Celebes has been discovered and described as *Paphiopedilum sangii.*

Schlüssel zu den Arten der Sektion BLEPHAROPETALUM

1. Staminodiumplatte verkehrt herzförmig, mehr oder weniger einfarbig hell gelblich grün; Schuh deutlich (*venustum*-ähnlich) von dunklen Adern durchzogen . . . . . *Paphiopedilum sangii*

1a. Staminodiumplatte hufeisen- bis halbmondförmig; Beaderung des Schuhs nicht „*venustum*-ähnlich" 2

2. Seitliche Petalen horizontal abstehend, an der Basis etwas schmaler als an der Spitze; Fahne undeutlich gestreift 3

2a. Seitliche Petalen mehr oder weniger nach unten gebogen, fast gleich breit über die ganze Länge, Fahne deutlich gestreift . . . . *Paphiopedilum violascens*

Key to the species belonging to section BLEPHAROPETALUM

1. Staminodal shield obcordate, more or less uniformly bright yellowish green; pouch with prominent veins (as in *Paphiopedilum venustum*) . . . . *Paphiopedilum sangii*

1a. Staminodal shield horse-shoe-shaped to lunate; pouch not with *venustum*-like venation 2

2. Lateral petals spreading horizontally, narrowest at the base, widening slightly towards the apex; dorsal sepal indistinctly striped . . . 3

2a. Lateral petals more or less drooping, nearly equally wide over their entire length; dorsal sepal distinctly striped . . . . *Paphiopedilum violascens*

3. Seitliche Petalen rötlich braun mit kleinen, schwärzlichen Flecken an den Rändern und entlang der Mittelader in der basalen Hälfte . . . . *Paphiopedilum mastersianum*

3a. Seitliche Petalen blaß karmesinrot mit dunkleren Flecken auf einem grünlich gelben Überzug im basalen Teil . *Paphiopedilum papuanum*

3. Lateral petals reddish-brown with blackish spots in the marginal and middle areas of the lower half . . . . *Paphiopedilum mastersianum*

3a. Lateral petals dull crimson dotted with darker spots in the lower half and tinged with greenish yellow . . . . . *Paphiopedilum papuanum*

Paphiopedilum mastersianum (Rchb. f.) Stein

Stein's Orchideenbuch: 477 (1892)

Basionym: *Cypripedium mastersianum* Rchb. f. **Gardeners' Chronicle,** 2. Ser., **12:** 102 (1879)

Synonym: *Cordula mastersiana* (Rchb. f.) Rolfe **Orchid Review 20** (1): 2 (1912)

Paphiopedilum mastersianum (Rchb. f.) Stein

Über die Entdeckungsgeschichte von *Paphiopedilum mastersianum* ist so gut wie nichts bekannt, außer daß der erste Import von den Sunda-Inseln gestammt haben soll.

Die Pflanze, die als Grundlage für die Beschreibung als *Cypripedium mastersianum* durch H. G. REICHENBACH diente, wurde ihm über die Fa. VEITCH aus dem Botanischen Garten in Kew übermittelt. Sie ist dem damaligen Herausgeber des **Gardeners' Chronicle,** Dr. Maxwell T. MASTERS gewidmet.

Paphiopedilum mastersianum ist eine kurzstämmige, mittelgroße, als Humusepiphyt wachsende krautige Pflanze mit bis zu 22 cm langen und bis ungefähr 4,5 cm breiten, schmal-länglichen, tief- und hellgrün marmorierten Blättern. Die abgerundeten Blattspitzen zeigen drei Zähnchen.

The only thing known about the discovery of this species is that the plants first imported to Europe originated from the Sunda Islands.

The plant on which REICHENBACH based his concept of *Cypripedium mastersianum* was sent to him by VEITCH who obtained it from the Royal Botanic Gardens at Kew. The species was named in honour of the editor of the **Gardeners' Chronicle,** Dr. Maxwell T. MASTERS.

Paphiopedilum mastersianum is a perennial herbaceous epiphyte with short stems. The leaves are narrowly oblong, up to about 22 cm long and more or less 4.5 cm wide, deep green with pale green tessellations, obtuse, the apex three-toothed.

The inflorescence is 30 to 40 cm high, deep purple, very hairy, usually one-flowered.

The bract is ovate, 2 to 2.5 cm long, acute.

Paphiopedilum mastersianum (Rchb. f.) Stein

Die tiefpurpurne, flaumig behaarte Infloreszenz wird bis 40 cm hoch.

Die eiförmige Braktee ist 2 bis 2,5 cm lang und spitz.

Die meist einzige Blüte ist 10 bis 12 cm breit und ungefähr 7,5 cm hoch.

Das mittlere Sepalum (die Fahne) ist 3 bis 4 cm lang und 3 bis 4,5 cm breit, breit eiförmig, stumpf, bewimpert, grün mit dunklerer undeutlicher Aderung. Die Ränder sind weiß bis cremefarben.

Das Synsepalum ist bis 3 cm lang, hellgrün, eiförmig, spitz, viel kleiner als die Fahne.

Die Petalen stehen horizontal ab. Sie sind 4,5 bis 6 cm lang und 1,2 bis 2 cm breit, spatelförmig mit bewimperten Rändern, bräunlich rot, zur Basis hin heller, mit vielen kleinen schwarzpurpurnen Wärzchen entlang dem oberen Rand und entlang der Mittelader.

Die sackförmige Lippe ist 4,5 bis 5,5 cm hoch und ungefähr 2,5 cm breit, hell kastanienbraun, am Öffnungsrand gelblich. Die nach innen gefalteten Seitenlappen sind grünlich braun mit purpurnen Flekken.

The flower is 10 to 12 cm wide and about 7.5 cm high.

The dorsal sepal is broadly ovate, obtuse, with ciliolate margins, bright green with a broad yellow-white border, and deep green veins; 3 to 4 cm long and 3 to 4.5 cm wide.

The synsepalum is up to 3 cm long, pale green, ovate, acute, much smaller than the dorsal one.

The lateral petals are spreading horizontally. They are 4.5 to 6 cm long and 1.2 to 2 cm wide, spathulate with ciliolate margins, brownish-red, paler towards the base where there are numerous small black-purple warts along the superior margin and along the mid-vein.

The lip is saccate, 4.5 to 5.5 cm high and about 2.5 cm wide, pale reddish-brown, yellowish around the rim, the infolded lobes greenish-brown spotted with dull purple.

The staminodal shield is horseshoe-shaped, with acute incurved cusps and a blunt middle tooth on the apical side; greenish-brown with yellow margin.

Die Staminodiumplatte ist grünlich braun mit gelbem Rand, hufeisenförmig mit lang nach innen gekrümmten Seitenzähnen und einem stumpfen Mittelzahn.

Paphiopedilum mastersianum wächst auf Borneo und den umliegenden Inseln in Höhenlagen bis zu 1200 m ü. NN.

Das Klima im Heimatareal ist gekennzeichnet durch anhaltende Luftfeuchtigkeit zwischen 80 und 90 % und Temperaturen von 20 bis 29 °C am Tage und 14 bis 20 °C in der Nacht.

Um *Paphiopedilum mastersianum* erfolgreich zu kultivieren, sollte man für die Pflanzen ein gut durchlässiges Substrat wählen, das aber viel Feuchtigkeit speichern kann. Reichliche Moosbeigaben sind also angebracht.

Es sollte vor allem darauf geachtet werden, die Pflanze immer zeitig umzutopfen, da sie kein saures Substrat vertragen. Auch reagieren sie negativ auf hohe Salzkonzentrationen im Gießwasser. *Paphiopedilum mastersianum* gedeiht am besten an einer schattigen Stelle im temperierten Gewächshaus mit ausreichender Luftzirkulation.

Diese Art blüht im allgemeinen von Juli bis September.

Varietäten:

Obwohl *Paphiopedilum papuanum* (Ridley ex Rendle) L. O. Williams nunmehr im allgemeinen als eigenständige Art betrachtet und deshalb auch in diesem Buch als solche abgehandelt wird, bin ich der Meinung, daß diese Pflanzen ebensogut als geographische Varianten von *Paphiopedilum mastersianum* betrachtet werden könnten.

Paphiopedilum mastersianum inhabits Borneo and the surrounding islands to the western part of New Guinea. The plants usually grow at altitudes of up to 1200 meters above sea-level.

The climate in the natural distribution area is characterized by lasting humidity between 80 and 90 % and temperatures of 20 to 29 °C during the day and 14 to 20 °C at night.

Paphiopedilum mastersianum can be grown successfully when potted in a well-drained open substrate capable of retaining plenty of moisture. This is best obtained by adding abundant moss.

One should take care to repot the specimens regularly as this species does not resist to a stale (acid) compost. The plants are also very sensitive to high salt-concentrations in the water. *Paphiopedilum mastersianum* does best in a shaded spot in the intermediate house with good air circulation.

This species usually flowers between July and September and between January and March, depending on the origin of the variant.

Varieties:

Although *Paphiopedilum papuanum* (Ridley ex Rendle) L. O. Williams is now generally considered an autonomous species and therefore treated as such in this book, those plants could easily be considered to be mere geographic varieties of *Paphiopedilum mastersianum.*

Paphiopedilum papuanum (Ridley ex Rendle) L. O. Williams

Bot. Mus. Leafl. Harvard Univ. 12 (5): 149 (1946)

Basionym: *Cypripedium papuanum* Ridley ex Rendle
Gardeners' Chronicle, 3. Ser., **58:** 131 (1915)

Synonym: *Paphiopedilum zieckianum* Schoser
Die Orchidee 18: 1–7 (1967)

Paphiopedilum papuanum (Ridley ex Rendle) L. O. Williams

Paphiopedilum papuanum wurde während der WOLLASTON-Expedition in Neu Guinea um 1913 bis 1914 durch C. B. KLOSS entdeckt. Die gesammelten Pflanzen gingen an RIDLEY, der für die Ausarbeitung zuständig war. Er benannte die Frauenschuhart als *Cypripedium papuanum* und katalogisierte das Material.

Während derselben Expedition wurde auch Lebendmaterial im Auftrag von N. Charles ROTHSCHILD gesammelt, und zwischen diesem Material befanden sich auch Pflanzen von *Cypripedium papuanum*, die im Sommer 1915 in England zur Blüte kamen.

Die offizielle Beschreibung der Art erfolgte durch A. B. RENDLE im **Gardeners' Chronicle** *(loc. cit.)*. RENDLE basierte seinen Beitrag auf den Angaben RIDLEYs, von dem auch die lateinische Diagnose stammte. Die Veröffentlichung war gepaart mit einem Foto der Pflanze und der Blüte. Der Transfer zu *Paphiopedilum* erfolgte 1946 durch L. O. WILLIAMS *(loc. cit.)*.

Während der letzten Jahre wurden eine Menge *Paphiopedilum*-Pflanzen als *Paphiopedilum papuanum* verkauft. Diese Pflanzen sind aber als geographische Variante von *Paphiopedilum violascens* Schlechter – einer anderen Art aus Neu-

Paphiopedilum papuanum was discovered by C. B. KLOSS, the botanical collector of the WOLLASTON-expedition to New Guinea around 1913 to 1914. The plants collected during the expedition were send to RIDLEY, who was responsible for working out the collection among which there was also a slipper orchid which he named *Cypripedium papuanum.*

During the same expedition, life material was collected for N. Charles ROTHSCHILD. Among those plants, the slipper orchid in question flowered in the summer of 1915.

The species was originally described as *Cypripedium papuanum* by A. B. RENDLE in the **Gardeners' Chronicle** *(loc. cit.)* together with a diagnosis rendered by H. N. RIDLEY and a photograph of the plant and flower.

The species was duly transferred to *Paphiopedilum* by L. O. WILLIAMS *(loc. cit.)*.

Lots of plants have been sold as *Paphiopedilum papuanum* over the last years. These plants, however, represent a geographical variety of *Paphiopedilum violascens* Schlechter, another species found in New Guinea. Until recently, the true *Paphiopedilum papuanum* was very rare in cultivation and usually grown under the synonymous

Guinea – anzusehen. Bis vor kurzem war das echte *Paphiopedilum papuanum* sehr selten in Kultur und wurde allgemein als *Paphiopedilum zieckianum* Schoser bezeichnet. Es besteht kein Zweifel daran, daß SCHOSERs Konzept von *Paphiopedilum zieckianum* identisch mit *Paphiopedilum papuanum* (Ridley ex Rendle) L. O. Williams und daher als Synonym zu betrachten ist.

Paphiopedilum papuanum ist ein bis 25 cm großer, krautiger Humusepiphyt. Die Laubblätter, meistens 4 bis 6 per Trieb, sind blaßgrün, oben mit dunkelgrüner Marmorierung, länglich, ledrig, im allgemeinen 7 bis 10 cm lang und bis zu 2,6 cm breit. Sie sind mehr oder weniger spitz, der Apex mit drei kleinen Zähnchen.

Die Infloreszenz ist 15 bis 25 cm hoch, dunkel karmesinrot, mit vielen abstehenden steifen Härchen.

Die Braktee ist eiförmig, behaart wie die Infloreszenz, bis 1,4 cm lang.

Der gestielte Fruchtknoten ist dunkelgrün, behaart, bis zu 5 cm lang.

Die Fahne ist breit eiförmig mit einer kurzen abrupten Spitze, grün, mit fünf oder sechs mehr oder weniger obskuren dunkel karmesinroten, unverzweigten Adern auf jeder Seite der Mittelader, nur 2,6 cm lang und ungefähr 2 cm breit, die Außenseite behaart, die Ränder bewimpert.

Die zu einem Synsepalum verwachsenen lateralen Sepalen sind elliptisch-spatelförmig, mehr oder weniger spitz, ungefähr 2 cm lang und bis ca. 1 cm breit, mit ungefähr sieben hervorstehenden Adern, die Rückseite behaart.

Die seitlichen Petalen sind länglich, zur Spitze hin etwas breiter werdend, gerundet, bis ungefähr 4,3 cm lang und bis 1,3 cm breit, unbehaart mit bewimperten Rändern, mehrfach geadert, blaß karmesinrot mit einem gelblich grünen Überzug und kleinen dunkel karmesinroten Flecken im basalen Teil.

Die Lippe ist insgesamt bis 4 cm lang, wobei die Klaue fast so groß ist wie der Schuh. Die Seitenlappen sind etwas gerundet und mit kleinen Härchen überzogen. Der Hauptlappen ist tief sackförmig, blaß karmesinrot wie die seitlichen Petalen, die Seitenlappen sind grünlich gelb überhaucht.

Die Staminodiumplatte ist halbmond- bis nierenförmig, am oberen Rand eingekerbt, am unteren Rand mit zwei gerundeten Seitenlappen und einem stumpfen Mittelzahn.

Paphiopedilum papuanum wurde an den Hängen von Mount Carstenz in einer Höhenlage von ungefähr 830 m ü. NN gefunden. Unabhängig voneinander fanden auch ZIECK und MATTES Pflanze

name *Paphiopedilum zieckianum* whereby SCHOSERs concept of *Paphiopedilum zieckianum* is doubtless identical with *Paphiopedilum papuanum* (Ridley ex Rendle) L. O. Williams.

Paphiopedilum papuanum is an herbaceous leaf humus epiphyte growing to an overall height of about 22 to 25 cm.

The leaves, usually 4 to 6 per growth, are oblong, leathery, generally 7 to 10 cm long and up to 2.6 cm wide, more or less acute, the extreme tip minutely tri-cuspidate; pale green, the upper surface with darker green tesselations.

The inflorescence is 15 to 25 cm high, deep cirmson, bearing numerous stiff hairs.

The floral bract is ovate, similarly hairy, up to about 1.4 cm long.

The pedicel and the deep green ovary are hairy like the scape and about 5 cm long.

The dorsal sepal is broadly ovate with a short, abrupt acute apex, green, with five to six somewhat obscure deep crimson unbranched nerves on either side of the mid-vein, barely 2.6 cm long and more or less 2 cm wide, hirsute on the back, the margins ciliate.

The lateral petals are united to a synsepalum which is elliptic-lanceolate, subacute, more or les 2 cm long and up to about 1 cm wide, hirsute on the back, with about seven raised nerves.

The lateral petals are oblong, broadening towards the apex, obtuse, up to about 4.3 cm long and up to 1.3 cm wide, glabrous except for the ciliolate margin, multiple-nerved, dull crimson dotted with darker spots in the lower half and tinged with greenish-yellow.

The lip is up to about 4 cm long, including the broad claw, which is about equal in length to the cup, the minutely pubescent lateral lobes somewhat obtuse, the main lobe deeply saccate, dull crimson as in the petals, the lobes tinged with greenish-yellow.

The staminodal shield is lunate-reniform, notched above, the lobes blunt, the median tooth obtuse.

Paphiopedilum papuanum was originally found on the slopes of Mount Carstenz at an elevation of about 830 m. Plants of this species were also found by ZIECK and MATTES, probably at the same site on the northern slopes of the Arfak-Mountains at an altitude of about 1050 m, above the Ngabret-lake along a tributary of the Warmolpi dieser Art, beide wahrscheinlich am selben Standort, an den nördlichen Hängen des Arfak-Gebirges auf ca. 1050 m Höhe an einem steilen Hang, entlang eines Seitenflusses des Warmolpi Marveni, ca. 30 km südlich von Manokwari, in der Nähe des Ngabret-Sees. Die Pflanzen wuchsen in der prallen Sonne.

Entsprechend seiner Herkunft sollte *Paphiopedilum papuanum* im warmen Gewächshaus in einem gut durchlässigen Substrat und bei konstant hoher Luftfeuchtigkeit gehalten werden. Die Pflanzen sollten immer feucht gehalten werden (siehe auch unter *Paphiopedilum violascens*).

Marveni about 30 km to the south of Manokwari. The plants are reported to grow in sunny spots.

According to its origin, *Paphiopedilum papuanum* does best in the warm house, potted in an open substrate under constant high humidity. The plants should be kept moist at all times. (See also under *Paphiopedilum violascens*).

Paphiopedilum sangii Braem

Die Orchidee 38 (4): 169–172 (1987)

Paphiopedilum sangii Braem

Paphiopedilum sangii Braem

Paphiopedilum sangii wurde auf Celebes (Sulawesi) entdeckt und durch Helmut SANG (Essen) nach Deutschland gebracht, wo im Frühjahr 1987 die ersten Pflanzen zur Blüte kamen.

Die Art gehört aufgrund ihrer Merkmalskombination (marmorierte Laubblätter, einblütige Infloreszenz, Aurikeln am Lippenhauptlappen, Synsepalum deutlich kleiner als die Fahne) zu der Untergattung SIGMATOPETALUM Hallier und aufgrund der Morphologie der lateralen Sepalen und der Fahne eindeutig in die Sektion BLEPHAROPETALUM Pfitzer. Die Pflanzen setzen sich aber deutlich von den Mitgliedern anderer bekannter Arten dieser Gruppe durch die „*venustum*-ähnliche" Beaderung des Lippenhauptlappens, und vor allem durch die verkehrt herzförmige Staminodiumplatte ab.

Paphiopedilum sangii was discovered on Celebes (Sulawesi) and introduced by Helmut SANG (Essen) to West Germany where the first specimens flowered in the spring of 1987.

The combination of the characteristics (tesselated leaves, uniflorous inflorescence, auriculate labellum, synsepalum distinctly smaller than the dorsal sepal) indicated that this species belongs within the subgenus SIGMATOPETALUM Hallier. Because of the morphology of the dorsal sepal and of the lateral petals, *Paphiopedilum sangii* is without any doubt to be positioned within the section BLEPHAROPETALUM Pfitzer. The plants however, differ distinctly from all known taxa of this group by their "*venustum*-like" venation of the labellum and especially by the obcordate staminodal shield.

Paphiopedilum sangii ist eine krautige, als Humusepiphyt lebende Pflanze, bis zu 50 cm breit (von Blattspitze zu Blattspitze) und bis zu 26 cm hoch (inkl. Blütenstand).

Pro Trieb werden bis zu 4 Laubblätter ausgebildet. Sie sind bis ca. 5,2 cm breit und bis ca. 28 cm lang, oberseitig deutlich graugrün-dunkelgrün marmoriert, unterseitig einfarbig grün mit leicht purpurbraun überhauchter Basis, länglich lanzettförmig, spitz, unterseitig deutlich und ganz scharf gekielt.

Der Blütenstand ist bis zu ca. 25 cm hoch. Der Blütenstandschaft ist rund, schwarz-braun gefärbt, intensiv von kurzen steifen elfenbeinfarbenen Härchen überzogen. Er trägt im allgemeinen eine einzelne Blüte und entspringt einer deutlich ausgebildeten Schutzscheide.

Die Blütenbraktee ist bis zu 2,4 cm lang und ca. 8 mm breit, spitz zulaufend, gelblich grün, an der Basis purpurbraun überzogen, an der Spitze dunkler grün.

Der Fruchtknoten ist bis zu 5,5 cm lang und mehr oder weniger 5 mm breit, schwarz-braun gefärbt, wie die Blütenstandsachse intensiv elfenbeinfarben behaart, die Härchen aber noch länger als an dem Infloreszensstiel.

Die Blüte ist bis 8,5 cm hoch und 5,4 cm breit (gemessen über den Sepalen an der lebenden Pflanze in natürlicher Haltung).

Das dorsale Sepalum (die Fahne) wird bis ungefähr 4,3 cm hoch und 3,3 cm breit. Es ist schmal eiförmig, spitz zulaufend und manchmal auf halber Höhe beidseitig bis ca. 8 mm tief eingeschnitten; apfelgrün, an der Basis purpurbraun, die Endregion hellgrün; mit einheitlich apfelgrünen Rändern und mit bis zu ca. 18 dunkelgrünen Längsstreifen. Die Fahne ist beiseitig elfenbeinfarbig behaart, wobei die Behaarung auf der Rückseite (Außenseite) am intensivsten ist.

Die seitlichen Sepalen sind, wie typisch für die Gattung *Paphiopedilum* zu einem Synsepalum verwachsen. Dieses ist bis ca. 3,2 cm lang und 1,6 cm breit, deutlich doppelspitzig, aber einfach gekielt, manchmal im basalen Teil beidseitig mit einer kleinen Kerbe am Rande; gelblich grün, mehr oder weniger undeutlich gestreift, wie die Fahne beidseitig behaart.

Die lateralen Petalen sind bis 4,6 cm lang und bis zu 1,4 cm breit, lanzettförmig, mit mehr oder weniger gerundeten Enden. Sie sind in einem Winkel von ungefähr 45° nach unten gerichtet, leicht nach hinten verdreht, und die oberen Ränder sind über etwa ⅔ der Länge intensiv gewellt. Sie sind dunkel englischrot gefärbt, zur Spitze hin etwas heller,

Paphiopedilum sangii is an herbaceous humus epiphyte attaining an overall width of 50 cm (measured from leaf tip to leaf tip). The overall height, including the inflorescence can be up to about 26 cm.

Per growth, up to four leaves are generated. These are up to about 5,2 cm in width and up to about 28 cm long. On the upper surface, they are grayish green, distinctly tesselated with darker green. The under side is unicoloured green with a light purple hue on the base. The leaves are oblong lanceolate, acute, with a distinct sharp keel underneath.

The inflorescence is up to about 25 cm high, terete, very dark brown, intensely covered by stiff, ivory-coloured hairs; usually bearing one single flower and generated from a distinct protective sheath.

The floral bract is up to 2.4 cm long and about 8 mm across, acuminate, yellowish green, the base covered with purple, the apex darker green.

The ovary is up to about 5.5 cm long and more or less 5 mm wide, blackish brown and just like the inflorescence covered with ivory-coloured hairs. These however, are longer than those covering the other parts of the plant.

The flower is up to about 8,5 cm high and usually about 5,4 cm wide (measured across the sepals on the living plant in its natural state).

The dorsal sepal is up to about 4.3 cm long and about 3.3 cm across, narrowly ovate, acuminate, sometimes bilaterally at half height with symmetrically incisions which are about 8 mm deep; apple green, the base covered with purplish brown, the apical region bright green, with uniform apple green margins and up to about 18 dark green longitudinal stripes, ventrally as well as dorsally covd with ivory-coloured hairs, whereby the hairs on the outside are longer than on the inside.

The lateral sepals are (typically for the genus *Paphiopedilum*) united to a synsepalum. This is about 3.2 cm long and about 1.6 cm in width, distinctly double-ended but with a single keel, sometimes bilaterally with a small incision at the margin; yellowish green, more or less indistinctly striped longitudinally, and just as the dorsal sepal covered with stiff hairs on both sides.

The lateral petals are up to about 4,6 cm long and up to about 1.4 cm wide, lanceolate, more or less rounded at the apex. The are directed downwards in an approximate angle of 45°, distorted slightly backwards, about ⅔ of the upper margins strongly undulate. They are deep englishred, the apical regions brighter, the apex and the narrow marginal regions apple-green; the upper half of the inside surface more or less intensely covered with ivory-coloured hairs.

Paphiopedilum sangii, Braem

Spitze und schmale Randregion apfelgrün; die obere Hälfte innenseitig mehr oder weniger intensiv elfenbeinfarben behaart.

Die Lippe (das Labellum) ist dreilappig, insgesamt bis 4,7 cm hoch und ca. 1,9 cm breit. Der Hauptlappen ist deutlich sackförmig, die Seitenlappen sind nach innen geschlagen und bilden eine fast geschlossene Röhre.

Die Seitenlappen sind hell purpur mit weißem, tiefpurpur geflecktem Rand, ca. 1,5 cm breit und bis 2,5 cm hoch.

Der Hauptlappen ist ca. 1,9 cm breit und bis 3,5 cm lang, mit gerundeter Spitze. Die Öffnung ist bis 2 cm breit, die Ränder der Öffnung sind nicht nach innen gebogen, sondern stehen deutlich nach außen gerichtet. Der Hauptlappen ist olivgrün, deutlich von dunkleren Adern durchzogen, der Öffnungsrand ist einfarbig grün.

Die Staminodiumplatte ist verkehrt herzförmig, bis ca. 1,4 cm breit und ungefähr ebenso hoch; hell gelblich grün, in der Mitte weißlich, die apikale Seite grün berandet.

Soweit bekannt, ist die Verbreitung von *Paphiopedilum sangii* auf Sulawesi (Celebes) begrenzt. Dort wächst die Art als Humusepiphyt am Waldboden in leicht humosen, stark mineralhaltigen Lehmböden.

Paphiopedilum sangii ist temperiert bis warm zu kultivieren. Die Art gedeiht am besten in einem gut durchlässigen, feuchtigkeitsspeichernden Substrat an einer halbschattigen bis schattigen Stelle.

Die Blütezeit ist im allgemeinen von April bis Juli.

The labellum is trilobate, overall up to 4.7 cm long and about 1.9 cm wide. The main lobe is distinctly saccate, the lateral lobes are folded inwards forming a nearly closed tube.

The lateral lobes are about 1.5 cm wide and up to 2.5 cm high, brightly purple with a white margin covered by deep purple spots. The main lobe is about 1.9 cm wide and up to about 3.5 cm long, the apex rounded, the aperture about 2 cm in width with borders that are not infolded but distinctly directed outwards. The main lobe of the labellum is olive green, distinctly veined with a darker green; at the borders of the aperture uniformly green.

The staminodal shield is obcordate, up to about 1.4 cm wide and about equally high, bright yellow green, the center part whitish, the apical margins green.

The distribution of *Paphiopedilum sangii* is probably restricted to Sulawesi (Celebes) where the plants grow as humusepiphytes in open woods.

Paphiopedilum sangii should be kept in a semi-shaded to shaded spot in the intermediate or warm greenhouse. The plants should be potted in an open, but moisture retaining substrate.

The flowers generally open between April and July.

Paphiopedilum violascens Schlechter

in **Die Orchidaceen von Deutsch-Neu-Guinea**

in F. FEDDE (Hrsg./edit.): **Reportorium specierum novarum regni vegetabilis,** Beiheft, **1:** 2–3 (1911)

Synonyma: *Cordula violascens* (Schlechter) Rolfe
Orchid Review 20 (1): 2 (1912)
Paphiopedilum bougainvilleanum Fowlie
Orchid Digest 35: 122 (1971)
Paphiopedilum violascens var. *gautierense* J. J. Smith
Nova Guinea 12: 175 (1915)
Paphiopedilum wentworthianum Schoser & Fowlie ex Fowlie
Orchid Digest 32: 285 (1968)

Paphiopedilum violascens Schlechter

Paphiopedilum violascens wurde 1907 durch den deutschen Botaniker Rudolf SCHLECHTER in Neu-Guinea entdeckt und wie oben zitiert einige Jahre später beschrieben.

Die durch FOWLIE veröffentlichten Konzepte von *Paphiopedilum wentworthianum* und *Paphiopedilum bougainvilleanum* sind meines Erachtens als identisch mit *Paphiopedilum violascens* zu betrachten. Sie können, wenn überhaupt, lediglich den Status einer geographisch bedingten, sich in der Blütenfärbung und der Marmorierung der Blätter unwesentlich unterscheidenden Variante in Anspruch nehmen.

Paphiopedilum violascens ist eine als Humusepiphyt wachsende krautige Pflanze, die per Trieb vier bis sechs bis 22 cm lange und 2,5 bis 4 cm breite, länglich lanzettförmige, vorne fein 3zähnige, graugrüne, auf einen dunkelgrünen Hintergrund marmorierte Laubblätter ausbildet.

Die fast immer einblütige, aufrecht stehende Infloreszenz ist bis zu 30 cm hoch, dunkelbraun bis fast schwarz, flaumig behaart.

Die Blüten werden bis zu 7 cm hoch.

Die Fahne ist weiß oder rosa mit magentaroten, grünen oder purpurnen Längsadern, ca. 3,2 cm

Paphiopedilum violascens was discovered in 1907 by the German botanist Rudolf SCHLECHTER who described it a few years later in his comprehensive treatise of the orchids from the German-controlled part of New Guinea.

The concepts published by FOWLIE as *Paphiopedilum bouquainvilleanum* and *Paphiopedilum wentworthianum* are, in my opinion, to be considered geographical variants of *Paphiopedilum violascens* differing from this species only in the colouration of the flower and in the tesselation of the leaves. They can, at the outmost, claim varietal status.

Paphiopedilum violascens is a perennial herbaceous plant, growing epiphytically in leaf-mould or debris. The stems are very short and bear up to 6 leaves.

The leaves are oblong lanceolate, the apex finely three-toothed; up to 22 cm long and 2.5 to 4 cm wide; deep green tesselated with grey-green.

The inflorescence is erect, up to 30 cm high, dark brown to blackish, downy; usually bearing one single, up to 7 cm large flower.

The dorsal sepal is broadly ovate, acute, symmetrically striped with magenta, green and black; margins pink to white; inner surface smooth, outer surlang und ungefähr 2,8 cm breit, eiförmig, spitz, bewimpert.

Das Synsepalum ist ungefähr 2,4 cm lang und 1,5 cm breit, länglich eiförmig, spitz, bleicher gefärbt als die Fahne.

Die Petalen sind 3,8 bis 4,6 cm lang und ca. 1,6 cm breit, länglich bis linealisch riemenförmig, sichelartig herabgebogen, bewimpert, an der Basis purpurn, zur Spitze hin weißlich und fast über die ganze Länge purpurn beadert.

Die Lippe ist olivgrün bis braun, glänzend, pantoffelförmig, ungefähr 4,5 cm lang und ca. 2,6 cm breit. Sie ist dreilappig mit nach innen gebogenen Seitenlappen. Die Beaderung ist vor allem auf der Außenseite sehr ausgeprägt.

Die Staminodiumplatte ist cremegelb mit grüner Zeichnung, ungefähr 1,2 cm lang und 1 cm breit, nieren- bis hufeisenförmig, oben etwas gespalten, mit drei apikalen Zähnen, die seitlichen länger als der mittlere.

Paphiopedilum violascens ist in den Küstengegenden Papua Neu Guineas, auf einigen der nahegelegenen Inseln und auf Bougainville (Typen von *Paphiopedilum bougainvilleanum* und *Paphiopedilum wentworthianum*) beheimatet.

Die Pflanzen wachsen mit Vorliebe im Halbschatten am Rande von Steilhängen im Gebirge als Humusepiphyt am Boden ziemlich offener Wälder oder auf humus- und moosbedeckten Felsen.

Das Heimatareal ist stets naß, und in diesen nahe am Äquator gelegenen Gebieten sind die Tage und die Nächte fast gleich lang. Die Regenperiode dauert von Juni bis September. Zu dieser Zeit herrschen Temperaturen um 30 °C am Tage und um 20 °C in der Nacht. In den trockeneren Monaten sind die Temperaturen um 3 bis 5 °C niedriger.

Am natürlichen Standort blühen die Pflanzen im allgemeinen von Mai bis Juli und oft nochmals von Oktober bis Dezember.

Paphiopedilum violascens gedeiht am besten im temperierten bis warmen Gewächshaus bei hoher Luftfeuchtigkeit und diffusem Licht. Die Pflanzen brauchen ein lockeres, durchlässiges Substrat und benötigen, vor allem während der Wachstumsperiode, viel Wasser.

Auch für eine gute Luftumwälzung sollte gesorgt werden. Zu hell stehende Pflanzen scheinen ihr Wachstum zu mindern.

face more green and brown, slightly keeled, ciliate; about 3.2 cm long and more or less 2.8 cm wide.

The synsepalum is about 2.4 cm long and more or less 1.5 cm wide, oblong-ovate, acute, coloured paler than the dorsal sepal.

The lateral petals are 3.8 to 4.6 cm long and about 1.6 cm wide, oblong linear to ligulate, drooping crescent like, more or less deflexed, ciliolate, the basal part purple, towards the apex whitish and veined purple over the whole length.

The saccate lip is inverted helm-shaped, olive-green to brown, shiny, about 4.5 cm long and about 2.6 cm wide; trilobate, the sidelobes infolded. The veining is very distinct on the outer surface.

The staminodal shield is cream-yellow with a green pattern in the center area; more or less reniform to lunar crescent shape, somewhat cleft at the top, with three apical teeth, the middle one distinctly shorter than the outer ones.

Paphiopedilum violascens occurs in the coastal regions of Papua New Guinea, on some of the adjacent islands and on the isle of Bougainville (Types of *Paphiopedilum bougainvilleanum* and *Paphiopedilum wentworthianum*).

The plants of this species prefer semi-shaded habitats on the border of steep mountain cliffs. They usually grow as humus epiphytes on the ground in open woods or on humus covered rock formations.

The habitats are constantly wet in this area, which is near the equator, day time and night time are of about equal length. The rainy season lasts from June to September. During that period of the year, the temperatures are about 30 °C during the day and about 20 °C at night. During the drier months, the analogous temperatures are about 3 to 5 °C lower.

In its natural habitat, *Paphiopedilum violascens* usually flowers between May and July and often a second time from October to December.

Paphiopedilum violascens grows best in the intermediate to warm greenhouse at constant high relative humidity and diffuse light. The plants should be potted in a well-drained open compost and should be extensively watered, especially during the growth period.

Good air circulation is a must. Plants grown in too bright light seem to loose substance.

Varietäten:

Einige Farbvarianten dieser Art werden auf dem Orchideenmarkt angeboten. Eine davon, eine hell gefärbte Variante (Abb. unten rechts) wird normalerweise als *Paphiopedilum papuanum* angeboten, hat aber mit dieser letzten Art nichts zu tun (siehe auch unter *Paphiopedilum papuanum*). Zwei andere wurden als eigenständige Arten beschrieben.

Die erste davon, beschrieben als *Paphiopedilum bougainvilleanum* (Abb. unten links) hat eine ziemlich einfarbig grüne Blüte mit einigen Spuren Purpur auf den seitlichen Petalen.

Die zweite, von FOWLIE und SCHOSER als *Paphiopedilum wentworthianum* beschrieben, unterscheidet sich von den typischen *Paphiopedilum violascens*-Formen nur durch die ziemlich einheitlich dunkelpurpur gefärbten lateralen Petalen.

Diese drei Konzepte sollten lediglich als geographische Varianten betrachtet werden und verdienen keineswegs den Artstatus.

Varieties:

A number of colour varieties of this species are being sold. One of them, a paler coloured variant (figure below, right), is usually sold a *Paphiopedilum papuanum* but has nothing to do with that species (see under *Paphiopedilum papuanum*). Two others have been described as distinct species.

The first of these, described as *Paphiopedilum bougainvilleanum* (figure below, left) has a fairly uniformly green coloured flower with some remnants of purple on the lateral petals.

The second one, described as *Paphiopedilum wentworthianum* by SCHOSER and FOWLIE, differs from the typical form of *Paphiopedilum violascens* only by its fairly uniform coloured dark violet lateral petals.

None of these concepts deserve to be upheld as autonomous species and should be regarded as mere geographical variants.

Paphiopedilum violascens Schltr.
„*Paphiopedilum bougainvilleanum*"

Paphiopedilum violascens Schltr.
„*Paphiopedilum papuanum* Hort."

Sektion PUNCTATUM
Karasawa & Saito

Diese Sektion umfaßt nur eine Art – *Paphiopedilum tonsum* – und unterscheidet sich von allen anderen Gruppen der Untergattung durch blaßbraune Blüten mit spatelförmigen, zur Spitze hin allmählich breiter werdenden, von tief purpurbraunen Flecken überzogenen seitlichen Petalen, deren Ränder kaum bewimpert sind.

Section PUNCTATUM
Karasawa & Saito

The section **Punctatum** comprises but one species – *Paphiopedilum tonsum.* This section differs from the others in the subgenus by the pale brown flowers with hardly ciliate lateral petals which are narrow spathulate, gradually widening towards the apex, and covered with large deep brownish-purple spots.

Paphiopedilum tonsum (Rchb. f.) Stein

Stein's Orchideenbuch: 488 (1892)

Basionym: *Cypripedium tonsum* Rchb. f.
Gardeners' Chronicle, 2. Ser., **20:** 262 (1883)

Synonym: *Cordula tonsa* (Rchb. f.) Rolfe
Orchid Review 20 (1): 2 (1912)

Paphiopedilum tonsum (Rchb. f.) Stein

Pflanzen dieser Art wurden erstmals durch S. CURTIS, Sammler im Auftrage der Firma VEITCH, zusammen mit *Paphiopedilum curtisii* auf Sumatra gesammelt. REICHENBACH beschrieb die Art *(loc. cit.)* auf Bitten von VEITCH als *Cypripedium tonsum*, bezugnehmend auf die kaum behaarten Wärzchen auf der Blüte.

Paphiopedilum tonsum ist eine bis ungefähr 40 cm hohe, als Humusepiphyt wachsende Pflanze mit verkürzten Sprossen, die 3 bis 8 Blätter tragen.

Diese Laubblätter sind bis 20 cm lang und bis ungefähr 5 cm breit. Sie sind länglich elliptisch, oberseitig hell- und dunkelgrün marmoriert, unterseitig intensiv purpurn gefleckt.

Die aufrecht stehende Infloreszenz ist bis zu 40 cm hoch, dunkelgrün mit purpurner Zeichnung und meistens einblütig.

Die behaarte Braktee ist bis 2 cm lang und somit ungefähr ein Drittel so lang wie der Fruchtknoten.

Die Blüte ist ungefähr 10 cm hoch.

The discovery of this species is credited to S. CURTIS who collected the plants together with *Paphiopedilum curtisii* in Sumatra whilst employed as a plant collector by VEITCH. REICHENBACH described the species and named it *Cypripedium tonsum* in reference to the hairless warts on the flower.

Paphiopedilum tonsum is an herbaceous humus epiphyte generating short stems, each bearing 3 to 8 leaves.

The leaves are elliptic oblong, up to about 20 cm long and more or less 5 cm wide, deep green, marbled and blotched with pale grayish green above, heavily mottled with dull purple beneath.

The inflorescence is terete, erect, up to 40 cm high, dark green, blotched purple, usually bearing one single flower.

The hairy bract is up to 2 cm long and sowith about one third as long as the ovary.

The flower is about 10 cm in diameter.

Die Fahne ist ca. 5 cm lang und bis 4 cm breit, breit herzförmig, spitz zulaufend, an den Rändern fein bewimpert, weißlich rosa, mit grünen und dunkelpurpurnen Längsstreifen.

Das Synsepalum ist ungefähr 4 cm lang und viel schmaler als die Fahne, länglich elliptisch, spitz.

Die Petalen sind ungefähr 6,5 cm lang und 2 cm breit, fast spatelförmig, am oberen Rand nach der Spitze hin fein bewimpert und mit kleinen, kaum behaarten Wärzchen besetzt, hellgrün mit dunkelgrünen Adern und manchmal purpurn befleckt. Die Mittelader ist mit Wärzchen besetzt, die, wie die Randwärzchen, nur im apikalen Teil einige spärliche Härchen aufweisen.

Die ausgeprägt sackförmige Lippe ist ungefähr 5,6 cm lang und 3,5 cm breit, mattgrün, mit leichtem braunem oder karminrotem Überzug. Die nach innen geschlagenen Seitenlappen sind mit Wärzchen bestückt.

Die fahlgrüne Staminodiumplatte ist nierenförmig, oben tief eingekerbt, unten mit einem hellgrünen abgerundeten Zähnchen.

Paphiopedilum tonsum wächst auf den westlichen Hängen des Barisangebirges auf Sumatra, einer Region in ziemlicher Nähe des Äquators, wodurch die Tage und die Nächte fast gleichlang sind.

Das Klima ist geprägt durch eine „Trockenzeit" von Juni bis Oktober und eine ausgesprochen regenreiche Periode von November bis März. Die jährliche Niederschlagsmenge beläuft sich insgesamt auf ungefähr 2500 mm.

The dorsal sepal is more or less 5 cm long and up to 4 cm wide, broadly cordate, acuminate, the margins ciliolate, whitish rose, with green and purple-brown longitudinal stripes.

The synsepalum is about 4 cm long and much smaller than the dorsal sepal, oblong elliptic, acuminate, pale green with dark green veins.

The lateral petals are spreading, sub-spathulate, margins ciliate, basal portion green with deep green veins and with some nearly hairless, blackish warts mainly near the margins and on the mid-vein, the apical portion dull purple toned with brown, about 6.5 cm long and 2 cm wide.

The lip is distinctly saccate, about 5.6 cm long and 3.5 cm wide, pale green tinged with rose-colour and veined and reticulated with green, the infolded lobes tawny yellow and covered with smallish warts.

The staminodal shield is reniform, deeply notched at the base and with a small, pale green, blunt tooth at the apex.

Paphiopedilum tonsum inhabits the western slopes of the Barisan Mountain Range in Sumatra. This region is near the equator leaving nights and days fairly equally long.

The climate in this region is characterized by a "dry period" from June to October and a distinct "rainy period" from November through March. The yearly rate of precipitation amounts to 2500 mm.

The differences in temperatures over the year are relatively small. The daytime temperatures are be-

Paphiopedilum tonsum
(Rchb. f.) Stein

Die Temperaturunterschiede über das Jahr hinweg sind nur geringfügig. Die Tagestemperaturen schwanken um 24 bis 28 °C, und die Nachttemperaturen liegen zwischen 11 und 16 °C.

Paphiopedilum tonsum ist ziemlich einfach zu kultivieren. Man stellt die Pflanzen am besten in ein temperiertes Gewächshaus. Sie werden in ein gut durchlässiges Substrat getopft.

Während der sommerlichen Wachstumsperiode muß viel gegossen werden und im Winter nur geringfügig weniger. Die Luftfeuchtigkeit sollte immer über 70 % gehalten werden.

Paphiopedilum tonsum blüht normalerweise von März bis Juni.

tween 24 and 28 °C and at night, the temperatures are between 11 and 16 °C.

Paphiopedilum tonsum is an easy to grow species. The plants do best in the intermediate greenhouse, potted in a well drained, open compost.

Water should be given abundantly during the growth period and only slightly less during the winter months. The relative humidity should be constantly above 70 %.

Paphiopedilum tonsum flowers in general between March and June.

Sektion PLANIPETALUM Karasawa & Saito

Die Sektion **Planipetalum** unterscheidet sich von allen anderen Sektionen der Untergattung **SIGMATOPETALUM** durch die Kombination folgender Merkmale: eine deutlich längsgestreifte Fahne, schmal längliche, schräg abstehende seitliche Petalen, deren Ränder mit langen Wimpern versehen sind und einer sichel- bis halbmondförmigen Staminodiumplatte.

Zu dieser Sektion setze ich *Paphiopedilum purpuratum* und *Paphiopedilum sukhakulii* sowie die Naturhybride *Paphiopedilum × wardii* wobei es ohne weiteres vertretbar wäre, das letztere Taxon in die Sektion **Sigmatopetalum** unterzubringen.

Section PLANIPETALUM Karasawa & Saito

The section **Planipetalum** is distinct from all other sections of the subgenus **SIGMATOPETALUM** by the combination of dorsal sepals distinctly longitudinally striped; the narrow oblong, transversely spreading plain lateral petals with margins covered by long cilia and the crescent-shaped or more or less lunate staminodal shield.

I include in this section besides *Paphiopedilum purpuratum* and *Paphiopedilum sukhakulii* also the natural hybrid *Paphiopedilum × wardii*. The latter taxon could however just as well be placed in section **Sigmatopetalum.**

Schlüssel zu den Taxa der Sektion PLANIPETALUM

1. Seitliche Petalen kurz, mehr oder weniger spitz, purpurn, im basalen Bereich mit kleinen schwarzen Flecken übersät; Fahne gerundet, meistens in der unteren Hälfte stark zurückgeschlagen, mit einer breiten Spitze, mit purpurner Längsstreifung . . . . *Paphiopedilum purpuratum*

1a. Seitliche Petalen lang, mit einer engen Spitze; Fahne mit grüner Längsstreifung 2

2. Seitliche Petalen horizontal abstehend, blaßgrün, fast einheitlich mit tief purpurnen bis schwarzen Flecken überzogen; Fahne breit eiförmig; Schuh lederig gelb bis grünlich gelb, außen rötlich überzogen, innen glänzend weinrot . . . . *Paphiopedilum sukhakulii*

2a. Seitliche Petalen schräg nach unten gerichtet, grün bis rötlich grün, mehr oder weniger einheitlich braun gefleckt, manchmal mit deutlich roten Spitzen; Fahne schmal eiförmig; Schuh grün, die Vorderseite mit braunen Adern und vielen braunen Flecken, innen einheitlich mit schokoladenbraunen Flecken überzogen . . . . *Paphiopedilum × wardii*

Key to the taxa belonging to section PLANIPETALUM

1. Lateral petals short, bluntly acute at the apex, purple, the basal portion finely spotted with minute blackish spots; dorsal sepal rounded, usually strongly recurved in the lower half, broadly acuminate at the apex, longitudinally striped purplish . . . . *Paphiopedilum purpuratum*

1a. Lateral petals long, sharply acute at the apex; dorsal sepal with green longitudinal stripes 2

2. Lateral petals spreading horizontally, pale green, almost uniformly spotted with deep purple dots; lateral sepal broadly ovate; pouch tawny to greenish-yellow, suffused red on the outside, the inner surface shiny, whinered . . . . *Paphiopedilum sukhakulii*

2a. Lateral petals pointed obliquely downwards, green to reddish-green, more or less regularly covered with brown spots, the tips sometimes distinctly red; dorsal sepal narrowly ovate; pouch green, the front covered by brown veins and multiple brown spots, the inside uniformly spotted chocolate-brown . . . . *Paphiopedilum × wardii*

4901

Paphiopedilum purpuratum (Lindley) Stein

Stein's Orchideenbuch: 481 (1892)

Basionym: *Cypripedium purpuratum* Lindley
Edwards's Botanical Register 23, t. 1991 (1837)

Synonyma: *Cordula purpurata* (Lindley) Rolfe
Orchid Review 20 (1): 2 (1912)

Cypripedium sinicum Hance ex Rchb. f.
Walp. Ann. 3: 602 (1853)

Paphiopedilum sinicum (Hance ex Rchb. f.) Stein
Stein's Orchideenbuch: 481 (1892)

Paphiopedilum purpuratum (Lindley) Stein

Über die Entdeckungsgeschichte von *Paphiopedilum purpuratum* ist relativ wenig bekannt. Pflanzen dieser Art wurden bereits um 1836 durch KNIGHT, den Vorgänger von VEITCH, in der englischen Gartenbaufirma ROYAL EXOTIC NURSERY kultiviert. Leider ist nicht überliefert, wo die Pflanzen herstammten.

Zur gleichen Zeit waren auch bei der Firma LODDIGES Pflanzen von *Paphiopedilum purpuratum* in Kultur, und eine dieser Pflanzen wurde im November 1836 im Hause LODDIGES gezeichnet und diente als Grundlage für die Beschreibung durch John LINDLEY im **Edwards's Botanical Register** als *Cypripedium purpuratum*.

Paphiopedilum purpuratum ist ein am Boden wachsender krautiger Humusepiphyt.

Die länglich elliptischen Laubblätter sind 7 bis 14 cm lang und ungefähr 4 cm breit. Sie enden unregelmäßig dreizähnig. Die Blattoberfläche variiert zwischen hell- und dunkelgrün marmoriert bis fast einfarbig. Es wurde beobachtet, daß deutlich marmorierte Neutriebe die Marmorierung nachträglich fast ganz verlieren.[1] Die Blattunterseite ist hellgrün und an der Basis bewimpert.

Die bis zu 20 cm hoch werdende Infloreszenz ist purpurn und flaumig weiß behaart, meistens einblütig.

[1] T. C. LEE, **Orchid Digest 39** (5): 192–196 (1975).

Little is known about the discovery of *Paphiopedilum purpuratum.* Plants of this species, of unknown origin, were grown by KNIGHT, the predecessor of VEITCH in the ROYAL EXOTIC NURSERY as early as 1836.

About the same time some specimens of the same species were under cultivation at the nursery of LODDIGES. One of these plants was drawn on the premises of LODDIGES and used to illustrate the description of *Cypripedium purpuratum* by John LINDLEY in **Edwards's Botanical Register.**

Paphiopedilum purpuratum is an herbaceous plant, usually growing as humus epiphyte on the ground of wet forests.

The leaves are oblong-elliptic, 7 to 14 cm long and about 4 cm wide, the apex irregularly three-toothed. The upper surface is variable between bright green of different shades, tesselated with dark green to apparent uni-coloured green. It was reported that distinctly tesselate new growths became nearly unicoloured with age.[1] The under surface is bright green, ciliate at the base.

[1] T. C. E. LEE, **Orchid Digest 39** (5): 192–196 (1975)

Paphiopedilum purpuratum (Lindley) Stein

Die hellgrüne, flaumig behaarte Braktee ist mit ca. 2,2 cm Länge ungefähr halb so lang wie der Fruchtknoten.

Die Blüte wird 7,5 bis 8,5 cm hoch.

Die Sepalen sind bewimpert. Das mittlere Sepalum (die Fahne) ist 3,8 bis 4,5 cm lang und 4 bis 4,6 cm breit, fast kreisrund, spitz, unten mit zurückgebogenen Rändern, weiß mit grün überlaufendem Mittelteil, mit 8 bis 10 symmetrisch gebogenen braunpurpurroten Streifen.

Die seitlichen Sepalen sind, typisch für die Gattung *Paphiopedilum*, zum Synsepalum verwachsen. Dieses ist eiförmig, spitz, ungefähr ein Drittel so groß wie die Fahne, grünlich.

Die fast spatelförmigen Petalen sind 5 bis 6 cm lang und 1,8 bis 2,2 cm breit. Sie stehen horizontal ab, sind bewimpert, purpurn karmesinrot mit tiefpurpurnen oder grünen Adern, an der Basis mit zahlreichen schwarzen Wärzchen besetzt.

Die Lippe ist 4,2 bis 4,4 cm lang und ungefähr 2,6 cm breit, fast zylindrisch und im unteren Drittel spitz zulaufend, bräunlich purpurn mit intensiver purpurner Aderung, dreilappig. Die nach innen ge-

The inflorescence is terete, erect, up to about 20 cm high, purple, covered by downy white hairs, generally bearing one single flower.

The floral bract is about 2.2 cm long, bright green, downy, about half as long as the ovary.

The flower is about 7.5 to 8.5 cm high.

The sepals are ciliate. The dorsal sepal is suborbicular, 3.8 to 4.5 cm long and 4 to 4.6 cm wide, acute, the basal portion of the margins reflexed, white suffused green on the central area, with 8 to 10 symmetrically curving, brown-purple-red longitudinal stripes.

The lateral sepals are united to a synsepalum as typical for the genus *Paphiopedilum*. This synsepalum is ovate, acute, about one third the size of the dorsal sepal, greenish.

The subspathulate lateral petals are 5 to 6 cm long and 1.8 to 2.2 cm wide, spreading horizontally, ciliolate, purple-crimson with deep purple or green veins, the base suffused with many blackish warts.

The lip is 4.2 to 4.4 cm long and about 2.6 cm wide, nearly cylindrical, ending in a rounded tip, brownish-purple with purple venation, trilobate. The falteten Seitenlappen sind purpurn und mit vielen schwarzen Wärzchen besetzt.

Die innere Struktur der Lippe ist faszinierend. Die Seitenwände sind mehr oder weniger glatt, aber der Boden und ein Teil der hinteren Wand ist warzig und dicht mit feinen nadelartigen Strukturen besetzt.

Die Staminodiumplatte ist halbmondförmig, oben eingekerbt, unten mit einem deutlich ausgeprägten Zähnchen, mattgrün, violettpurpurn überhaucht.

Obwohl die LINDLEYsche Beschreibung von einer Art aus dem Malaiischen Archipel spricht, beruht diese Herkunftsangabe auf einer Fehlinformation.

Paphiopedilum purpuratum ist in Hongkong beheimatet und wahrscheinlich auch in der gegenüberliegenden Gebirgsregion der chinesischen Provinz Guangdong. Auch auf einigen kleinen Inseln zwischen Hongkong und dem Festland soll die Art vorkommen.

Die Pflanze wächst epiphytisch auf humus- und moosbedeckten Felsen in Flußnähe von Meereshöhe bis 700 m ü. NN, an Nord- und Nordwesthängen. *Paphiopedilum purpuratum* wächst auch zwischen Bambus an den Flußufern. Die sehr schattigen Standorte sind immer den kühlen, trokkenen Nordwinden ausgesetzt.

Die Sommertemperaturen in diesem Heimatareal schwanken auf Meereshöhe zwischen 28 und 36 °C. Im Winter ist die mittlere Temperatur um 13 °C, wobei Temperaturstürze bis auf 0 °C vorkommen. An den höher gelegenen Standorten sind die Temperaturen entsprechend niedriger, und die Pflanzen müssen manchmal sogar Frost ertragen.

Die Luftfeuchtigkeit liegt, trotz der trockenen Winde, aufgrund der Flußnähe immer bei 70 bis 80 %.

Die Blütezeit im Heimatareal ist zwischen Oktober und Januar.

Paphiopedilum purpuratum ist eine relativ einfach zu kultivierende Art. Im Sommer sollen die Pflanzen relativ warm und im Winter ziemlich kühl stehen. Es sollte aber für immer gleichbleibende hohe Luftfeuchtigkeit gesorgt werden. Das Substrat sollte durchlässig, aber wasserspeichernd sein.

Varietäten:

Beide als *Cypripedium purpuratum* var. beschriebenen Varietäten von *Paphiopedilum purpuratum* weichen nur geringfügig von der Normalform ab und sind innerhalb der natürlichen Bandbreite der Art anzusiedeln.

infolded lateral lobes are purple suffused with numerous small blackish warts. The inner structure of the lip is fascinating. The sidewalls are more or less smooth but the bottom and part of the rear wall is warty and densely covered by fine needle-like structures.

The staminodal shield is semi-lunate, notched at the top; at the bottom with a distinct tooth in the middle; dull green suffused violet-purple.

Although Lindley indicated this species to originate from the Malaysian Archipelago, we now know that *Paphiopedilum purpuratum* inhabits Hong Kong and most probably also the montane regions of the Chinese province of Guangdong. Plants of this species have also been said to grow on the small islands between Hong Kong and the Chinese mainland.

The plants grow as epiphytes in humus and leafy debris as well as on moss-covered rocks near rivers in altitudes up to about 700 meters above sea-level, on north and northwestern slopes. *Paphiopedilum purpuratum* has also been found between patches of bamboo along rivers. The habitats are always very shaded and open to the cool, dry northerly winds.

The summer temperatures in the natural distribution area of *Paphiopedilum purpuratum* are between 28 and 36 °C. In winter, the mean temperature is about 13 °C, but temperature drops to about 0 °C are not seldom. In higher habitats, the temperatures are somewhat lower and the plants sometimes must endure some degrees below the freezing point.

In spite of the dry cool winds, the relative humidity is always between 70 and 80 % due of the vicinity of water.

Paphiopedilum purpuratum flowers in its natural habitat between October and January.

Paphiopedilum purpuratum is a relative easy species to grow. During the summer, the plants should be kept relatively warm and during the winter, they should be moved to the cold house. Relative humidity however, should be kept high over the entire year. The substrate should be well-drained, open and should be capable of retaining a lot of moisture.

Varieties:

Two varieties have been described as *Cypripedium purpuratum* var. Neither of them differs from the typical form of *Paphiopedilum purpuratum* to such an extent as that they could be considered separate taxa.

Paphiopedilum sukhakulii Schoser & Senghas

Die Orchidee 16 (Sonderheft Juni / extra edition June): 109–110 (1965)

Paphiopedilum sukhakulii Schoser & Senghas

Paphiopedilum sukhakulii wurde angeblich durch den thailändischen Orchideenhändler Prasong SUKHAKUL im Norden seiner Heimat als *Paphiopedilum callosum* gesammelt. Die Pflanzen wuchsen zwischen den Importen von *Paphiopedilum callosum* in verschiedenen Gärtnereien in Belgien und Deutschland, bis der Unterschied in der Blütenmorphologie auffiel und die anfängliche Vermutung, es handle sich um eine Wiederentdeckung des 1932 erstmals beschriebenen *Paphiopedilum × wardii* sich nicht bestätigte. SCHOSER und SENGHAS beschrieben die Pflanze als autonome Art *(loc. cit.)*.

Paphiopedilum sukhakulii hat 4 bis 8 schmale, längliche, bis 25 cm lange und bis 5 cm breite, an der Spitze kurz dreizähnige Blätter. Deren Oberfläche ist mehr oder weniger rauh und zeigt ein undeutliches, verwachsenes Zeichnungsmuster. Die Unterseite ist nur an der Basis mit einigen kleinen roten Punkten versehen.

Paphiopedilum sukhakulii is said to have been discovered by the Thai horticulturalist Prasong SUKHAKUL in the northern part of his home country. The plants obviously were collected when not in flower and were sold as *Paphiopedilum callosum* to several orchid firms in Belgium and Germany.

Because of the leaf morphology, it was first suggested that the plants were specimens of *Paphiopedilum × wardii,* lost since 1932. Upon flowering, it became clear that they represented a new species which was duly described as *Paphiopedilum sukhakulii* by SCHOSER and SENGHAS *(loc. cit.)*.

Paphiopedilum sukhakulii has 4 to 8, narrow elliptic leaves that are up to about 25 cm long and up to 5 cm wide, the apex shortly three-toothed, the upper surface more or less rough, and more or less distinctly, tesselate; the under surface only at the base with small reddish spots.

Die Infloreszenz ist aufrecht, bis 25 cm hoch, dicht und kurz weiß behaart, purpurbraun. Sie trägt meist eine (selten zwei) bis 14 cm breite Blüte(n).

Die Braktee ist 1,5 bis 2,5 cm lang, 1,2 bis 2 cm breit, oval, spitz, gefaltet, am Rücken deutlich gekielt, unterseitig kurz behaart, hellgrün, meist mit bis zu 7 dunkleren Nerven.

Der ca. 4 bis 6 cm lange, aufrechte, stumpf dreikantige Fruchtknoten ist an der Spitze deutlich nach vorne gebogen und wie der Infloreszenzschaft gleichmäßig dicht und kurz behaart.

Die Fahne ist breit oval, lang zugespitzt, 3,2 bis 4,2 cm lang und 2,5 bis 3,2 cm breit, weiß mit 10 bis 13 großen und 2 bis 6 kürzeren intensiv grünen Adern.

Das kleinere Synsepalum ist gleichgestaltet mit 9 dunklen Adern.

Die horizontal abstehenden Petalen sind 6 bis 7 cm lang und 1,4 bis 2 cm breit mit zum größten Teil parallel verlaufenden Rändern, die erst im Enddrittel in einem ziemlich spitzen Winkel zusammenlaufen. Sie sind hellgrün mit 6 bis 8 intensiv gefärbten Adern, übersät mit vielen mehr oder weniger gleichgroßen braunroten Flecken, die jedoch niemals

The inflorescence is erect, up to 25 cm high, densely covered by white hairs, purple-brown, usually bearing one (seldomly two), up to 14 cm broad flower(s).

The floral bract is 1.5 to 2.5 cm long and 1.2 to 2 cm across, oval, acute, folded, distinctly keeled on the back, the under surface covered by short hears, bright green, usually with up to 7 darker veins.

The ovary is about 4 to 6 cm long, erect, blunt triangular, the apex distinctly incurved, uniformly dense pilose.

The dorsal sepal is broadly ovate, acuminate; 3.2 to 4.2 cm long and 2.5 to 3.2 cm wide, white with 10 to 13 large and 2 to 6 shorter deep green veins.

The synsepalum is similar but smaller, with 9 dark veins.

The lateral petals are spreading, 6 to 7 cm long and 1.2 to 1.4 cm wide, the margins running parallel over nearly the extire length, becoming acuminate only in the apical one-third; bright green with 6 to 8 darker longitudinal veins and densely covered with regular brownish red spots never having a wartlike structure; glossy, the margins ciliate, the hairs

Paphiopedilum sukhakulii Schoser & Senghas

eine warzenartige Struktur annehmen. Die Petalen sind glänzend und an den Rändern behaart, wobei die Härchen an der Basis am längsten sind und in der Zahl und Länge in Richtung Spitze abnehmen.

Die in einem etwa 30- bis 40°-Winkel zu der Blütenachse stehende Lippe ist deutlich schuhförmig, insgesamt 4 bis 5 cm lang (wobei der Schuh selbst 3 bis 3,5 cm lang ist), ziemlich spitz endend und dadurch in der Vorderansicht mehr oder weniger dreieckig. Die Spitze ist gelblichgrün. In Richtung der Öffnung geht die Färbung in einen bräunlichen bis rotbraunen Ton über. Das Adernetz hebt sich durch seine intensivere Färbung deutlich hervor.

Außen ist der Schuh gleichmäßig von sehr kleinen Härchen besetzt. Innen ist er deutlich glänzend, intensiv weinrot, vorne kahl, am Boden und auf der Rückseite mit zahlreichen langen weißen Drüsenhaaren besetzt.

Die Staminodiumplatte ist halbmond- bis hufeisenförmig, 11 bis 13 mm im Durchmesser, weißlich grün mit dunkelgrüner Zeichnung.

Paphiopedilum sukhakulii wächst epiphytisch in Humus auf sandigem Lehm entlang schattiger Wasserläufe, oft mit *Paphiopedilum callosum* zusammen. Die Standorte liegen in Höhenlagen bis 1000 m ü. NN im Nordosten Thailands, wobei es nicht auszuschließen ist, daß das Verbreitungsgebiet sich bis nach Oberbirma und in andere Richtungen ausdehnt.

Paphiopedilum sukhakulii wird wie *Paphiopedilum callosum* kultiviert.

Die Hauptblütezeit von *Paphiopedilum sukhakulii* ist von September bis Dezember, aber die Pflanzen blühen oft nochmals im Frühjahr. Die einzelnen Blüten halten mehrere Wochen an der Pflanze.

longer and more numerous at the base than towards the apex.

The lip stands in an angle of about 30 to 40° to the shaft of the inflorescence. It is overall about 4 to 5 cm long, the saccate midlobe about 3 to 3.5 cm long, fairly pointed and when looked at frontally, fairly triangular, the apex yellowish-green, towards the aperture fading into a brownish to reddish-brown tone, with deeper venation; on the outside regularly covered with minuscule hairs; the inner surface glossy, intensively wine-red, the front glabrous, the back and the bottom densely covered with long white glandular hairs.

The staminodal shield is semi-lunate (horseshoe-shaped), 11 to 13 mm in diameter, whitish-green with a deep green pattern.

Paphiopedilum sukhakulii grows along with *Paphiopedilum callosum* epiphytically in leafy humus in shaded spots along rivers. The habitats are at altitudes of up to 1000 m above sea-level in the north-eastern parts of Thailand. It is quite possible that the natural distribution is much larger.

Paphiopedilum sukhakulii should be grown like *Paphiopedilum callosum.* The plants do not require special care and are rather floriferous. *Paphiopedilum sukhakulii* does best in the intermediate or warm greenhouse, potted in an open substrate.

Relative humidity should be high at all times. A resting period at lower temperatures and with lesser water upon completion of the new growth will induce good flowering.

Paphiopedilum sukhakulii flowers generally from September to December but many well grown specimens will also flower in Spring. The flowers last several weeks on the plant.

Paphiopedilum × *wardii* Summerhayes (pro sp.)

Die Orchidee 37 (1), 4. Umschlagseite / back cover (1986)

(Hybrid. nat. Paphiopedilum venustum × Paphiopedilum sukhakulii)

Basionym: *Paphiopedilum wardii* Summerhayes
Gardeners' Chronicle, 3. Ser., **92:** 446 (1932)

Synonyma: *Cypripedium wardii* Anon., non Rolfe[1)]
Orchid Review 41: 2, Abb. / Figure 9 (1933)
Orchid Review 43: 30 (1935)

Paphiopedilum × wardii Summerhayes (pro sp.)

Paphiopedilum × wardii wurde von dem englischen Armeeoffizier F. Kingdon WARD auf einem Marsch von China in die Gebirgsgegend Nord-Birmas im November des Jahres 1922 erstmals gefunden. Die Blüte, die er mitnahm, ging jedoch verloren. Obwohl Hauptmann WARD die Gegend in den nachfolgenden Jahren mehrmals durchstreifte, dauerte es bis in den Winter 1930/31, bis er die Pflanzen an derselben Stelle wie im Jahre 1922, zwischen Fort Hertz und Nam Tamai, wiederfand.

Hauptmann WARD veröffentlichte eine kurze Entdeckungsgeschichte,[2)] worin die Pflanze als *Cypripedium sp.* bezeichnet wurde.

SUMMERHAYES griff die Notiz WARDs auf und veröffentlichte die Art als *Paphiopedilum wardii* ungefähr ein halbes Jahr später in derselben Zeitung.

Seit die Pflanzen bekannt sind, haben Autoren festgestellt, daß es sich um eine Naturhybride handelt. Inzwischen sind eine Reihe von Untersuchungen möglich geworden. Abgesehen von elektro-

[1)] Dieses Taxon sollte nicht verwechselt werden mit *Cypripedium wardii* Rolfe, ein echtes *Cypripedium* vom oberen Salween-Fluß in Tibet.

[2)] F. K. WARD in **Gardeners' Chronicle,** 3. Ser., **91:** 366, Abbildung 163 (1932).

Paphiopedilum × wardii was discovered by the British army officer F. Kingdon WARD on a march back from China in the montane region of northern Burma but the flower he took on that occasion was lost. Although Captain WARD searched the same area during the next years, it was not until the winter of 1930/1931 that he managed to rediscover some plants of the same species, actually on the same location of his find of 1922 between Fort Hertz and Nam Tamai.

WARD published a short account of the discovery referring to the plant as *Cypripedium sp.*[2)]

SUMMERHAYES described the species in detail *(loc. cit.),* about half a year later, using the notes of WARD and dedicating the species to him.

Ever since plants of this taxon have been known, its hybrid origin has been suggested. In the mean time, new research methods have become available. Besides electron microscopy used to examine the surface structure of pollen and seeds, cytogenetic examinations haven become more

[1)] This taxon is not to be confused with *Cypripedium wardii* Rolfe, which is a true *Cypripedium* originating from the upper Salween River in Tibet.

[2)] F. K. WARD in **Gardeners' Chronicle,** 3. Ser., **91:** 366, Figure 163 (1932).

nenmikroskopischen Untersuchungen an Pollen und Samenoberflächenstrukturen ist die Cytogenetik weit vorangekommen.

Es ist inzwischen nicht nur möglich, die Chromosomenzahlen einzelner Pflanzen zu bestimmen, sondern wir sind heute auch in der Lage, einzelne Chromosome miteinander zu vergleichen.

Auf diesem Wege ist abschließend bewiesen worden, daß *Paphiopedilum* × *wardii* eine Hybride zwischen *Paphiopedilum venustum* und *Paphiopedilum sukhakulii* ist.

Paphiopedilum × *wardii* ist ein am Boden in Humus wachsender krautiger Epiphyt mit vielen kurzen Sprossen, die von einem kurzen Rhizom ausgehen.

Jeder Sproß trägt 3 bis 4 Blätter. Diese stehen zweireihig. Sie sind länglich lanzettförmig, undeutlich dreilappig am Ende, bis 17 cm lang und bis ungefähr 5,5 cm breit. Oberseitig sind sie unregelmäßig hell- und dunkelgrün marmoriert. Die Unterseite ist überzogen mit einer Vielzahl von kleinen purpurnen Flecken.

Die aufrecht stehende Infloreszenz ist bis 20 cm hoch, rötlich, dicht behaart, im allgemeinen einblütig.

Die Braktee ist lanzettförmig, spitz, kurz dreilappig an der Spitze, 2 bis 3 cm lang, sehr zusammengedrückt und deutlich gekielt. Die Ränder und der Kiel sind behaart.

Der Fruchtknoten ist 4 bis 6 cm lang, grün oder rötlich, behaart.

Die Fahne ist schmal eiförmig, spitz zulaufend, nach vorne gebogen, 4 bis 5 cm lang und 2,5 bis 3 cm breit, konkav und leicht kapuzenförmig am Ende, leicht behaart, bewimpert, weiß mit vielen parallellaufenden grünen Adern und rötlichen Wimpern am Rande.

Das Synsepalum ist spitz, 3,5 bis 4,5 cm lang, schmaler als die Fahne, aber ähnlich gefärbt.

Die Petalen sind abstehend, etwas nach unten geneigt, länglich lanzettförmig, spitz, schmaler im Endteil, 5 bis 6,5 cm lang und ungefähr 1,5 cm breit; grünlich bis rötlich grün in der Grundfärbung, mehr oder weniger regelmäßig braun befleckt; mit 7 oder 8 parallel verlaufenden dunkelgrünen Adern. Die Ränder sind regelmäßig mit steifen Härchen besetzt.

Die Lippe ist deutlich sackförmig, dreilappig, 4 bis 5 cm lang. Die nach innen geschlagenen Seitenlappen formen eine 2 bis 3 cm lange Röhre, die grün bis gelblich und mit roten Flecken besetzt ist. Der sackförmige Hauptlappen ist an der Basis ge-

common and it is now not only possible to count the chromosome sets of single plants, but we are capable of comparing the single chromosomes of different plants.

In this way, it has been conclusively proven, that *Paphiopedilum* × *wardii* is a natural hybrid between *Paphiopedilum venustum* and *Paphiopedilum sukhakulii.*

Paphiopedilum × *wardii* is an herbaceous plant, usually growing on the ground in leaf-mould. Short stems of about 5 cm height are generated from a fairly short rhizome.

Each stem bears about 3 to 4 leaves. These are spreading in two opposite rows, oblong lanceolate, obtusely three-lobed at the apex, up to 17 cm long and up to about 5.5 cm wide, the upper surface tesselated irregularly bright- and deep-purple, the under surface covered with multiple small purple spots.

The inflorescence is erect, up to 20 cm high, reddish, densely pilose, generally bearing one single flower.

The bract is lanceolate, acute, shortly trilobate at the apex, 2 to 3 cm long, compressed and distinctly keeled. The margins and the keel are hairy.

The ovary is 4 to 6 cm long, greenish to reddish, pilose.

The dorsal sepal is narrowly ovate, acuminate, inclined forwards, 4 to 5 cm long and 2.5 to 3 cm broad, concave, slightly hooded at the apex, slightly pubescent, ciliate, the whole with numerous parallel green nerves and reddish marginal cilia.

The synsepalum is lanceolate, acute, 3.5 to 4.5 cm long, narrower than the dorsal sepal but similarly coloured.

The lateral petals are spreading and somewhat pendulous, oblong-lanceolate or oblong, acute, narrowed in the apical part, 5 to 6.5 cm long and about 1.5 cm wide, with the margins regularly ciliate, more or less regularly spotted brown on a green or pinkish ground, the reddish-pink colour increasing towards the apex, with 2 or 8 dark green parallel veins.

The lip is deeply saccate, obtuse, trilobed, 4 to 5 cm long, the lateral lobes folded in to form an almost closed, 2 to 3 cm long tube which is green to yellowish with red pustulate spots; the terminal lobe at the base with two more or less erect auricles, the whole densely papillose-puberulous outside, the outside of the pouch green, veined brown, in front with numerous small brown spots, inside evenly spotted chocolate.

Paphiopedilum × *wardii* Summerhayes (pro sp.)

ohrt, außen grün, vorne mit braunen Adern und mit vielen braunen Flecken, innen gleichmäßig schokoladenbraun gefleckt.

Die Staminodiumplatte ist breit halbmondförmig, oben eingekerbt und unten dreilappig, bis 1 cm lang und 1,2 bis 1,5 cm breit. Die unteren Seitenlappen sind breit, gerundet, leicht nach innen gebogen; der mittlere Lappen ist viel kürzer und dreieckig, spitz. Die Staminodiumplatte ist grünlich mit dunkleren grünen Adern, manchmal mit einem lila Überzug. Der untere Mittellappen ist dunkelrot bis fast schwarz.

Paphiopedilum × *wardii* ist immer noch eine Seltenheit in unseren Sammlungen. Dies ist sicherlich der Fall, weil das Heimatareal in einem Land liegt, das nicht frei zu bereisen ist.

Paphiopedilum × *wardii* wurde auf humusbedeckten Felsen und an trockenen Hängen, mehr oder weniger der direkten Sonne ausgesetzt, gefunden. In Höhenlagen von 1200 bis 1500 m ü. NN, in Nord-Birma (siehe oben). Die Monate Mai bis Oktober sind am wärmsten, im Januar ist es am kältesten.

The staminodal shield is stalked, broadly lunate, with a short split at the back, in front 3-lobed, scarsely 1 cm long and 1.2 to 1.4 cm wide, the sidelobes are broad, scarsely incurved; middle lobe much shorter, triangular, acute, curved downwards. The staminode is greenish with darker veins, sometimes deeply tinged pink, the middle lobe deep red or nearly black.

Paphiopedilum × *wardii* is still a rarity in our collections. The reason for this being most certainly that travelling in Burma is very restricted.

Paphiopedilum × *wardii* was found growing on humus-covered rocks and dry slopes, more or less in direct sun at altitudes between 1200 and 1500 meters above sea-level in the northern parts of Burma. In this region the warmest season is between May and October and the coolest month is January.

Paphiopedilum × *wardii* does well when grown in the warm or intermediate greenhouse. The plants should be potted in a well-drained, open substrate with plenty of moss to ascertain the retaining of abundant moisture. The plants prefer semi-shaded spots with plenty of air circulation.

Paphiopedilum × wardii gedeiht gut im warmen oder temperierten Gewächshaus. Man sollte ein gut durchlässiges Substrat mit Moosbeimischung verwenden und für ausreichende Drainage sorgen. Die Pflanzen bevorzugen einen halbschattigen Standort mit guter Luftzirkulation.

Die Pflanzen blühen willig, wenn sie eine Ruheperiode mit weniger Wasser und etwas niedrigeren Temperaturen im Dezember durchleben. Im allgemeinen blüht *Paphiopedilum × wardii* zwischen Ende Dezember und März.

Paphiopedilum × wardii flowers freely after a resting period with somewhat lower temperatures and less water after the completion of the new growth in December. *Paphiopedilum × wardii* generally flowers between the end of December and March.

Anmerkung:

Seitdem ich dieses Taxon entsprechend seinem natürlichen Status zur Naturhybride umkombinierte, ist Kritik an dieser Entscheidung laut geworden.

Die Kritik basiert vor allem auf der vermeintlichen Tatsache, daß das Verbreitungsgebiet von *Paphiopedilum × wardii* weit abgelegen ist von den Verbreitungsgebieten von *Paphiopedilum venustum* und *Paphiopedilum sukhakulii*.

Dieses Argument ist unhaltbar. Sicherlich gibt es eine Reihe von Gebieten, wo diese Arten vorkommen, die uns bisher nicht bekannt sind. Jeder, der selbst einmal Standorte gesucht hat, weiß, wie schwierig das ist und wie schnell man an einer Population nichtblühender Orchideen vorbeiläuft, ohne sie zu sehen.

Abgesehen davon können Standorte existiert haben, die inzwischen durch Naturkatastrophen oder aus anderen Gründen vernichtet wurden.

Daß immer wieder neue oder als verschollen betrachtete *Paphiopedilum*-Arten aus China oder anderen bisher unzugänglichen Gebieten auftauchen beweist, daß unsere Kenntnis über die tatsächliche Verbreitung von Orchideen-Arten sehr zu wünschen übrig läßt.

Note:

Since I transferred the taxon to the hybrid status *(loc. cit.)* this decision has been repeatedly criticized.

The criticism is mainly based on the argument that the natural habitats of *Paphiopedilum × wardii* are far off from the natural habitats of *Paphiopedilum venustum* and *Paphiopedilum sukhakulii.*

That argument cannot be upheld. There are certainly a number of habitats of these species not known to us at the time. Anyone having searched for orchids in the wild knows how difficult it is to discover a population when the plants are not in flower.

Besides, it is quite possible that habitats have been destroyed in the course of deforestation or for other reasons or by other means.

The late discovery of species like *Paphiopedilum sukhakulii* and *Paphiopedilum emersonii* (to name only two) and the rediscovery of "lost" species from more or less inaccessable areas proves that our knowledge about the natural distribution of orchids is quite unsatisfactory.

Sektion BARBATA
Krzl.

Die Sektion **Barbata** ist mit 10 Arten und einer Vielzahl von intra- und interspezifischen Merkmalsvariationen, welche primär blütenmorphologischer Natur sind, eine der größten Gruppen innerhalb der Gattung *Paphiopedilum*.

Trotz der immensen Variationsbreite einiger Charakteristika, kann die Sektion **Barbata** anhand der folgenden Merkmalskombinationen eindeutig erkannt werden:

1. Fahne mit deutlicher Längsstreifung
2. Deutlich bewimperte Ränder der seitlichen Petalen
3. Seitliche Petalen länglich-linealisch, im apikalen Bereich mehr oder weniger gedreht
4. Staminodiumplatte halbkreisförmig bis länglich
5. Unterseite der Laubblätter einfarbig grün, kaum purpurn überlaufen

Die Arten dieser Sektion können auf der Basis der blütenmorphologischen Charakteristika, in drei Gruppen aufgeteilt werden. Obwohl wiederholt behauptet wurde, daß diese drei Gruppen natürliche phylogenetische Einheiten darstellen, kann ich nicht umhin, dies zu bezweifeln.

Section BARBATA
Krzl.

Section **Barbata** is one of the largest sections of the genus *Paphiopedilum* including a total of 10 species with many intra- and interspecific variations of numerous characters, especially the floral ones.

In spite of the vast variations in some of the morphological characters, Section **Barbata** is easily recognized by the following combination of characteristics:

1. dorsal sepal with distinct longitudinal stripes
2. margins of the lateral petals with well developed cilia
3. lateral petals linear-oblong, more or less twisted near the apex
4. staminodal shield semi-orbicular to oblong
5. under surface of leaves pale green throughout, hardly tinged with purple.

The species within this section can be classified into three groups on the basis of floral characteristics. Although it has repeatedly been suggested that these three groups represent natural phylogenetic entities, I do have my doubts in this respect.

Schlüssel zu den Untersektionen der Sektion BARBATA

1\. Fahne deutlich ausgeprägt, groß, mehr oder weniger gerundet, die Ränder im apikalen Bereich nicht nach innen, sondern leicht nach außen gebogen; Ränder der seitlichen Petalen mit großen, dunkelpurpurnen und leicht verdickten warzenartigen, mit langen schwarzen Wimpern versehenen Flecken . . . . Untersektion BARBATA

1a. Fahne eiförmig, sehr spitz und mit deutlich kahnförmigem Apex; Ränder der seitlichen Petalen nur wenig befleckt 2

2\. Ränder der seitlichen Petalen einheitlich über die ganze Länge mit langen, schwarzen Wimpern versehen; Staminodiumplatte halbkreisförmig, die Oberfläche dicht kurz behaart . . . . Untersektion LORIPETALUM

2a. Ränder der seitlichen Petalen nur an den schwarzen Flecken bewimpert oder leicht behaart; Oberfläche der Staminodiumplatte fast glatt . . . . Untersektion CHLORONEURA

Key to the subsections of section BARBATA

1\. Dorsal sepal well-developed, more or less rounded, the apical portion not incurved but gently recurved; the margins of the lateral petals with large, deep purple, slightly swollen wart-like blotches with long blackish cilia . . . . subsection BARBATA

1a. Dorsal sepal ovate, sharply pointed and strongly carinate at the apex; margins of the lateral petals with hardly any blotches . . . 2

2\. Margins of the lateral petals with black, regularly arranged long cilia throughout; staminodal shield semi-orbicular, surface densely covered with short cilia . . . . subsection LORIPETALUM

2a. Lateral petal margins ciliate only from the wart-like blotches, or faintly hairy; surface of the staminodal shield almost smooth . . . . subsection CHLORONEURA

Untersektion BARBATA

Diese Untersektion umfaßt vier Arten. Sie sind gekennzeichnet durch die warzenartigen, lang bewimperten Flecken auf den Rändern der seitlichen Petalen und durch eine große, ziemlich runde Fahne, die im oberen Teil am breitesten ist und im apikalen Bereich leicht zurückgebogene Ränder aufweist.

Schlüssel zu den Arten der Untersektion BARBATA

1. Warzenartige Flecken nur am oberen Rand der seitlichen Petalen 2

1a. Warzenartige Flecken nicht auf den oberen Rand der seitlichen Petalen begrenzt . . 3

2. Seitliche Petalen fast waagerecht abstehend oder nur leicht nach unten gerichtet; Fahne 3 bis 4,5 cm breit . *Paphiopedilum barbatum*

2a. Seitliche Petalen stark herabgezogen, Fahne 5 bis 6 cm breit . . *Paphiopedilum callosum*

3. Seitliche Petalen schräg nach unten gerichtet, Flecken über die ganze Petalenoberfläche verteilt . . . . *Paphiopedilum argus*

3a. Seitliche Petalen horizontal abstehend, Flecken nur entlang der Ränder . . . . *Paphiopedilum lawrenceanum*

Subsection BARBATA

This subsection comprises 4 species and is characterized by lateral petals with warty, long ciliate spots at the margins, and a large, nearly rounded, dorsal sepal which is widest above the middle, with margins that are gently recurved in the upper half.

Key to the species of subsection BARBATA

1. Warty blotches on lateral petals only on the upper margins 2

1a. Warty blotches on lateral petals not restricted to the upper margins 3

2. Lateral petals spreading almost horizontally or directed downwards only slightly; dorsal sepal 3 to 4.5 cm wide . *Paphiopedilum barbatum*

2a. Lateral petals strongly declivitous, dorsal sepal 5 to 6 cm wide . . *Paphiopedilum callosum*

3. Lateral petals declivitous, blotches scattered on the whole surface . . . . *Paphiopedilum argus*

3a. Lateral petals spreading horizontally, blotches restricted to the marginal areas . . . . *Paphiopedilum lawrenceanum*

Paphiopedilum argus (Rchb. f.) Stein

Stein's Orchideenbuch: 453 (1892)

Basionym: *Cypripedium argus* Rchb. f.
Gardeners' Chronicle, 1. Ser., **33:** 608 (1873)

Synonyma: *Cordula argus* (Rchb. f.) Rolfe
Orchid Review 20 (1): 2 (1912)
Cypripedium pitcherianum W. A. Manda
Amer. Flor. 3: 178 (1887)
Paphiopedilum barbatum var. *argus* Hort.
Belg. Horticole 25: 57 (1875)

Paphiopedilum argus (Rchb. f.) Stein

Paphiopedilum argus wurde 1872 durch Gustav WALLIS auf der philippinischen Hauptinsel Luzon entdeckt und kurz darauf von der englischen Orchideenhandlung VEITCH nach Europa eingeführt. Die ersten Pflanzen blühten bereits im April 1873.

Paphiopedilum argus ist eine ziemlich variable Art. Die augenförmigen, warzenartigen Flecken auf den Petalen sind das auffallendste Merkmal und waren auch der Anlaß für die Namensgebung.[1)]

Paphiopedilum argus hat elliptische bis länglich lanzettförmige, spitz zulaufende, 12 bis 20 cm lange und 2,5 bis 4 cm breite, hellgrüne, von länglichen dunklen Flecken überzogene Blätter. Die Unterseite ist einfarbig grün mit einem basalen schwach purpurnen Schimmer. Die Blattränder sind fein gekerbt.

Die aufrecht stehende Infloreszenz wird 30 bis 40 cm hoch und trägt meistens eine, selten auch zwei, 6 bis 7 cm große Blüten.

Die Fahne ist breit ei- bis herzförmig, spitz, an den Rändern fein behaart, weiß mit sich abwechselnden langen und kurzen Längsadern. Die Streifen sind manchmal allesamt grün, manchmal sind aber nur die kürzeren grün und die längeren purpurn

[1)] REICHENBACH entnahm den Namen der griechischen Mytologie. Dort ist **Argus** ein Monster mit vielen Augen.

Paphiopedilum argus was discovered in 1872 by Gustav WALLIS who found the plants on the Philippine island of Luzon. They were shortly after imported to Europe by VEITCH.

The first plants of the shipment flowered in April 1873.

Paphiopedilum argus is a rather variable species. The eyeshaped, wart-like blotches on the lateral petals are very characteristic and responsible for the name the plant was given by REICHENBACH.[1)]

Paphiopedilum argus has elliptic to oblong-lanceolate, acute leaves. They are 12 to 20 cm long and 2.5 to 4 cm wide, pale green, tesselated with oblong dark green spots. The under surface is unicoloured green, suffused slightly purple in the basal areas, the margins finely serrate.

The inflorescence is erect, 30 to 40 cm high, usually bearing one, seldomly two, 6 to 7 cm large flower(s).

The dorsal sepal is ovate to cordate, acute, the margins minutely ciliolate, white with alternately longer and shorter longitudinal veins that are sometimes all green, sometimes the longer ones

[1)] REICHENBACH borrowed the name from Greek Mythology were **Argus** is a multi-eyed monster.

gefärbt. Der basale Fahnenteil ist manchmal von dunkelpurpurnen Flecken überzogen.

Die lateralen Sepalen sind zu einem Synsepalum verwachsen. Dieses ist der Fahne sehr ähnlich, aber kleiner. Es hat eine eingekerbte Spitze und schwach gefärbte einheitliche Adern.

Die seitlichen Petalen sind nach hinten gebogen. Sie sind feinwellig, bandförmig, spitz endend. Die Ränder sind wie bei der Fahne fein behaart. Die Petalen werden von Adern durchzogen, die von der Basis bis zu zwei Drittel der Länge schwachgrün und dann bis zur Spitze schwach purpurn gefärbt sind. Die ganze innere Oberfläche der Sepalen ist mehr oder weniger intensiv mit dunklen Warzen überzogen.

Die Lippe ist breit pantoffelförmig, im oberen Teil braunpurpurn, unten grünbraun. Die nach innen gefalteten Seitenlappen sind schmal, schwach purpurn mit dunklen purpurnen Flecken.

Die Staminodiumplatte ist fast hufeisenförmig, mit zwei nach innen gebogenen halbmondförmigen Hörnern.

Paphiopedilum argus ist auf der südwestlichen und östlichen Seite des zentralen Gebirgszuges der Insel Luzon beheimatet. Die Pflanzen wachsen in Humus und Moos auf verwittertem Quarzgestein in einer Höhe von 1300 bis 1800 m ü. NN zusammen mit *Phajus, Calanthe, Acanthephippium* etc.

Das Klima in diesem Heimatareal ist gekennzeichnet durch starke, oft tagelang anhaltende Monsunregenfälle während der Monate Dezember bis Februar. Ab März wird es etwas wärmer und trokkener. Die Temperaturen bewegen sich dann zwischen 30 °C am Tage und 17 °C in der Nacht. Während der Monate Mai und Juni gibt es wenig Regen. Dieser Periode folgt eine nasse Zeit mit heftigen Regenfällen. Im August gibt es oftmals orkanartige Stürme. Von Oktober bis November wird es wieder etwas trockener. Die Wintertemperaturen liegen zwischen 18 °C am Tage und 4 °C in der Nacht.

Die durchschnittliche Jahresniederschlagsmenge liegt bei 3500 mm.

Paphiopedilum argus benötigt entsprechend ihrer Herkunft auch in der Kultur eine hohe Luftfeuchtigkeit. Nur während der Ruheperiode, nach Vollendung des Neutriebes, können die Pflanzen etwas trockener gehalten werden.

Paphiopedilum argus liebt viel Moos um seine Wurzeln und ein verhältnismäßig saures Substrat.

Für eine gute Luftumwälzung ist zu sorgen.

Die Blüten halten mehrere Wochen.

purple and the shorter ones green, the basal area occasionally spotted with blackish purple.

The lateral sepals are united to a synsepalum. This is similar to the dorsal sepal but smaller, notched at the tip, the veins paler and uniform.

The lateral petals are deflexed and undulate, ligulate, acute, margins ciliate; white with pale green veins to two-thirds of their length, the apical third pale purple, the whole inner surface more or less densely spotted with blackish warts.

The lip is broadly calceiform, dull brownish purple above, pale greenish brown beneath; the infolded lateral lobes are narrow, pale purple spotted with deep purple.

The staminodal shield is nearly horseshoe-shaped, with two incurved cusps.

Paphiopedilum argus inhabits the southwestern and eastern slopes of the central mountain range on the island of Luzon. The plants grow in humus and moss on eroded quartz-rock in altitudes of about 1300 to 1800 meters above sea-level, together with *Phajus, Calanthe, Acanthephippium* etc.

The climate in this natural distribution area is characterized by strong monsoon rains, sometimes lasting for days at a time from December through February. Beginning in March, it becomes somewhat warmer and drier. During this period, the temperatures are between 30 °C during the day and 17 °C at night. From May to June, there is little rainfall. This period is followed by a wet season with rainstorms and in August typhoons are not uncommon. From October to November it is drier again. The winter temperatures are between 18 °C during the day and 4 °C at night.

The average yearly rainfall is about 3500 mm.

Paphiopedilum argus needs high humidity in cultivation. The plants should be watered abundantly and should be kept a little drier only during the resting period.

This species does well when potted in an well-drained open substrate with plenty of moss.

Air circulation should be assured.

The flowers last several weeks on the living plants.

Paphiopedilum barbatum (Lindley) Pfitzer

in PRINGSHEIM: **Jahrb. Wiss. Botanik 19:** 163 (1888)

Basionym: *Cypripedium barbatum* Lindley
Edwards's Botanical Register 27,
Misc. No. 110: 53 (1841)

Synonyma: *Cordula barbata* (Lindley) Rolfe
Orchid Review 20 (1): 2 (1912)

Cordula nigrita (Rchb. f.) Merrill
Journ. Str. Br. Roy. As. Soc. Spec.
Nr. 135 (1921)

Cypripedium barbatum Blume,
non Reinwardt
Bl. Cat.: 98 (1823)

Cypripedium biflorum Williams
Orch. Grower's Man. ed. 3: 109 (1868)

Cypripedium Crossi Morren
Belg. Horticole 15: 227 (1865)

Cypripedium nigritum Rchb. f.
Gardener's Chronicle, 3. Ser., **16:** 102 (1882)

Cypripedium orbum Rchb. f.
Gardeners' Chronicle, 3. Ser., **2:** 778 (1887)

Paphiopedilum nigritum (Rchb. f.) Pfitzer
in ENGLER, A.: **Bot. Jahrb. 19:** 40 (1894)

Paphiopedilum barbatum (Lindley) Pfitzer

Paphiopedilum barbatum wurde durch John LINDLEY 1841 als *Cypripedium barbatum* beschrieben. Die Grundlage war eine Pflanze, die H. CUMING am Mount Ophir, etwa 160 km nordwestlich von Singapur, gefunden hat und die im Jahre 1841 in der englischen Gärtnerei LODDIGES zur Blüte kam. LINDLEYs Beschreibung war sehr spartanisch; aber ein Jahr später erschien in derselben Zeitschrift[1] eine Farbtafel mit einem etwas ausführlicheren Begleittext. Auch im **Curtis's Botanical Magazine**[2] erschien eine von einer Besprechung durch W. J. HOOKER begleitete Farbtafel. HOOKER schrieb, daß er keine eindeutigen Abgrenzungsmerkmale zu *Paphiopedilum*

[1] **Edwards's Botanical Register 28,** t. 17 (1842).
[2] **Curtis's Botanical Magazine 72,** t. 4234 (1846).

This species was described as *Cypripedium barbatum* by Lindley in 1841. His concept was based on a plant collected by H. CUMING on the slopes of Mount Ophir, about 160 km to the west of Singapore, which flowered at the LODDIGES nursery in 1841. LINDLEYs description was very brief but one year later, a coloured plate of the species was published in the same journal[1] together with a more detailed description.
Also in **Curtis's Botanical Magazine**[2] a coloured plate of the species was published, together with a discussion of the species by W. J. HOOKER. HOOKER wrote: "It is with no small reluctance that I represent this as a species distinct from *Cypripe-*

[1] **Edwards's Botanical Register 28,** t. 17 (1842)
[2] **Curtis's Botanical Magazine 72,** t. 4234 (1846)

tum und zu *Paphiopedilum purpuratum* zu erkennen vermochte.

Diese Einschätzung war sicherlich auf die mangelhafte Erfahrung mit der Gattung zurückzuführen; damals waren nur einige wenige Arten bekannt.

Es steht außer Zweifel, daß *Paphiopedilum barbatum* nach modernen taxonomischen Gesichtspunkten als eigenständige Art zu betrachten ist.

Paphiopedilum barbatum wächst als Humusepiphyt am Boden oder auf Felsen. Die einzelnen Pflanzen werden bis ungefähr 40 cm hoch.

Die Laubblätter sind schmal-elliptisch, 10 bis 20 cm lang und 2 bis 3 cm breit, dünn, mattgrün mit dunklerer Marmorierung.

Die meist einblütige Infloreszenz wird ungefähr 25 cm hoch.

Die Braktee ist ca. 1,5 cm lang, der Fruchtknoten dagegen bis 6 cm lang.

Die Blüte erreicht einen Durchmesser von ungefähr 8 cm.

Die Fahne ist fast kreisrund, ungefähr 4 cm hoch. Die Ränder im basalen Bereich sind zurückgebogen. Die Fahne ist hellgrün, mit ca. 15 an der Basis grünen und oben purpurnen Längsstreifen. Die Fahne ist längs des Mittelstreifens etwas gefaltet.

Das Synsepalum ist kleiner, ca. 3 cm lang und ungefähr 2 cm breit, schmal elliptisch, mehr oder weniger stumpf, ähnlich wie die Fahne gefärbt.

Die Petalen sind linealisch-länglich, 4 bis 5 cm lang und 1 bis 1,4 cm breit. Sie sind am Ende etwas breiter als an der Basis und stehen meist etwas nach unten weisend. Die Ränder der Petalen sind bewimpert, die oberen Ränder außerdem mit einigen schwarzen behaarten Warzen besetzt. Die Petalen sind an der Basis bräunlich grün, zu der Spitze hin purpurn gefärbt.

Die Lippe ist deutlich pantoffelförmig, ungefähr 4 cm lang und ungefähr 2,5 cm breit, dunkel braunpurpurn. Die nach innen geschlagenen Seitenlappen sind mit einigen purpurnen Punkten und kleinen warzenartigen Flecken besetzt.

Die Staminodiumplatte ist ungefähr 1,5 cm hoch und 1 cm breit, hufeisenförmig, grün mit rotbrauner bis purpurner Zeichnung,[1] am unteren Rand beidseitig mit einem ausgeprägten Anhängsel.

Paphiopedilum barbatum ist in Malaysia und auf der thailändischen Halbinsel beheimatet. Die Pflanzen bewohnen feuchte, schattige Gebirgstäler und sind auf moosbewachsenen Felsen und

[1] Einige Autoren gehen hier von einer rotbraunen bis purpurnen Grundfärbung aus und betrachten den grünen Teil als die Zeichnung.

dium venustum . . . or even from the *Cypripedium purpuratum* . . . the only distinguishing character being in the 'purple hairy warts which border the upper edge of the petals' in our present plant."

This view was most certainly due to the relatively little experience with the genus of which only few species were known at that time.

In the mean time, there is no doubt that *Paphiopedilum barbatum* is to be regarded as an autonomous species according to modern taxonomic views.

Paphiopedilum barbatum is an epiphyte in leaf-mould on the ground or on moss covered rocks. The single plants can attain an overall height of about 40 cm.

The leaves are narrow-elliptic, 10 to 20 cm long and up to 3 cm wide, thin, usually pale dull green above, tesselated with oblong deep green spots.

The inflorescence is terete, erect, up to 25 cm high, usually bearing one single flower.

The floral bract is up to about 1.5 cm long, therefore much smaller than the 6 cm long ovary.

The flower attains a diameter of up to 8 cm.

The dorsal sepal is sub-orbicular, about 4 cm high, folded at the mid-vein, pointed, green at the base, the remainder white, more or less stained with vinous purple; the veins prominent, deep purple, the central ones green at the base.

The synsepalum is much smaller, about 3 cm long and about 2 cm wide, ovate, obtuse to acute, pale green with deep green and purple veins.

The lateral petals are linear-oblong, spreading, slightly deflexed, with ciliate margins, the superior one fringed with small blackish warts; 4 to 5 cm long and 1 to 1.4 cm wide; brownish green towards the base, purplish towards the apex.

The lip is distinctly saccate, inverted helmet-shaped, about 4 cm long and more or less 2.5 cm wide, deep brown-purple, paler beneath; the infolded sidelobes are purple spotted with deeper purple and covered with small warty spots.

The staminodal shield is horseshoe-shaped, about 1.5 cm high and more or less 1 cm wide, green with a reddish-brown or purple pattern,[1] with a deep cleft in the back convex edge and a tooth at the front concave side.

Paphiopedilum barbatum is known from Malaysia and Thailand. The plants inhabit moist, shaded mountain valleys. They generally grow on moss-

[1] Some authors consider the ground colour to be reddish-brown to purple and consider the green part to be the pattern.

zwischen Moos auf sandigen und moorigen Böden in Höhenlagen von 700 bis 1200 m ü. NN zu finden.

Das Klima im Heimatareal ist gekennzeichnet durch eine heiße trockene Periode mit hoher Luftfeuchtigkeit durch nächtliche Taubildung von Ende November bis Februar, gefolgt von einer etwas nasseren Periode von März bis Mai. Die Monate Juni und Juli sind wieder etwas trockener, und die Hauptregenzeit fällt zwischen August und November mit einer Niederschlagsmenge von über 3300 mm. Die Temperaturen schwanken in der Trokkenzeit zwischen 24 und 30 °C, in der Regenzeit zwischen 15 und 21 °C.

Die Hauptblütezeiten sind Dezember bis Januar sowie Juni bis Juli.

Paphiopedilum barbatum gehört im warmen bis temperierten Gewächshaus an einen schattigen, gut belüfteten Platz.

Die Pflanzen mögen ein offenes Substrat mit viel Moos und eine ständig hohe Luftfeuchtigkeit sowie ein regelmäßiges Wässern außerhalb der Ruhezeit. *Paphiopedilum barbatum* ist äußerst dankbar in der Kultur, blüht regelmäßig, mag aber weder hohe Salzkonzentrationen im Gießwasser noch ein vermodertes Substrat.

Varietäten:

Ungefähr 50 Taxa sind als Varietäten von *Paphiopedilum barbatum* (Lindley) Pfitzer beschrieben worden. Drei davon werden inzwischen allgemein als selbständige Art anerkannt[1)] und eine[2)] ist als Synonym von *Paphiopedilum superbiens* (Rchb. f.) Stein zu betrachten.

Die übrigen sind mehr oder weniger stark abweichende Farbvarianten des Nominaltyps. Diese „Varianten“ basierten meistens auf einer längst nicht mehr existierenden Pflanze. Viele dieser „Varietäten“ verdienen kaum den ihnen zugeteilten taxonomischen Status und einige sind als solche noch nicht einmal gültig veröffentlicht. Obwohl die meisten dieser „Varietäten“ als *Cypripedium barbatum* var. beschrieben wurden, halte ich es nicht für angebracht, sie durch eine gültige Umkombination zu der Gattung *Paphiopedilum* aufzuwerten.

---

1) *Paphiopedilum argus* (Rchb. f.) Stein
Paphiopedilum hookerae (Rchb. f.) Stein
Paphiopedilum superbiens (Rchb. f.) Stein

2) var. Veitchii Van Houtte

covered rock on sandy and moorlike underground at altitudes of 700 to 1200 meters above sea-level.

The climate in this area is characterized by a hot, dry period from the end of November to February with high humidity due to nighttime dews, followed by a wetter period from March to May. The months of June and July are somewhat drier again and the main rainy season is between August and November with over 3300 mm of rainfall. The summer temperatures are between 24 and 30 °C the temperatures during the rainy season are between 15 and 21 °C.

Paphiopedilum barbatum usually flowers from December through January and from June through July.

Paphiopedilum barbatum should be grown in the warm or intermediate greenhouse in shaded conditions and under good air-circulation.

The plants of this species do best in an well-drained open substrate with plenty of moss. The relative humidity should be kept high at all times and waterings should be abundant. During the resting period after the completion of the new growth, the plants should be kept somewhat cooler and drier.

Paphiopedilum barbatum reacts very sensitive to high concentrations of salts in the water and should be repotted regularly.

Varieties:

About 50 varieties of *Paphiopedilum barbatum* (Lindley) Pfitzer have been described. Three of them are now generally recognized as autonomous species[1)] and one[2)] is to be considered synonymous with *Paphiopedilum superbiens* (Rchb. f.) Stein.

The remaining differ more or less from the type in the colouration of their flower(s). These "variants" usually were described on the basis of one single plant. Many of these "varieties" hardly deserve to be considered as separate taxa and some of them are not even been published validly. Although most of these "varieties" have been described as *Cypripedium barbatum* var. I do not consider them worthwhile enough to transfer them to *Paphiopedilum.*

---

1) *Paphiopedilum argus* (Rchb. f.) Stein
Paphiopedilum hookerae (Rchb. f.) Stein
Paphiopedilum superbiens (Rchb. f.) Stein

2) var. *veitchii* Van Houtte

A

B

Paphiopedilum callosum (Rchb. f.) Stein – Varianten / Variants

C

D

Paphiopedilum callosum (Rchb. f.) Stein

Stein's Orchideenbuch: 457 (1892)

Basionym: *Cypripedium callosum* Rchb. f.
Gardeners' Chronicle, 2. Ser., **26:** 326 (1886)

Synonyma: *Cordula callosa* (Rchb. f.) Rolfe
Orchid Review 20 (1): 2 (1912)
Cypripedium barbatum var. *crossii* Hort. ex Veitch
A Manual of Orchidaceous Plants 4: 12 (1889)
Cypripedium barbatum var. *warneri* Hort.
Journ. of Hortic. 4: 115 (1882)
Cypripedium barbatum var. *warnerianum* Moore
in WARNER: **Select. Orchidaceous Plants** (1862–1891)
Cypripedium schmidtianum Krzl.
Botan. Tidsskr. 24: 13 (1901)
Paphiopedilum birkii Fowlie
nomen nudum
Paphiopedilum reflexum Hort. ex Stein
Stein's Orchideenbuch: 457 (1892)
Paphiopedilum regnieri Hort. ex Stein
Stein's Orchideenbuch: 457 (1892)
Paphiopedilum sublaeve (Rchb. f.) Fowlie
Orchid Digest 43 (6): 224 (1979)
Paphiopedilum thailandense Fowlie
nomen nudum

Diese mit *Paphiopedilum barbatum* und *Paphiopedilum lawrenceanum* eng verwandte Art wurde durch den französischen Pflanzensammler Alexandre REGNIER in Siam (heute Thailand) gefunden und nach Europa gebracht. VEITCH[1] mutmaßte, daß es sich um eine „Zwischenform" von *Paphiopedilum barbatum* und *Paphiopedilum lawrenceanum* handeln könnte.

H. G. REICHENBACH beschrieb die Pflanze als selbständige Art unter dem Namen *Cypripedium callosum (loc. cit.)*.

Die drei Taxa werden hier als eigenständige Arten abgehandelt, obwohl ich keineswegs auszuschließen vermag, daß *Paphiopedilum barbatum, Paphiopedilum callosum* und *Paphiopedilum lawrenceanum* „nur" Varietäten von ein und derselben Art sind.

Im Frühjahr 1900 wurden durch den dänischen Botaniker Joh. SCHMIDT in Thailand auf der Insel Koh Chang Pflanzen gefunden, die KRÄNZLIN als *Cypripedium schmidtianum* beschrieb *(loc. cit.)*.

Mehrere Autoren und einige Orchideenhändler betrachten „*Paphiopedilum schmidtianum*" als eine eigenständige Art innerhalb der *Paphiopedilum*

[1] **A Manual of Orchidaceous Plants 4:** 16 (1889).

This species was discovered in Siam (now Thailand) and brought to Europe by the French collector Alexandre REGNIER. VEITCH[1] suggested that *Paphiopedilum callosum* might be a natural hybrid between *Paphiopedilum barbatum* and *Paphiopedilum lawrenceanum*, but H. G. REICHENBACH described the plant as an autonomous species as *Cypripedium callosum (loc. cit.)*.

In that sense, the three taxa are treated here, although I can, by no means rule out the possibility that *Paphiopedilum callosum, Paphiopedilum barbatum,* and *Paphiopedilum lawrenceanum,* are "mere" varieties of one single species.

In the spring of 1900, the Danish Botanist Joh. SCHMIDT found a number of plants on the island Koh Chang off the coast of Thailand that were described as *Cypripedium schmidtianum* by KRÄNZLIN *(loc. cit.)*.

Several authors and some orchid dealers consider "*Paphiopedilum schmidtianum*" (as they incorrectly call it) to be an autonomous species of the *Paphiopedilum appletonianum*-complex or as a natural hybrid of *Paphiopedilum callosum* with *Paphiopedilum appletonianum.* If that were so,

[1] **A Manual of Orchidaceous Plants 4:** 16 (1889).

appletonianum-Verwandtschaft oder als eine Naturhybride zwischen *Paphiopedilum callosum* und *Paphiopedilum appletonianum*. „*Paphiopedilum schmidtianum*" wäre somit identisch mit *Paphiopedilum × siamense*. Beide Interpretationen sind eindeutig falsch und beruhen auf mangelhaften Recherchen. Die Beschreibung KRÄNZLINs und der im Botanischen Museum der Universität Kopenhagen aufbewahrte Typus[1] lassen keinerlei Zweifel daran, daß *Paphiopedilum schmidtianum* identisch ist mit *Paphiopedilum callosum*.

Paphiopedilum callosum ist eine als Humusepiphyt wachsende krautige Pflanze mit länglich ovalen, bis ungefähr 25 cm langen und 5 cm breiten, meist hellgrünen Blättern. Auf der Oberseite zeigen sie eine dunkle Marmorierung. Vorne sind sie eingeschnitten. Die Färbung und Marmorierung der Laubblätter ist auch bei dieser Art sehr variabel. Die Blattunterseite ist einfarbig, an der Basis spärlich behaart.

Die aufrecht stehende Infloreszenz wird bis 40 cm hoch. Sie ist tief violett, flaumig behaart, meist einblütig, nicht selten aber zweiblütig.

Die Braktee ist ungefähr 2,2 cm lang und somit etwa ein Drittel so lang wie der Fruchtknoten.

Die Blüte wird etwa 10 cm breit.

Die Fahne ist breit herzförmig, bis 7,5 cm breit und ungefähr 5 cm hoch. Die Ränder der oberen Hälfte sind zurückgebogen. Die Fahne ist weiß mit einem grünen Basisteil und vielen abwechselnd, kurzen und langen purpurnen Längsstreifen.

Das Synsepalum ist viel kleiner als die Fahne, ungefähr 3,5 cm lang, ei- bis lanzettförmig, stumpf, weiß mit grünen Längsadern.

Die zungenförmigen bis ca. 5,5 cm langen und ungefähr 1 cm breiten Petalen sind um ca. 45° (manchmal mehr) nach unten abgewinkelt. Sie sind an den Rändern bewimpert und am oberen Rand mit schwarzen Wärzchen besetzt; grün mit dunkleren Adern und im Enddrittel rosapurpurn.

Die umgekehrt helmförmige Lippe wird ca. 4,7 cm lang und 2,2 cm breit. Sie ist dunkelbronze gefärbt und hat, von großen schwarzen Warzen besetzte, nach innen geschlagene grüne Seitenlappen.

Die Staminodiumplatte ist sehr variabel, rundlich hufeisenförmig bis fast quadratisch, oben deutlich gespalten, unten mit drei Zähnen versehen, wovon die seitlichen deutlich länger sind als der mittlere.

Paphiopedilum callosum ist in Thailand und Indochina weit verbreitet. Die Pflanzen leben epiphy-

[1] Eine in Alkohol eingelegte Blüte.

then "*Paphiopedilum schmidtianum*" would be identical with *Paphiopedilum × siamense*.

Both interpretations are clearly erroneous. The description rendered by KRÄNZLIN and the type of *Cypripedium schmidtianum*[1] leave no doubt that *Cypripedium schmidtianum* is identical with *Paphiopedilum callosum*.

Paphiopedilum callosum is an herbaceous perennial plant, usually growing epiphytically in humus. The leaves are oval-oblong, acute, up to about 25 cm long and up to ca. 5 cm wide, variable in colour, usually bright green, with blackish green hieroglyphic markings and tesselations above, but often much paler; grey-green and keeled beneath, the basal part slightly pilose.

The inflorescence is erect, up to 40 cm high, deep violet, downy, usually bearing one single flower, not seldomly two-flowered.

The floral bracts are about 2.2 cm long, covering only one-third of the ovary.

The flower is up to 10 cm wide.

The dorsal sepal is broadly cordate, up to 7.5 cm wide and more or less 5 cm high, folded at the midvein and undulate in the apical half; white with numerous alternately longer and shorter veins that are green at the base and deep vinous purple upwards.

The synsepalum is much smaller than the dorsal sepal, about 3.5 cm long, ovate to lanceolate, more or less blunt, white with pale green longitudinal veins.

The lateral petals are ligulate, up to about 5.5 cm long and about 1 cm wide, drooping at an angle of 45 degrees or even closer to the lip, the margins ciliate with up to about 5 blackish warts on the upper one, pale green with darker veins and tinted with pale rose-purple in their apical one-third.

The lip is deeply saccate, inverted helmet-shaped, up to about 4.7 cm long and about 2.2 cm wide, deep bronze-brown, greenish beneath; the infolded lobes green, spotted with deep purple.

The staminodal shield is very variable, rounded horseshoe-shaped to nearly quadrangular, clearly cleft at the top, at the lower end with three teeth of which the lateral two are much longer than the middle one.

Paphiopedilum callosum is rather common in Thailand and the whole of Indochina. The plants are generally found in humus on the forest ground and

[1] A pickled flower.

tisch in Blatthumus am Waldboden oder in humusgefüllten Felsspalten bis 1400 m ü. NN.

Paphiopedilum callosum ist sehr einfach zu kultivieren. Die Pflanzen sind anspruchslos und ausgesprochen blühwillig.

Die Art gedeiht am besten im temperierten bis warmen Gewächshaus in einem lockeren Substrat mit ausreichender Drainage.

Die Luftfeuchtigkeit sollte möglichst hoch gehalten werden. Eine Ruheperiode im Dezember mit weniger Wassergaben und etwas niedrigeren Temperaturen bewirkt eine gute Blüteninduktion.

Varietäten:

Auch von *Paphiopedilum callosum* sind eine Unmenge Varietäten beschrieben worden. Mit Ausnahme der Albinoform basieren diese „Varianten" auf zweiblütigen oder „besonders schönen" oder „besonders großen" Exemplaren des Normaltyps. Alle sind innerhalb der natürlichen Bandbreite einer Art anzusiedeln.

Obwohl die meisten dieser „Varietäten" als *Cypripedium callosum* var. beschrieben wurden, halte ich es nicht für angebracht, sie durch eine gültige Umkombination zu der Gattung *Paphiopedilum* aufzuwerten.

Paphiopedilum callosum var. ***viridiflorum*** (Hort.) Braem *comb. nov.*

Basionym: *Cypripedium callosum* var. *viridiflorum* Hort.
Journ. des Orchid., 15. März 1893

Synonyma: *Cypripedium callosum* var. *sanderae* Hort.
Gardeners' Chronicle, 3. Ser., **15:** 663 (1894)

Paphiopedilum callosum var. *sanderae* (Hort.) Pfitzer
in ENGLER, A.: **Das Pflanzenr. IV** (50) Heft 12: 93 (1903)

Blüten ohne jegliche rote Pigmente. Blütenblätter grünlich weiß; somit eine richtige Albinoform.

Obwohl das Albino von *Paphiopedilum callosum* im allgemeinen als „var. *sanderae*" bekannt ist, wurde ein Jahr vor der Veröffentlichung der var. *sanderae,* die gleiche Varietät als var. *viridiflorum* beschrieben. Dieser Benennung ist somit aus Prioritätsgründen zu folgen, und sie wird hier deshalb zu *Paphiopedilum callosum* var. umkombiniert.

in rockcracks filled with leafy debris at altitudes of up to 1400 meters above sea-level.

Paphiopedilum callosum is an easy to grow plant which does not require special conditions and is very floriferous.

The plants do best in the intermediate or warm greenhouse potted in a well-drained open compost.

The relative humidity should be kept high at all times and the plants flower best after a rest in December with lesser water and lower temperatures.

Varieties:

A vast quantity of varieties of *Paphiopedilum callosum* have been described (all as *Cypripedium callosum* var.). With the exception of the *albino*-form these "variants" are either based on two-flowered or on "very beautiful" or "very large" specimens which do not differ enough from the type to be recognized as autonomous taxa.

Paphiopedilum callosum var. ***viridiflorum*** (Hort.) Braem *comb. nov.*

Basionym: *Cypripedium callosum* var. *viridiflorum* Hort.
Journ. des Orchid., 15. March 1893

Synonyma: *Cypripedium callosum* var. *sanderae* Hort.
Gardeners' Chronicle, 3. Ser., **15:** 663 (1894)

Paphiopedilum callosum var. *sanderae* (Hort.) Pfitzer
in ENGLER, A.: **Das Pflanzenr. 4** (50) Heft 12: 93 (1903)

Flower greenish-white without any red pigmentation, therefore a real albino.

Although the albino of *Paphiopedilum callosum* is generally referred to as var. *sanderae,* the publication as var. *viridiflorum* is the oldest and therefore the valid one.

Paphiopedilum lawrenceanum (Rchb. f.) Pfitzer

in PRINGSHEIM: **Jahrb. Wiss. Botanik 19:** 163 (1888)

Basionym: *Cypripedium lawrenceanum* Rchb. f. **Gardeners' Chronicle,** 2. Ser., **10:** 748 (1878)

Synonyma: *Cordula lawrenceana* (Rchb. f.) Rolfe **Orchid Review 20** (1): 2 (1912)

Paphiopedilum barbatum subspecies *lawrenceanum* (Rchb. f.) M. W. Wood **Orchid Review 84** (1001): 352 (1976)

Paphiopedilum barbatum subspecies *lawrenceanum* var. *hennisianum* M. W. Wood **Orchid Review 84** (1001): 352 (1976)

Paphiopedilum fowliei Birk **Orchid Digest 45** (2): 63 (1981)

Paphiopedilum hennisianum (Schoser ex Wood) Fowlie **Orchid Digest 41** (2): 60 (1977)

Paphiopedilum hennissianum var. *fowliei* (Birk) Cribb **The Genus Paphiopedilum:** 190 (1987)

Paphiopedilum lawrenceanum (Rchb. f.) Pfitzer

Die Art wurde 1878 in Nord-Borneo durch F. W. BURBIDGE auf einer Sammelreise für die Firma VEITCH entdeckt. BURBIDGE fand die Pflanzenpopulation am linken Ufer des Lawas-Flusses, in der Nähe von Merengit, in 300 bis 500 m Höhe.

Die Firma VEITCH überließ H. G. REICHENBACH frisches Material, und er beschrieb die neue Art zu Ehren von Sir Trevor LAWRENCE.

Der Grund für die noch immer andauernde Diskussion über den taxonomischen Status dieser Art wurde von REICHENBACH selbst in seiner Erstbeschreibung gelegt. Er schrieb u. a.:

„Die Blüte ist identisch mit der von *Cypripedium barbatum majus*".

In der Tat ist die Ansicht einiger Autoren, *Paphiopedilum lawrenceanum* (und auch *Paphiopedilum callosum*) sei in Wirklichkeit nur eine Varietät von *Paphiopedilum barbatum*, unschwer nachzuvollziehen (siehe unten).

Paphiopedilum lawrenceanum was discovered in the northern part of Borneo by F. W. BURBIDGE when collecting for VEITCH. BURBIDGE found the plants on the left bank of the river Lawa, near Merengit between 300 and 500 meters above sea-level.

Life material was sent by VEITCH to REICHENBACH who described the species *(loc. cit.)* in honour of Sir Trevor LAWRENCE.

There has been some confusion about the taxonomic status of this species. The reason for this originates in REICHENBACHs description in which he wrote:

"The flower is equal to those of *Cypripedium barbatum majus*".

And to this day, some authors consider *Paphiopedilum lawrenceanum* (as well as *Paphiopedilum callosum*) to be a variety of *Paphiopedilum bar-*

Das von BIRK beschriebene *Paphiopedilum fowliei* sowie das von FOWLIE als autonome Art beschriebene *Paphiopedilum hennisianum* sind eindeutig geographische Varianten von *Paphiopedilum lawrenceanum*. Sie können, wenn überhaupt, nur den Varietätsstatus beanspruchen.

Paphiopedilum lawrenceanum ist eine bis 50 cm hoch werdende, als Humusepiphyt wachsende krautige Pflanze mit bis zu 27 cm langen und ca. 5 cm breiten Laubblättern. Diese sind stark dunkel- und olivgrün marmoriert, vorne spitz und zweilappig.

Die aufrecht stehende Infloreszenz wird bis 45 cm hoch und ist im allgemeinen einblütig, selten aber auch zweiblütig.

Die Braktee ist ungefähr 2 cm lang und somit ungefähr ein Viertel so lang wie der Fruchtknoten.

Die Blüte ist bis 12 cm breit.

Die Fahne ist 6 cm lang und ebenso breit, fast kreisrund, in der Mitte gefaltet, kurz gespitzt; weiß, an der Basis grün überlaufend, mit einer Vielzahl sich abwechselnder kurzer und langer grüner und purpurner Längsstreifen.

Das Synsepalum ist 3,7 cm lang, länglich eiförmig.

batum – an opinion not difficult to agree with (see below).

The taxon described as *Paphiopedilum fowliei* by BIRK and *Paphiopedilum hennisianum*, a variety transferred to the species status by FOWLIE, represent mere geographical variants and can, at the outmost, claim the varietal status.

Paphiopedilum lawrenceanum is an herbaceous, perennial humusepiphyte attaining an overall height of about 50 cm. The leaves are up to 27 cm long and up to about 5 cm wide, oval-oblong, tesselated with olive green and deep grass green (variable), acute, bilobed.

The inflorescence is erect, up to 45 cm high, generally one- and sometimes two-flowered.

The floral bract is about 2 cm long, covering about one-fourth of the ovary.

The flower is up to about 12 cm across.

The dorsal sepal is about 6 cm long and wide, nearly orbicular, folded at the middle, white, with broad, alternately longer and shorter veins, the central ones usually green at the base, the others deep vinous purple.

The synsepalum is ovate-oblong, about 3.7 cm long.

Paphiopedilum lawrenceanum (Rchb. f.) Pfitzer

Die Petalen sind bis 6,4 cm lang und ungefähr 1,2 cm breit, fast waagerecht abstehend, im apikalen Teil gerade bis deutlich nach hinten gebogen, gewimpert, an jedem Rand mit 5 bis 10 schwarzen Warzen („Unterschied" zu *Paph. barbatum*), an der Basis grün, zur Spitze hin allmählich in purpurbraun übergehend.

Die Lippe ist deutlich sackförmig, bis 5,5 cm lang, mehr oder weniger dunkel purpurbraun, am oberen Teil oft dunkler. Die nach innen geschlagenen Seitenlappen sind mit kleinen Wärzchen besetzt.

Die Staminodiumplatte ist halbmondförmig, oben tief eingekerbt und unten mit fünf Zähnchen in der Ausbuchtung, wobei das mittlere am stärksten ausgebildet ist. Die Staminodiumplatte ist wachsartig gelb, mit einem purpurnen Rand und grüner Zeichnung.

Paphiopedilum lawrenceanum ist beheimatet in Nord-Borneo (Sabah). Die Pflanzen wachsen in Höhenlagen von 300 bis 500 m ü. NN in Humusansammlungen am Waldboden oder auf humusbedeckten Felsen, immer an schattigen Stellen.

Das Klima in der Heimatregion ist geprägt von einem nordöstlichen Monsun von Anfang November bis Mitte Februar. In dieser Zeit kommen oft tagelange Regenfälle vor. Ab März wird es allmählich trockener. April ist der trockendste und heißeste Monat mit Tagestemperaturen um 30 °C und um 18 °C in der Nacht.

Ab Mai wird es wieder etwas feuchter, und ab Anfang Juni bläst der südwestliche Monsun, der bis Mitte Oktober gelegentliche Regenfälle bringt.

Juni und Juli sind die kühlsten Monate mit Durchschnittstemperaturen um 25 °C am Tage und 15 °C in der Nacht.

Die Gegend ist niemals richtig trocken. Der jährliche Niederschlag beläuft sich auf ungefähr 4000 mm. Die Luftfeuchtigkeit ist das ganze Jahr über sehr hoch.

Paphiopedilum lawrenceanum blüht im allgemeinen von März bis Mai.

Paphiopedilum lawrenceanum gehört in das warme oder temperierte Gewächshaus. Die Pflanzen sollten in ein gut durchlässiges Substrat getopft werden, am besten unter Beimischung von Moos. Die Kultur sollte halbschattig erfolgen, und die Pflanzen sollten niemals richtig austrocknen. Von Februar bis Anfang März ist es für eine gute Blüteninduktion angebracht, die Pflanzen kühler zu stellen und weniger zu gießen.

The lateral petals are up to 6.4 cm long and about 1.2 cm wide, spreading nearly horizontally, straight to distinctly recurved in the apical portion, ligulate, ciliolate, with 5 to 10 blackish warts on each margin ("difference" to *Paphiopedilum barbatum*), green with purple tips.

The lip is deeply saccate, up to 5.5 cm long, more or less deep purple-brown, the upper parts usually darker than the lower areas, the infolded lateral lobes covered with small wart-like structures.

The staminodal shield is semi-lunate, the upper margin deeply notched, the lower one with five small teeth in the sinus, the middle tooth being more distinct than the others; yellowish with a purple margin and green pattern.

Paphiopedilum lawrenceanum is an endemic of Borneo where it is reported from the northern parts only. The plants grow at altitudes of 300 to 500 meters above sea-level in humus on the ground of the forest or on humus-covered rocks, always in shady places.

In this area, the northeast monsoon blows from the end of November to mid-February, bringing warm moist winds and periods of heavy rainfall, often lasting for days at the time. Starting in March, it gradually gets drier and April is the hottest and driest months with temperatures averaging 30 °C during the day and 18 °C at night.

Convectional rains start in May and by the end of that month, the southwest monsoon arrives, bringing more heavy rainfall until mid-October.

June and July are the coolest months with average temperatures of 25 °C during daytime and 15 °C at night.

The distribution area of *Paphiopedilum lawrenceanum* is continuously wet with only short periods of dryness. The annual rainfall averages 4000 mm. The humidity remains very high over the entire year.

Paphiopedilum lawrenceanum should be kept in the intermediate to warm greenhouse, potted in a well-drained open compost retaining plenty of moisture. The plants should be grown under semi-shaded conditions and should never be allowed to dry out completely.

A short resting period in February and at the beginning of March with lower temperatures and less abundant waterings is indicated.

Paphiopedilum lawrenceanum generally flowers between March and May.

Varietäten:

Paphiopedilum lawrenceanum var. ***hyeanum***
(L. Linden & Rodigas) Braem, *comb. nov.*

Basionym: *Cypripedium lawrenceanum*
var. *hyeanum* L. Linden & Rodigas
Lindenia 1: 89. Tafel 42 (1885)

Dies ist die richtige Bezeichnung für die Albinoform von *Paphiopedilum lawrenceanum*. Die Fahne und das Synsepalum sind reinweiß mit grünen Streifen. Die Petalen sind gelblich grün mit dunkelgrünen Adern. Die Lippe ist ebenso glänzend grün, mit dunkelgrüner Zeichnung.

Dies ist die einzige echte Varietät von *Paphiopedilum lawrenceanum*, bisher aber noch nicht ordnungsgemäß zu der Gattung *Paphiopedilum* gestellt. Dies sei hiermit nachgeholt.

Anmerkung:

Es sind ungefähr 30 weitere „Varietäten", alle als *Cypripedium lawrenceanum* var. beschrieben worden. Keine davon ist ausreichend vom Nominaltyp verschieden, um den Varietätsstatus in Anspruch nehmen zu können.

Varieties:

Paphiopedilum lawrenceanum var. ***hyeanum***
(L. Linden & Rodigas) Braem, *comb. nov.*

Basionym: *Cypripedium lawrenceanum*
var. *hyeanum* L. Linden & Rodigas
Lindenia 1: 89, plate 42 (1885)

This is the valid designation of the *albino* of *Paphiopedilum lawrenceanum*. The dorsal sepal and the synsepalum are pure white, striped green. The lateral petals are yellowish-green with deep green veins. The lip is equally glossy green with deep-green tesselation.

This is the only real variety of the species, up til now lacking the transfer to *Paphiopedilum*.

Note:

About 30 more "varieties" of *Paphiopedilum lawrenceanum* have been described (all as *Cypripedium lawrenceanum* var.). None of these differ distinctly from the type and cannot be upheld as separate taxa.

Untersektion LORIPETALUM
Karasawa & Saito

Die Untersektion **Loripetalum** umfaßt drei Arten. Die Pflanzengruppe ist der Untersektion **Barbata** ähnlich, unterscheidet sich jedoch deutlich durch folgende Merkmale:

1. relativ kleine eiförmige Fahne
2. Rand der Fahne im apikalen Bereich leicht nach innen gebogen und somit ein spitzes Ende bildend
3. Randgegenden der seitlichen Petalen ohne große, bewimperte, warzenähnliche Flecke
4. Ränder der seitlichen Petalen einheitlich und regelmäßig mit langen schwarzen Wimpern besetzt
5. Staminodiumplatte halbkreisförmig, kurz behaart

Schlüssel zu den Arten der Untersektion LORIPETALUM

1. Seitliche Petalen länglich lanzettförmig, unbefleckt . . . . . . *Paphiopedilum dayanum*

1a. Seitliche Petalen lanzettförmig, zumindest teilweise mit kleinen Flecken überzogen . . . 2

2. Apikaler Teil der seitlichen Petalen kaum gebogen, unbefleckt . . *Paphiopedilum ciliolare*

2a. Apikaler Teil der seitlichen Petalen mehr oder weniger sinusartig gebogen, bis in die Endregion mit kleinen Flecken überzogen . . . . *Paphiopedilum superbiens*

Subsection LORIPETALUM
Karasawa & Saito

The subsection **Loripetalum** comprises three species. This section is somewhat similar to subsection **Barbata** but it is well characterized by the following combination of characters:

1. dorsal sepal relatively small, ovate
2. apical margin of the dorsal sepal slightly incurved and forming a sharply pointed apex
3. marginal portion of the lateral petals without large, warty, ciliate spots
4. margins of lateral petals uniformly and regularly covered with long black cilia
5. staminodal shield semi-orbicular, covered with minute hairs.

Key to the species of subsection LORIPETALUM

1. Lateral petals without spots, linear lanceolate . . . . *Paphiopedilum dayanum*

1a. Lateral petals lanceolate, at least in part minutely spotted 2

2. Apical portion of the lateral petals hardly sinuate, not spotted *Paphiopedilum ciliolare*

2a. Apical portion of the lateral petals more or less sinuate, finely spotted from the middle to the apex . . . . *Paphiopedilum superbiens*

Paphiopedilum ciliolare (Rchb. f.) Stein

Stein's Orchideenbuch: 462 (1892)

Basionym: *Cypripedium ciliolare* Rchb. f.
Gardeners' Chronicle, 2. Ser., **18:** 488 (1882)

Synonyma: *Cordula ciliolaris* (Rchb. f.) Rolfe
Orchid Review 20 (1): 2 (1912)

Cypripedium miteauanum Linden & Rodigas
Lindenia 3: 77 (1887)

Paphiopedilum superbiens subsp. *ciliolare* M. W. Wood
Curtis's Botanical Magazine 183 (2), t. 803, S. 52 (1981)

Paphiopedilum ciliolare var. *miteauanum* (Linden) Pfitzer in ENGLER, A.: **Pflanzenr. IV** (50): 89 (1903)

Paphiopedilum ciliolare (Rchb. f.) Stein

Paphiopedilum ciliolare wurde 1882 durch den im Auftrage der Firma LOW sammelnden BOXALL auf den Philippinen gefunden und nach England gesandt. Im selben Jahr wurde die Pflanze durch REICHENBACH, der auch die enge Verwandtschaft dieser Art zu *Paphiopedilum superbiens* vermerkte, als *Cypripedium ciliolare* beschrieben *(loc. cit.)*. In der Tat gehen einige Autoren davon aus, daß *Paphiopedilum ciliolare* und *Paphiopedilum superbiens* als zwei Unterarten derselben Art zu betrachten sind, eine Ansicht, die unschwer nachzuvollziehen ist.

Paphiopedilum ciliolare ist eine, bis zu 45 cm hoch werdende, als Humusepiphyt wachsende Pflanze mit schmal länglichen bis elliptisch länglichen, stumpfen Laubblätter, die bis über 20 cm lang und bis ca. 4 cm breit werden. Sie sind hell- und dunkelgrün marmoriert.

Die intensiv flaumig behaarte Infloreszenz wird bis ca. 40 cm hoch und trägt meistens eine einzelne Blüte.

Paphiopedilum ciliolare was discovered in 1882 by BOXALL when collecting on the Philippines for the British horticulture firm LOW. Within the same year, the plants were described as *Cypripedium ciliolare* by REICHENBACH *(loc. cit.)* who also noted the close relationship of this species with *Paphiopedilum superbiens.* Indeed, some authors do consider *Paphiopedilum ciliolare* and *Paphiopedilum superbiens* to be two subspecies of one single taxon.

Paphiopedilum ciliolare is an herbaceous plant, attaining an overall height of about 45 cm, usually growing epiphytically in leaf-mould.

The leaves are narrowly oblong to elliptic-oblong, obtuse, up to about 20 cm long and up to 4 cm wide, tesselated deep and pale green.

The inflorescence is erect, terete, densely downy, up to about 40 cm high, usually bearing one single flower.

Die Braktee ist etwa 1,7 cm lang und somit ungefähr ein Viertel so lang wie der Fruchtknoten.

Die Blüte hat einen Durchmesser von ca. 10 cm.

Die Fahne wird 4,5 bis 5 cm lang und 3,5 bis 4,5 cm breit. Sie ist breit eiförmig, lang zugespitzt, an den Rändern bewimpert, an der Basis purpurn, sonst weiß mit alternierenden, kurzen, grünen und purpurnen Längsadern.

Das Synsepalum ist ca. 2,9 cm lang, eiförmig, langspitzig, weißlich mit grünen Adern.

Die 5,5 bis 6,5 cm langen und 1,5 bis 2,5 cm breiten, zungenförmigen, heruntergeschlagenen oder nach unten gebogenen Petalen weisen an den Rändern lange schwarze Haare auf. Im Bereich der unteren zwei Drittel sind sie grün und intensiv mit schwarzen Wärzchen überzogen. Der obere (apikale) Teil ist hellpurpurn.

Die umgekehrt helmförmige Lippe ist ungefähr 5,5 cm lang, matt purpurbraun mit hell gelbgrünen nach innen gebogenen und mit purpurnen Wärzchen besetzten Seitenlappen.

Die Staminodiumplatte ist nierenförmig, oben eingekerbt und unten mit drei Zähnchen versehen, wovon das mittlere die seitlichen überragt. Die Staminodiumplatte ist grünlich, mehr oder weniger rosa überhaucht.

Paphiopedilum ciliolare unterscheidet sich in der Blütenmorphologie von *Paphiopedilum superbiens* durch die nur über zwei Drittel ihrer Länge befleckten Petalen, einen kleineren Schuh und durch den hervorstehenden Mittelzahn der Staminodiumplatte.

Die Art ist auf den philippinischen Inseln Luzon, Mindanao und Dinagat in Höhenlagen von 300 bis 1800 m ü. NN anzutreffen. Dieses Gebiet hat ein immer feuchtes Klima mit einer Periode relativer Trockenheit von März bis Juni.

Die Temperaturen variieren zwischen 22 und 30 °C am Tage und 15 bis 18 °C in der Nacht, je nach Jahreszeit.

Paphiopedilum ciliolare kann man im temperierten oder im warmen Gewächshaus kultivieren. Im Monat Januar sollten die Wassergaben reduziert und die Pflanze kühler gestellt werden, um eine gute Blüteninduktion zu erreichen.

Paphiopedilum ciliolare blüht im allgemeinen zwischen April und Juni.

The bract is about 1.7 cm long, about one-third the length of the ovary.

The flower is up to about 10 cm across.

The dorsal sepal is 4.5 to 5 cm long and 3.5 to 4.5 cm wide, broadly ovate, acuminate, ciliolate at the margin; purple at the base; the remainder white, with closely set, alternately longer and shorter veins which are green; sometimes purple towards the lateral margins.

The synsepalum is about 2.9 cm long, ovate, acuminate, whitish with green veins.

The lateral petals are ligulate, 5.5 to 6.5 cm long and 1.5 to 2.5 cm wide, deflexed and recurved, margins with long black hairs, the basal portion green, densely spotted with blackish warts to two-thirds of the length, the apical portion pale purple.

The lip is deeply saccate, inverted helm-shaped, about 5.5 cm long, dull purplish brown, the infolded side-lobes pale yellow-green spotted with purple warts.

The staminodal shield is reniform, greenish, somewhat suffused with pale rose, the basal edge with a notch, the distal margin obscurely five-toothed.

Paphiopedilum ciliolare differs from *Paphiopedilum superbiens* by the lateral petals, spotted only over two-thirds of their length, by a smaller pouch and by a protruding center tooth of the staminodal shield.

Paphiopedilum ciliolare is a native of the Philippine islands of Luzon, Mindanao and Dinagat. The plants grow at altitudes between 300 and 1800 m above sea-level. This area always remains fairly wet with a drier period between March and July. The average temperatures are, depending on the season, between 22 and 30 °C during the day and 15 to 18 °C at night.

Paphiopedilum ciliolare does well in the intermediate or in the warm greenhouse. The plants are easy to grow. Putting the specimens somewhat cooler and giving them less water during January will insure good blooming.

Paphiopedilum ciliolare usually flowers between April and June.

Paphiopedilum dayanum (Stone ex Lindley) Stein

Stein's Orchideenbuch: 464 (1892)

Basionym: *Cypripedium spectabile* var. *dayanum* Stone ex Lindley
Gardeners' Chronicle, 1. Ser., **20:** 693 (1860)

Synonyma: *Cordula dayana* (Stone ex Lindley) Rolfe
Orchid Review 20 (1): 2 (1912)

Cordula petri (Rchb. f.) Rolfe
Orchid Review 20 (1): 2 (1912)

Cypripedium burbidgei Rchb. f.
Gardeners' Chronicle, 2. Ser., **16:** 38 (1881)

Cypripedium dayanum (Stone ex Lindley) Lindley
Proc. Royal. Hort. Soc. 1: 245 (1860)

Cypripedium dayi Stone, MSS
Gardeners' Chronicle, 1. Ser., **20:** 693 (1860)

Cypripedium ernestianum Castle
J. Hort., 3. Ser., **14:** 375, t. 67 (1887)

Cypripedium peteri D. Dean
Gardeners' Chronicle, 3. Ser., **1:** 577 (1887)

Cypripedium peteri De Vos
Belg. Hort. 31: 242 (1881)

Cypripedium petri Rchb. f.
Gardeners' Chronicle, 2. Ser., **13:** 680 (1880)

Cypripedium superbiens var. *dayanum* (Stone ex Lindley) Rchb. f.
Xenia Orchidacea 2: 10 (1862)

Cypripedium × *petri* (Rchb. f.) Rolfe
Orchid Review 4: 247–248 (1896)

Cypripedium × *petri* var. *burbidgei* (Rchb. f.) Rolfe
Orchid Review 4: 247–248 (1912)

Paphiopedilum burbidgei (Rchb. f.) Pfitzer
in ENGLER, A.: **Botan. Jahrb. 19:** 40 (1894)

Paphiopedilum dayanum var. *petri* (Rchb. f.) Pfitzer
in ENGLER, A.: **Das Pflanzenr. IV** (50): 86 (1903)

Paphiopedilum petri (Rchb. f.) Pfitzer
in ENGLER, A.: **Botan. Jahrb. 19:** 40 (1894)

Paphiopedilum dayanum (Stone ex Lindley) Stein

Paphiopedilum dayanum wurde am Mount Kinabalu im Norden Borneos durch Hugh LOW entdeckt. Die Pflanzen, die er nach England sandte, wurden allesamt durch John DAY aufgekauft. John DAYs Gärtner STONE gelang es bereits kurz danach, eine Pflanze zur Blüte zu bringen. Er stellte sie unter dem Namen *Cypripedium dayi* aus.

In einer kurzen Beschreibung, die aller Wahrscheinlichkeit nach von John LINDLEY stammte, wurden die Pflanzen als Varietät zu *Cypripedium spectabile* gestellt. LINDLEY korrigierte seinen Irrtum bereits kurz darauf, indem er das Taxon als eigenständige Art als *Cypripedium dayanum* beschrieb.

Paphiopedilum dayanum war sehr selten in Kultur, bis sie 1879 durch Peter VEITCH und F. W. BURBIDGE wiederentdeckt wurde und eine „beträchtliche" Zahl von Pflanzen nach England importiert wurden.

REICHENBACH erhielt eine der Pflanzen dieser Aufsammlung und beschrieb sie prompt als *Cypripedium petri (loc. cit.).* Ein Jahr später wurde REICHENBACH nochmals eine Pflanze aus diesem Import überlassen, und auch diesmal hielt er es für angebracht, sie als neue Art, diesmal zu Ehren von BURBIDGE, dem Begleiter und Freund des Peter VEITCH, zu beschreiben.

Die Unterschiede zwischen *Paphiopedilum dayanum, Paphiopedilum petri* und *Paphiopedilum burbidgei* bestehen in der unterschiedlichen Blattmarmorierung, der unterschiedlichen Intensität der Blütenfärbung und der unterschiedlichen Blütengröße. Diese Variabilität läßt sich ohne weiteres auf die unterschiedlichen Höhenlagen der Fundorte zurückführen. Wie wir bereits besprochen haben, ist auch bei mehreren anderen *Paphiopedilum*-Arten eine große Variabilität in Blattfärbung, Blattmarmorierung, Blütengröße und Blütenfärbung zu beobachten.[1] Auch bei genauerer Analyse des Staminodiums ist kein eindeutiger Unterschied zu erkennen.

Die Verwirrung wurde noch erheblich gesteigert durch ROLFE, der 1896 die Meinung vertrat, *Paphiopedilum petri* sei eine Naturhybride zwischen *Paphiopedilum dayanum* und *Paphiopedilum virens,* und *Paphiopedilum burbidgei* sei eine „Varietät" dieser Hybride.[2]

Eine ausführliche Studie der Originalbeschreibungen führte zu der Feststellung, daß die drei Taxa

---

[1] Siehe z. B. unter *Paphiopedilum hookerae, Paphiopedilum purpuratum* und *Paphiopedilum sukhakulii.*

[2] Vgl. R. A. ROLFE, **Orchid Review 4:** 247–248 (1886).

Paphiopedilum dayanum was discovered by High LOW on the slopes of Mount Kinabalu in northern Borneo. The whole shipment which he sent back to England was purchased by John DAY. DAYs gardener STONE managed to flower one of the plants shortly after and exhibited the flowering specimen as *Cypripedium dayi.* In a short description, John LINDLEY erroneously described the taxon as a variety of *Cypripedium spectabile,* but shortly after redescribed the concept as *Cypripedium dayanum.*

Paphiopedilum dayanum remained rare in cultivation until it was rediscovered at the foot of the Marie-Parie Spur of Mount Kinabalu by Peter VEITCH and F. W. BURBIDGE in 1879. They brought a "considerable number" of the plants back to England. REICHENBACH was sent a plant of this collection and promptly described it as *Cypripedium petri (loc. cit.).* A year later, another plant from the same collection was sent to REICHENBACH who described it as a new species again, this time under the name of *Cypripedium burbidgei.*

The differences between the concepts of *Paphiopedilum dayanum, Paphiopedilum petri,* and *Paphiopedilum burbidgei* are solely to be found in different tesselations of the leaves, differences in the intensity of the colouration of the flower, and some difference in the size of the flower. This variability can, without doubt, be explained with the different altitudes of the habitats. As we have already discussed, a number of other species of the genus *Paphiopedilum*[1] demonstrate a distinct variability in leaf tesselation, flower size and flower colour. Neither can a difference in the morphology of the staminode be observed.

The chaos was considerably added to by ROLFE, who, in 1896, expressed the opinion that *Paphiopedilum petri* is an natural hybrid between *Paphiopedilum dayanum* and *Paphiopedilum virens,* and that *Paphiopedilum burbidgei* is a variety of this hybrid.[2]

A detailed study of the original descriptions has shown the three taxa *(Paphiopedilum dayanum, petri* and *burbidgei)* to be identical.

Paphiopedilum dayanum is an herbaceous perennial, usually growing epiphytically in leaf-mould. The stems are short and erect, each of them carrying up to 6 leaves.

---

[1] See also under *Paphiopedilum hookerae, Paphiopedilum purpuratum* and *Paphiopedilum sukhakulii.*

[2] See R. A. ROLFE, **Orchid Review 4:** 247–248 (1886).

(Paphiopedilum dayanum, petri und *burbidgei)* identisch sind.

Paphiopedilum dayanum ist ein meist am Boden in Humusansammlungen wachsender, krautiger Epiphyt mit kurzen, bis zu 5 cm langen, aufrecht stehenden Sprossen, die einem kurzen Rhizom entspringen.

Jeder Sproß trägt 4 bis 6 zweireihig angeordnete, bis zu 4 Jahre überdauernde Laubblätter.

Diese sind länglich lanzettförmig, bis ca. 21 cm lang und 5 cm breit, im basalen Teil gefaltet. Sie haben eine Spitze, die kaum sichtbar in zwei Lappen geteilt und mit einem kurzen Mittelzahn versehen ist. Zur Spitze hin sind die Ränder leicht gesägt. Die Blattfärbung und Blattmarmorierung sind unterschiedlich. Sie variieren von blaßgrün mit dunkler, bläulicher Marmorierung über hellgrün mit dunkleren grünen Flecken, bis grau-blaugrün mit entsprechend dunklerer Marmorierung. Die Marmorierung ist auch an der Unterseite des Blattes noch sichtbar.

Die aufrecht stehende Infloreszenz wird bis ungefähr 25 cm hoch. Sie ist dicht behaart, blaß braunrot und trägt eine bis ca. 13 cm breite Blüte.

Die spitz zulaufende, lanzettförmige Braktee ist bis 2,5 cm lang, hellgrün mit rötlicher Behaarung.

Der ähnlich behaarte Fruchtknoten ist bis 7 cm lang, mattgrün, blaß rotbraun an den Kanten.

Die an den Spitzen leicht konkaven Sepalen (Fahne und Synsepalum) sind leicht kapuzenförmig, weiß, mit vielen parallel verlaufenden grünen Adern. Die Ränder der äußeren Seite sind kurz behaart.

Die Fahne ist schmal eiförmig, bis 6 cm hoch und 3 cm breit.

Das Synsepalum ist lanzettförmig, bis 5 cm lang und 2 cm breit.

Die abstehenden Petalen sind elliptisch spatelförmig, mehr oder weniger spitz zulaufend, bis 8 cm lang und 1,5 cm breit, fahl purpurrosa mit vielen parallel verlaufenden grünen und rotbraunen Adern und gleichmäßig behaarten Rändern.

Die sackförmige Lippe ist bis 5 cm lang, dreilappig. Die Seitenlappen sind nach innen gefaltet und bilden somit eine fast geschlossene, 2 cm lange, blaß rötlichgrüne Röhre, die von glänzenden rotpurpurnen Wärzchen übersät ist. Der Mittellappen ist purpurbraun mit dunkleren rötlichbraunen Adern, außen kahl, innen mit steifen Härchen besetzt.

Die Staminodiumplatte ist nierenförmig mit ein bis drei Zähnen am unteren Rand, mehr oder weniger zweilappig am oberen Ende, bis 1,3 cm breit, blaßgrün und mit dunkelgrüner Marmorierung.

The leaves are distichous, oblong-lanceolate, up to 21 cm long and up to about 5 cm wide, folded in the basal area, the margins slightly serrate in the apical part and obtusely bilobed at the end, variable in colour, sometimes pale green with some oblong spots of deep green scattered over the upper surface, sometimes tesselated with deep and light green. The tesselation goes through to the under surface. The leaves last up to 4 years on the plants.

The inflorescence is erect, up to about 25 cm high, densely pilose, dull brown-red, bearing a single flower which is up to about 13 cm across.

The floral bract is up to about 2.5 cm long, acuminate lanceolate, pale green covered with reddish hairs.

The ovary is similarly hairy, up to about 7 cm long, buff green, ridged pale reddish-brown.

The sepals are concave at the apex, slightly cucullate, white with numerous parallel green veins, the margins and the outer surface shortly ciliate.

The dorsal sepal is narrowly ovate, up to 6 cm long and 3 cm broad.

The synsepalum is lanceolate, up to 5 cm long and about 2 cm wide.

The lateral petals are spreading, elliptic-spathulate, more or less acute, up to about 8 cm long and 1.5 cm wide, dull maroon pink, with numerous parallel green and maroon-brown veins, the margins evenly ciliate.

The lip is deeply saccate, up to 5 cm long, trilobed, the infolded lateral lobes forming an almost closed tube, 2 cm long, dull reddish-green with shiny maroon warts; the mid-lobe is auriculate at the base, purplish brown, veined deep maroon, glabrous on the outer surface, the inside covered with stiff hairs.

The staminodal shield is semi-ovate to reniform, 1 to 3-toothed on the lower edge, more or less bilobed on the upper one, up to 1.3 cm wide, pale green with deeper green marblings towards the central tooth.

Paphiopedilum dayanum is an endemic of Borneo. The plants grow in humus on the ground or on humus-covered serpentine cliffs, often in full sun, on northeastern slopes of Mount Kinabalu at altitudes of up to about 1200 meters above sea-level. This area is never dry and has a distinct rainy season between August and October. The hottest and driest month is April with average temperatures of 30 °C during daytime and 18 °C at night. June and July are the coldest months with average

Paphiopedilum dayanum ist auf der Insel Borneo beheimatet. Die Pflanzen wachsen in Wäldern und auf Serpentin-Gestein, oft im hellen Licht, auf nordöstlichen Berghängen des Mount Kinabalu bis 1200 m ü. NN. Dieses Gebiet ist niemals trocken und hat eine ausgeprägte Nässeperiode von August bis Oktober. Der trockendste und wärmste Monat des Jahres mit Durchschnittstemperaturen um 30 °C am Tage und um 18 °C in der Nacht ist der April. Juni und Juli sind mit 25 °C am Tage und 10 °C in der Nacht am kältesten.

Paphiopedilum dayanum läßt sich im warmen und temperierten Gewächshaus gut kultivieren. Im Winter sollten die Pflanzen zwecks Blüteninduktion etwas kühler gestellt werden. Man sollte reichlich gießen, aber für ein durchlässiges, mit Moos vermischtes Substrat sorgen. Pflanzen, die schattig kultiviert werden, haben deutlichere Blattmarmorierung und farbigere Blüten.

Die Blütezeit ist im allgemeinen von März bis Mai.

Varietäten:

Von *Paphiopedilum dayanum* sind 5 Varietäten beschrieben worden, 3 oder 4 davon als *Cypripedium dayanum* var. Sie sind ohne Ausnahme als „normale" Exemplare zu betrachten.

temperatures of about 25 °C during the day and 10 °C at night.

Paphiopedilum dayanum should be kept in the intermediate or warm house. During the winter, the plants should be kept somewhat cooler to obtain flower induction. They should be potted in a well-drained open substrate and watered abundantly. Plants grown under shaded conditions have more clearly tesselate leaves and darker flowers.

Paphiopedilum dayanum usually flowers between March and May.

Varieties:

Five varieties of this species have been describes, four of them as *Cypripedium dayanum* var. None of them differ from the type.

Paphiopedilum superbiens (Rchb. f.) Stein

Stein's Orchideenbuch: 487 (1892)

Basionym: *Cypripedium superbiens* Rchb. f.
Bonplandia 3: 227 (1855)

Synonyma: *Cordula superbiens* (Rchb. f.) Rolfe
Orchid Review 20 (1): 2 (1912)
Cypripedium veitchianum Hort. ex Lemaire
Ill. Horticole 12, t. 429 (1865)

Paphiopedilum superbiens (Rchb. f.) Stein „CURTISII"

Paphiopedilum superbiens (Rchb. f.) Stein

Pflanzen dieser Art wurden erstmals durch die Beschreibung REICHENBACHs in der **Bonplandia** erwähnt. Die Pflanze, die als Grundlage für die Veröffentlichung diente, wurde durch ROLLISON nach England importiert und später an den Hamburger Konsul SCHILLER verkauft. Eine zweite Pflanze wurde 1857 bei der Firma VEITCH zwischen einem Import von *Paphiopedilum barbatum*

Paphiopedilum superbiens was first described by REICHENBACH. His concept was based on a plant brought to England by ROLLISON and later purchased by Consul SCHILLER of Hamburg. A second plant was found at VEITCHs in 1857 among a shipment of *Paphiopedilum barbatum.* Up to 1889, all plants known to exist in cultivation originated from divisions of the two specimens gefunden. Bis 1889 kamen alle sich in Kultur befindlichen Pflanzen von *Paphiopedilum superbiens* aus Teilungen dieser zwei Exemplare,[1] obwohl kaum auszuschließen ist, daß andere Exemplare als *Paphiopedilum barbatum* var. in den Sammlungen versteckt waren.

Die Herkunft der von ROLLISON importierten Pflanze ist unbekannt geblieben. Es ist aber anzunehmen, daß diese Pflanze genauso wie das durch VEITCH importierte Exemplar aus den Gebirgen Sumatras stammte.

Inzwischen hatte aber Charles CURTIS, ein Sammler für die Firma VEITCH, im Jahre 1882 eine größere *Paphiopedilum*-Population in West-Sumatra gefunden und eine große Menge Pflanzen nach Hause gesandt. Die enge Affinität dieser Pflanzen zu *Paphiopedilum superbiens* wurde zwar bemerkt, die von CURTIS gefundenen Pflanzen hatten aber etwas kleinere Blüten und etwas kürzere Petalen. Dies brachte den unermüdlichen REICHENBACH dazu, eine neue Art mit dem Namen *Cypripedium curtisii* zu beschreiben *(loc. cit.)*.

Es steht aber ohne Zweifel, daß *Paphiopedilum curtisii* lediglich als Varietät von *Paphiopedilum superbiens* zu betrachten ist.

Auch das 1882 als *Cypripedium ciliolare* beschriebene Taxon gehört in diesen engen Komplex und wird von einigen Autoren als eine Unterart von *Paphiopedilum superbiens* behandelt, eine Ansicht, die ohne weiteres vertretbar ist.

Die meisten Exemplare von *Paphiopedilum superbiens*, die sich heute in Kultur befinden, stammen aus der Tobasee-Gegend Nord-Sumatras. Einige Exemplare mit besonders schöner grüner Fahne stammen aus dem Gebirge um Gunang Kurinski, ungefähr in der Mitte des entlang der Westküste Sumatras verlaufenden Gebirgszuges.

Paphiopedilum superbiens ist ein bis ca. 40 cm hoher, krautiger Humusepiphyt. Pro Trieb werden 4 bis 6, bis 20 cm lange und bis zu 8 cm breite, elliptische bis länglich elliptische, spitze, vorne fein dreizähnige Laubblätter ausgebildet. Ihre Oberseite ist hell- und dunkelgrün marmoriert, mit dunklerer Aderung. Die Unterseite ist deutlich gekielt, blaßgrün, purpurn überhaucht.

Die Infloreszenz ist dunkelpurpurn, flaumig behaart, bis ungefähr 30 cm hoch, meistens einblütig.

Die Braktee ist ungefähr 2 cm lang und somit ungefähr ein Drittel so lang wie der Fruchtknoten.

[1] Vgl. VEITCH, **A Manual of Orchidaceous Plants 4:** 52 (1889).

mentioned,[1] although it cannot be ruled out that other specimens were being grown as *Cypripedium barbatum*.

The origin of the plant imported to England by ROLLISON remained unknown. It is however, rather probable that it, just like the one discovered among the cultures at VEITCH originated from the mountainous regions of Sumatra.

In 1882, Charles CURTIS, collecting for VEITCH discovered a population of *Paphiopedilum*-plants in West Sumatra and the plants became quite plentiful in European collections. The close affinity of these plants to *Paphiopedilum superbiens* was noted but the plants found by CURTIS happened to have slightly smaller flowers and somewhat shorter lateral petals which led REICHENBACH *(loc. cit.)* to describe them as a separate species under the name of *Cypripedium curtisii*.

There is no doubt that *Paphiopedilum curtisii* (Rchb. f.) Stein is merely a geographical variety of *Paphiopedilum superbiens* (Rchb. f.) Stein.

Also *Cypripedium ciliolare,* described in 1882, belongs to this complex of closely allied species and is treated as a subspecies of *Paphiopedilum superbiens* by some authors.

Most of the specimens of *Paphiopedilum superbiens* now in cultivation originate from the area around Lake Toba in the northern part of Sumatra. Some clones with a particular nice green dorsal sepal were collected in the mountains around Gunang Kurinchi at about the center of the mountain chain running parallel to the west coast of Sumatra.

Paphiopedilum superbiens is an herbaceous perennial, usually growing in leafy moulder on the ground in forest areas. The plants have short, up to 3 cm long, erect stems bearing 4 to 6 leaves. The leaves are persisting for up to three years. They are up to 20 cm long and up to 8 cm wide, elliptic to oblong elliptic, minutely tridentate at the acute apex, sparsely ciliate on the basal margins, ventrally keeled, pale green tinged with purple below and distinctly tesselated with light and darker green above.

The inflorescence is deep purple, shortly pubescent, up to about 30 cm high, usually bearing one single flower.

The floral bract is about 2 cm long, covering about a third of the ovary.

[1] See VEITCH, **A Manual of Orchidaceous Plants 4:** 52 (1889).

Paphiopedilum superbiens (Rchb. f.) Stein

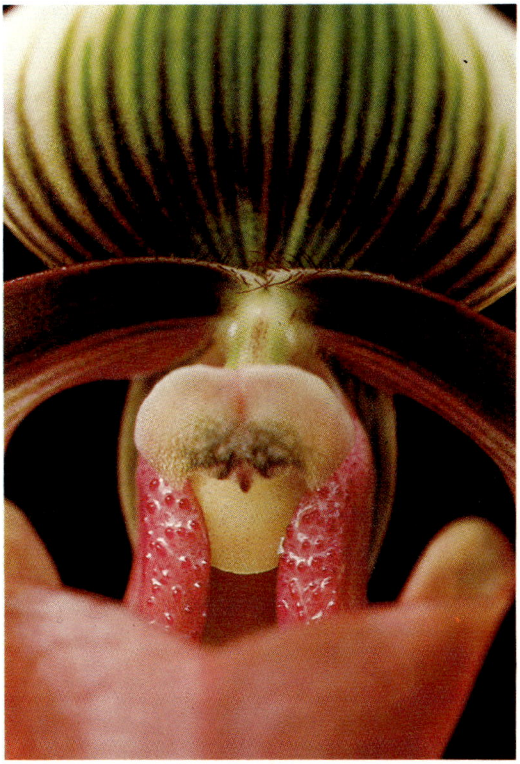

Die Fahne ist bis 6 cm hoch und 5 cm breit, fast herzförmig, langspitzig, bewimpert, in der Mitte hellgrün, mit grünen und purpurnen Längsstreifen.

Das Synsepalum ist bis 3,5 cm lang und schmaler als die Fahne, eiförmig, langspitzig, bewimpert, weiß mit grünen Längslinien.

Die riemenförmigen, leicht herabgeschlagenen und nach unten gekrümmten Petalen werden bis ungefähr 7 cm lang und bis 2 cm breit. Sie sind an den Rändern bewimpert, rosaweiß, mit purpurnen Adern, mit schwärzlichen Wärzchen besetzt, die am oberen Rand am größten und fast bis zur Spitze mit dunklen Flecken überzogen sind.

Die große, sackförmige Lippe wird bis 6 cm lang und bis ca. 3,6 cm breit. Sie ist dunkelpurpurn gefärbt, dreilappig. Die nach innen geschlagenen Seitenlappen sind blaßpurpurn und mit purpurnen Wärzchen besetzt.

Die Staminodiumplatte ist ca. 1,2 cm hoch und bis 2 cm breit, breit hufeisenförmig mit einer deutlichen Einkerbung am oberen Rand und drei kleinen Zähnen an der unteren Seite.

Paphiopedilum superbiens ist auf Sumatra beheimatet und wächst epiphytisch in Humus in Höhenlagen von 300 bis 1400 m ü. NN. Die Heimatge-

The dorsal sepal is up to 6 cm long and more or less 5 cm wide, subcordate, acuminate, ciliate, green at the center, with green and purple longitudinal stripes.

The synsepalum is up to about 3.5 cm long and narrower than the dorsal sepal, ovate, acuminate, ciliate, white, with longitudinal green veins.

The lateral petals are ligulate, falcate-deflexed, acute, up to about 7 cm long and up to 2 cm wide, whitish-rose, with purple veins and covered with blackish warts that are largest at the upper margin and suffused with dark blotches nearly over the whole length, the margins ciliate.

The large lip is deeply saccate, up to 6 cm long and up to about 3.6 cm across, deep purple, trilobate; the infolded sidelobes pale purple covered with deep purple warts.

The staminodal shield is about 1.2 cm high and more or less 2 cm across, broadly horseshoe-shaped with a distinct indentation at the upper margin and three small teeth at the lower end.

Paphiopedilum superbiens is a native of (and endemic to) Sumatra. The plants grow epiphytically in leafy debris and humus at altitudes between 300 and 1400 meters above sea-level. These areas remain constantly wet.

biete von *Paphiopedilum superbiens* werden niemals richtig trocken.

Im November und Dezember gibt es viel Regen und wärmere Winde, die vom Südpazifik kommen. Von Januar bis März gibt es nordöstliche, regenärmere Winde. Diese Periode geht in eine „Trokkenzeit" über, die von April bis Juni herrscht, aber immer noch ziemlich feucht ist. Ab Ende Juni bringen die südwestlichen Monsune wieder hohe Niederschlagsmengen.

Die Luftfeuchtigkeit bleibt somit das ganze Jahr über sehr hoch.

Die Tagestemperaturen der Region variieren zwischen 23 und 30 °C, die Nachttemperaturen zwischen 12 und 18 °C.

Paphiopedilum superbiens gedeiht gut im warmen oder im temperierten Gewächshaus. Man sollte ein durchlässiges Substrat verwenden und für eine gute Drainage sorgen. Die Pflanzen sollten im Sommer schattig stehen und niemals völlig austrocknen.

Eine kühle Ruheperiode mit verminderten Wassergaben induziert die Blüte.

In unseren Breiten kultivierte Exemplare blühen im allgemeinen von März bis Juni.

In November and December warm, moist winds bring large quantities of rain from the South Pacific. From January until March the northeast monsoon blows regularly but there is somewhat less precipitation. It becomes somewhat drier during the months of April through June but by the end of June, the south-west monsoon brings more rain again.

The relative humidity is high over the entire year. The daytime temperatures of this area average between 23 and 30 °C, the temperatures at night are between 12 and 18 °C.

Paphiopedilum suberbiens does best in the warm or intermediate greenhouse potted in an open, well-drained substrate. The plants should be shaded during the summer and never be allowed to dry out completely.

There should be a cooler resting period with less water after completion of the new growth.

In our latitudes, *Paphiopedilum superbiens* generally flowers between March and June.

Varietäten:

Paphiopedilum superbiens var. ***curtisii*** (Rchb. f.) Braem, *comb. et stat. nov.*

Basionym: *Cypripedium curtisii* Rchb. f.
Gardeners' Chronicle, 2. Ser., **20:** 8 (1883)

Synonyma: *Cordula curtisii* (Rchb. f.) Rolfe
Orchid Review 20 (1): 2 (1912)
Paphiopedilum curtisii (Rchb. f.) Stein
Stein's Orchideenbuch: 464 (1892)

Unterscheidet sich von der Nominalform durch eine kürzere Fahne, kürzere, nach hinten geschlagene Petalen, eine etwas längere Braktee und leichte Farbunterschiede in Blüte und Blatt. Alles spricht dafür, daß dies die Gebirgsvariante von *Paphiopedilum superbiens* ist.

Varieties:

Paphiopedilum superbiens var. ***curtisii*** (Rchb. f.) Braem, *comb. et stat. nov.*

Basionym: *Cypripedium curtisii* Rchb. f.
Gardeners' Chronicle, 2. Ser., **20:** 8 (1883)

Synonyma: *Cordula curtisii* (Rchb. f.) Rolfe
Orchid Review 20 (1): 2 (1912)
Paphiopedilum curtisii (Rchb. f.) Stein
Stein's Orchideenbuch: 464, 1892

This variety differs from the nominal type of *Paphiopedilum superbiens* by its shorter dorsal sepal, the shorter, reflexed lateral petals, a somewhat longer floral bract, minor differences in the tesselation of the leaves and in the colouration of the flower. This is the high altitude variant of the species.

Paphiopedilum superbiens var. ***sanderae*** (Curtis) Braem, *comb. nov.*

Basionym: *Cypripedium curtisii* var. *sanderae* Curtis
Gard. Mag., neue Serie, **58:** 303 (1915)

Paphiopedilum superbiens var. ***sanderae*** (Curtis) Braem, *comb. nov.*

Basionym: *Cypripedium curtisii* var. *sanderae* Curtis
Gard. Mag., new series **58:** 303 (1915)

Synonyme: *Cypripedium curtisii* forma *sanderae* (Curtis) Pearson
Gardeners' Chronicle, 3. Ser., **57:** 338 (1915)
Paphiopedilum superbiens forma *sanderae* (Curtis) M. W. Wood
Orchid Review 83 (990): 399 (1975)

Pflanzen ohne rote Farbpigmente. Blattunterseite rein grün. Blütenstandschaft grün. Blütenblätter grün und weiß.

Diese Varietät ist sehr selten, und beim Kauf ist Vorsicht geboten, da man diese Varietät ohne weiteres mit Hybriden von Albinoformen verwandter Arten verwechseln kann.

Synonyma: *Cypripedium curtisii* forma *sanderae* (Curtis) Pearson
Gardeners' Chronicle, 3. Ser., **57:** 338 (1915)
Paphiopedilum superbiens forma *sanderae* (Curtis) M. W. Wood
Orchid Review 83 (990): 399 (1975)

Plants without any red pigmentation. Under surface of the leaves uni-coloured green. Scape green, corolla green and white.

This variety is extremely rare and often hybrids of albinos of related species are offered for sale under its name.

Untersektion CHLORONEURA Hallier

Diese Untersektion kann problemlos von allen anderen anhand der glatten Oberfläche der Staminodiumplatte und der mehr oder weniger rötlich purpurnen Endregion der seitlichen Petalen unterschieden werden. Ich bezweifle sehr stark, daß dies eine natürliche Gruppe ist.

Subsection CHLORONEURA Hallier

This subsection is readily distinguished from all others by the entirely smooth surface of the staminodal shield and by the reddish purple apical portion of the lateral petals. I have severe doubts whether this is a true phylogenetic group.

Schlüssel zu den Arten der Untersektion CHLORONEURA

1. Fahne grünlich mit dunkelgrünen Längsstreifen, die Endregion mehr oder weniger nach innen gebogen und ein schmales, spitzes Ende bildend; Staminodiumplatte halbkreisförmig . . . . *Paphiopedilum javanicum*

1a. Fahne weißlich mit dunkel purpurnen Längsstreifen, oberer Teil kaum nach innen gefaltet, eine breite Spitze bildend; Staminodiumplatte länglich rautenförmig 2

2. Schuh in der Mitte und an der Öffnung fast gleich breit, Mittelregion mit einem Längskiel an der Öffnung; seitliche Petalen schmal länglich, die Spitze mehr oder weniger gerundet; unterer Teil der Fahne oft rötlich purpurn überhaucht . . . *Paphiopedilum acmodontum*

2a. Schuh viel breiter an der Öffnung als in der Mitte; seitliche Petalen länglich linealisch mit sehr spitzem Ende; Fahne gänzlich weißlich . . . . *Paphiopedilum urbanianum*

Key to the species of subsection CHLORONEURA

1. Dorsal sepal greenish, longitudinally striped dark green, apical portion more or less incurved, forming a sharply pointed apex; staminodal shield semi-circular . . . . *Paphiopedilum javanicum*

1a. Dorsal sepal whitish, longitudinally striped deep purple, apical portion hardly incurved, broadly acute at the apex; staminodal shield oblong-rhomboid 2

2. Lip almost equal in width at middle and aperture, middle portion longitudinally keeled at the aperture; lateral petals narrow-oblong, almost rounded or broadly acute at the apex; basal portion of the dorsal sepal often tinged with reddish purple *Paphiopedilum acmodontum*

2a. Lip much wider at aperture than at the hardly carinate middle; lateral petal linear-oblong, sharply acute at the apex; dorsal sepal whitish throughout . . *Paphiopedilum urbanianum*

Paphiopedilum acmodontum Schoser ex M. W. Wood

Orchid Review 84 (1001): 350 (1976)

Pflanzen dieser Art erreichten den europäischen Markt um 1968 aus den Philippinen. In seinem Vortrag anläßlich des 6. Weltorchideenkongresses erwähnte Gustav SCHOSER, daß er die Pflanzen als neue Art unter dem Namen *Paphiopedilum acmodontum* beschreiben wolle.

Als dies bis 1976 noch nicht erfolgt war, nahm ihm der englische Gartenbauunternehmer Mark W. WOOD die Aufgabe ab.

Paphiopedilum acmodontum ist eng mit *Paphiopedilum argus* verwandt, unterscheidet sich aber durch kleinere Blüten, unterschiedliche Lippen- und Staminodiummorphologie sowie durch eine unterschiedliche Blütenfärbung.

Paphiopedilum acmodontum ist eine ziemlich kurzstämmige, als Humusepiphyt wachsende Pflanze. Sie hat bis zu 4 cm breite und bis zu 18 cm lange, elliptisch-längliche, vorn mit drei winzigen Zähnchen versehene Laubblätter, die auf der Oberseite

This species was first recognized when plants reached Europe from the Philippines in 1968. Gustav SCHOSER, director of the Palmengarten at Frankfurt announced his intention to describe the new taxon as *Paphiopedilum acmodontum* in a speech delivered at the 6th World Orchid Conference in 1969 in Sydney.

When this had not been done by 1976, the British horticulturalist Mark W. WOOD published the concept in the **Orchid Review.**

Paphiopedilum acmodontum is closely related to *Paphiopedilum argus* and differs by smaller, differently-coloured flowers and different structures of the lip and the staminodal shield.

Paphiopedilum acmodontum is a fairly short-stemmed epiphytic plant.

The leaves are up to 18 cm long and 4 cm wide, distichous, oblong-elliptic, with a minutely tridenhell- und dunkelgrün marmoriert sind und teilweise bewimperte Ränder aufweisen.

Die aufrecht stehende Infloreszenz ist bis zu 25 cm hoch und meistens einblütig. Der Schaft ist grün mit violettpurpurner Zeichnung und flaumig behaart.

Die bis 3 cm lange Braktee ist eiförmig und zottig behaart.

Die Fahne ist weiß, deutlich rosapurpurn überlaufen und mit grünlichpurpurnen Längslinien versehen, 4 cm hoch und 3 cm breit. Sie ist langspitzig eiförmig.

Das Synsepalum ist ca. 3,2 cm lang und bis zu 1,5 cm breit, spitz eiförmig, weiß, mehr oder weniger purpurn überhaucht und mit grünen und purpurnen Streifen versehen.

Die Petalen sind in der apikalen Hälfte purpurn, in der unteren (basalen) Hälfte grünlich mit dunkleren Linien und deutlichen purpurnen Flecken. Sie sind bis 4,5 cm lang und bis 1,5 cm breit, umgekehrt lanzettförmig, fast waagerecht abstehend und etwas nach hinten gebogen, stumpf, am Rande mit einigen haarigen Wärzchen und schwach bewimpert, innen mehr oder weniger flaumig behaart.

Die Lippe ist bis zu 4 cm lang und ungefähr 2,3 cm breit, umgekehrt helmförmig, dreilappig, geöhrt, sehr fein flaumig behaart, bleich olivgrün, mehr oder weniger bräunlich purpurn überhaucht, vorne an der Öffnung mit einem spitzen Längskiel.

Die verkehrt eiförmige Staminodiumplatte ist ungefähr 1 cm lang und unwesentlich schmaler, bleich grün mit dunkelgrüner Zeichnung, purpurn überhaucht, am Vorderrand zweiteilig, hinten beidseitig mit einem Zahn und mit einem dritten Zahn in der Mitte. Auch das Staminodium ist ein wenig flaumig behaart.

Die genauen natürlichen Standortverhältnisse dieser Art sind unbekannt. Die Pflanzen sollen aus der Visayangegend auf den Philippinen stammen.

Das Klima in dieser Gegend ist gekennzeichnet durch eine hohe, andauernde Luftfeuchtigkeit und „Sommertemperaturen" um 29 °C am Tage und um 16 °C in der Nacht. Im „Winter" sind die Temperaturen um etwa 5 °C niedriger.

Paphiopedilum acmodontum blüht normalerweise zwischen Februar und April.

Die Pflanzen dieser Art gedeihen am besten im warmen Gewächshaus in einem gut durchlässigen Substrat bei hoher Luftfeuchtigkeit und guter Umwälzung. Die Pflanzen mögen kein direktes Sonnenlicht.

ticulate apex, with sparsely ciliate basal margins, tesselate pale and dark green above.

The inflorescence is erect, up to about 25 cm high, usually bearing one single flower. The scape is sparsely pubescent, green, marked violet-purple, downy.

The floral bract is about 3 cm long, ovate, pubescent.

The dorsal sepal is white, distinctly suffused with rose-purple and longitudinally striped greenish-purple, up to 4 cm long and about 3 cm wide, ovate, acute.

The synsepalum is about 3.2 cm long and up to 1.5 cm wide, acutely ovate, white, more or less suffused purple, longitudinally striped green.

The lateral petals are up to about 4.5 cm long and 1.5 cm wide, the basal portion greenish veined with and spotted deep purple, their apical half uniformly purple; spreading-reflexed, ligulate-oblanceolate, obtuse, the margins with some hairy warts and sparsely ciliate, the inner surface more or less downy.

The lip is up to about 4 cm long and more or less 2.3 cm across, deeply saccate, inverted helmet-shaped, trilobate, prominently auriculate, dentate or apiculate at the apex, minutely pubescent, olive-green, more or less suffused purple-brown.

The staminodal shield is abovate, about 1 cm long and nearly as wide, pale green with deep green pattern, more or less tinged bronze, minutely pubescent, the distal margin with 3 teeth, the middle tooth longer than the two lateral ones.

The exact natural distribution of this species is not known up to date. The plants are said to originate from the Visayan area in Central Philippines.

The climate of that region is characterized by high humidity over the entire year. The "summer" temperatures average between 29 °C during the day and 16 °C at night. In the cooler months, the temperatures are about 5 °C lower.

The normal flowering period of *Paphiopedilum acmodontum* is between February and April.

This species does best when grown in the warm house, potted in a well-drained open compost at high humidity and a high rate of air exchange. The plants should be kept semi-shaded.

Paphiopedilum javanicum (Reinwardt ex Blume) Pfitzer

in Pringsh.: **Jahrb. Wiss. Bot. 19:** 165 (1888)

Basionym: *Cypripedium javanicum* Reinwardt ex Blume
Catal. Gew. Buitenzorg: 98 (1823)

Synonyma: *Cordula javanica* (Reinwardt ex Blume) Rolfe
Orchid Review 20 (1): 2 (1912)
Cordula virens (Rchb. f.) Rolfe
Orchid Review 20 (1): 2 (1912)
Cypripedium virens Rchb. f.
Bot. Zeitung 21: 128 (1863)
Paphiopedilum virens (Rchb. f.) Pfitzer in ENGLER, A.: **Botan. Jahrb. 19:** 40 (1894)
Paphiopedilum purpurascens Fowlie
Orchid Digest 38: 153–157 (1974)

Paphiopedilum javanicum (Reinw. ex Blume) Pfitzer

Diese Art wurde durch den niederländischen Botaniker REINWARDT im Jahre 1823 in den östlichen Gebirgszügen Javas entdeckt und durch BLUME bereits im selben Jahr als kultivierte Art in den Unterlagen des Botanischen Gartens Buitenzorg aufgenommen.

Erst 1858 erfolgte die erste Einfuhr nach Europa zusammen mit anderen *Paphiopedilum*-Pflanzen, die aus Nord-Borneo stammten. Eine dieser Pflanzen wurde durch John DAY aufgekauft und war später die Vorlage für REICHENBACHs Konzept als *Cypripedium virens.*

REICHENBACH selber schrieb,[1] daß sein *Cypripedium virens* von *Cypripedium javanicum* nur durch die waagerecht stehenden Petalen und das Staminodium zu unterscheiden sei. Ein Unter-

[1] **Xenia Orchidacea 2:** 154–155 (1874).

This species was discovered by the Dutch botanist REINWARDT in 1823 in the mountains of the eastern part of Java. In the same year, they were registered as cultivated plants at the Botanical Gardens Buitenzorg by BLUME.

The first plants of this species reached Europe in 1858 together with other *Paphiopedilum*-species originating from Borneo. One of these specimens was purchased by John DAY and later on the basis for REICHENBACHs concept of *Paphiopedilum virens.*

REICHENBACH wrote[1] that *Paphiopedilum virens* differs from *Paphiopedilum javanicum* by the horizontally spreading petals and the staminodal shield. A difference in the morphology or structure of the staminodal shield of both concepts cannot

[1] **Xenia Orchidacea 2:** 154–155 (1874).

schied in der Staminodiumstruktur ist aber weder anhand der REICHENBACHschen Zeichnungen und Herbarbelege[1)] noch anhand von lebenden Pflanzen festzustellen. Die etwas hängende Stellung der Petalen bei *Paphiopedilum javanicum* und die mehr oder weniger deutlichen Farbunterschiede einzelner Blütenteile können nicht das Aufrechterhalten von *Paphiopedilum virens* als eigenständige Art begründen, ebensowenig der Anlaß für die Aufrechterhaltung einer Blütenvariante als autonomen Art als *Paphiopedilum purpurascens* sein.

Es steht außer Zweifel, daß die drei hier aufgeführten Taxa als Varianten ein und derselben Art anzusehen sind.

Paphiopedilum javanicum ist eine bis 40 cm hoch werdende, als Humusepiphyt wachsende krautige Pflanze mit bis 20 cm langen und 5 cm breiten, schmal-elliptischen oder länglich-elliptischen, hell- und dunkelgrün marmorierten, vorn fein eingeschnittenen Laubblättern.

Die bis 25 cm hoch werdende, aufrecht stehende Infloreszenz ist hellgrün und karminrot marmoriert. Sie ist meistens ein-, selten zweiblütig.

Die Braktee ist ungefähr 2 cm lang und somit ungefähr ein Drittel so lang wie der Fruchtknoten.

Bei der bis 10 cm hohen Blüte sind außer der Lippe alle Teile bewimpert.

Die Fahne ist ca. 4,3 cm lang und ca. 2,5 cm breit, herzförmig, lang gespitzt, hellgrün mit tiefgrüner Aderung, zur Spitze hin weißlich.

Das Synsepalum ist ungefähr 3,1 cm lang, schmaler als die Fahne, länglich eiförmig.

Die Petalen sind ungefähr 5 cm lang, 1,5 cm breit, breit riemenförmig, etwas herabgeschlagen, im oberen Drittel (apikaler Teil) mattpurpurn, darunter hellgrün und mit winzigen schwarzen Wärzchen getupft.

Die fast zylindrische, sackförmige Lippe ist bräunlich grün mit heller Spitze. Die Seitenlappen sind hellgrün mit purpurner Tupfung.

Die Staminodiumplatte ist nierenfömig, oben eingekerbt und unten etwas ausgebuchtet.

Paphiopedilum javanicum ist auf Java, Bali, Flores, Sumatra und Borneo beheimatet und besiedelt humusreiche, nasse Gebirgshänge. Die Pflanzen wachsen mit Vorliebe an sehr hellen Stellen zwischen 1000 und 2000 m ü. NN.

Das Verbreitungsgebiet von *Paphiopedilum javanicum* ist sehr groß, und eine allgemein gültige Klimaangabe ist kaum möglich.

[1)] Im Naturhistorischen Museum in Wien.

be ascertained, neither by the original descriptions nor by the specimens in the REICHENBACH-Herbarium,[1)] and the somewhat drooping lateral petals of *Paphiopedilum javanicum*, nor the more or less distinct differences in the colouration of flower parts are evidence enough to retain *Paphiopedilum virens* or to maintain a colour variant as an autonomous species as *Paphiopedilum purpurascens.*

In my opinion, there can be no doubt that the three taxa noted here are varieties of one single species.

Paphiopedilum javanicum is an humus epiphyte attaining an overall height of about 40 cm.

The leaves are up to about 20 cm long and 5 cm wide, narrowly to oblong elliptic, tesselate pale and dark green, at the apex deeply notched.

The inflorescence is up to 25 cm high, green, spotted purple, usually bearing one, seldomly two flower(s).

The floral bract is about 2 cm long, covering one third of the ovary.

The flower is up to about 10 cm across, all parts except the lip ciliate.

The dorsal sepal is up to about 4.3 cm long and more or less 2.5 cm wide, cordate, acuminate, green with deep green longitudinal veins, the apical area whitish.

The synsepalum is about 3.1 cm long, narrower than the dorsal sepal, oblong ovate.

The lateral petals are about 5 cm long and 1.5 cm wide, broadly strap-shaped, more or less deflexed, the apical one-third purple, the basal 2 thirds green, spotted with small blackish warts.

The lip is deeply saccate, greenish-brown with a brighter apex. The infolded sidelobes are green, spotted purple.

The staminodal shield is reniform, the upper margin notched, the lower margin with a distinct sinus.

Paphiopedilum javanicum is known from Java, Bali, Flores, Sumatra, and Borneo. The plants grow on humus-covered wet slopes in bright places at altitudes between 1000 and 2000 meters above sea-level.

The natural distribution area of this species is too large as to allow generally valid remarks about the climate.

Paphiopedilum javanicum generally flowers from November through December, sometimes from April through May.

This species is fairly easy to grow as *Paphiopedilum javanicum* is easily adapted to different condi-

[1)] Deposited at the Museum for Natural History at Vienna.

Paphiopedilum javanicum var. *virens* (Rchb. f.) Veitch

Paphiopedilum javanicum blüht meistens von November bis Dezember, manchmal aber auch im April und Mai.

Diese Art ist relativ einfach in der Kultur, da die Pflanzen sich als sehr anpassungsfähig erwiesen haben. *Paphiopedilum javanicum* gedeiht am besten im temperierten Gewächshaus, getopft in ein gut durchlässiges Substrat und bei konstant hoher Luftfeuchtigkeit. Eine kurze, kühle Ruheperiode mit geringeren Wassergaben ist für die Blüteninduktion von Nutzen.

Varietäten:

Paphiopedilum javanicum var. *virens* (Rchb. f.) Veitch
A Manual of Orchidaceous Plants 4: 35 (1889)

Hat im Vergleich zum typischen *Paphiopedilum javanicum* kleinere Blüten mit dunkelgrüner, glänzender Grundfärbung der Sepalen und Petalen. Die Lippe ist glänzend dunkelbraun.

tions. The plants do best when grown in the intermediate greenhouse, potted in a well-drained open substrate and kept at constant high humidity. After the completion of the new growths, the plants should be allowed a short rest at lower temperatures and somewhat drier conditions.

Varieties:

Paphiopedilum javanicum var. *virens* (Rchb. f.) Veitch
A Manual of Orchidaceous Plants 4: 35 (1889)

This variety differs from the typical *Paphiopedilum javanicum* by its smaller flowers, the glossy dark green ground colour paired with a glossy deep-brown pouch.

Paphiopedilum urbanianum Fowlie

Orchid Digest 45 (4): 131 (1981)

Paphiopedilum urbanianum Fowlie

Diese Art wurde entdeckt zwischen einer Sendung *Paphiopedilum argus*-Pflanzen, die von der philippinischen Insel Mindoro stammten, und von Frau Jacinta T. URBAN nach USA verkauft wurden.

Paphiopedilum urbanianum ist eine als Humusepiphyt am Waldboden oder auf moosbedeckten Felsen wachsende krautige Pflanze.

Die Laubblätter sind bis 20 cm lang und bis ungefähr 4 cm breit, schmal elliptisch, grasgrün, marmoriert mit dunklerem Grün.

Die aufrecht stehende Infloreszenz ist bis 25 cm hoch, ungefähr 0,5 cm im Durchmesser, dicht behaart, ein- oder zweiblütig.

Die blattartige Braktee ist 2 bis 2,2 cm lang und 1 cm breit, dunkelgrün, behaart.

Der Fruchtknoten ist 5 bis 7,5 cm lang und bis 0,5 cm im Durchmesser.

Die Fahne ist leicht nach vorne gebogen, 2,6 bis 3,4 cm lang und bis 3,2 cm breit, breit elliptisch, spitz, weißlich mit bis zu 13 grünen Adern manchmal mit einer Spur Lila.

Das Synsepalum ist bis 3,2 cm lang, schmal elliptisch, spitz, bis 1,8 cm breit, weiß mit bis zu 9 grünen Adern.

Die Petalen sind schmal bandförmig, mehr oder weniger gerundet an den Enden, 5,5 bis 6 cm lang und 1,2 bis 1,5 cm breit. Die basale Hälfte ist grün

Paphiopedilum urbanianum was discovered between a shipment of *Paphiopedilum argus* originating from the Philippine island of Mindoro and sold to the American orchid dealer RANDS by Mrs. Jacinta T. URBAN.

Paphiopedilum urbanianum is an herbaceous perennial, growing epiphytically in leaf-mould.

The leaves are up to about 20 cm long and up to about 4 cm wide, narrow elliptic, grass-green, tesselate with darker green.

The inflorescence is erect, up to about 25 cm high and about 5 mm in diameter, densely pilose, bearing one or two flowers.

The leafy bract is 2 to 2.2 cm long and about 1 cm wide, deep green, pubescent.

The ovary is 5 to 7.5 cm long and about 5 mm across.

The dorsal sepal is slightly bent forwards, 2.6 to 3.4 cm long and up to 3.2 cm wide, broad elliptic, acute, whitish, with up to 13 green, longitudinal veins, sometimes slightly lilac.

The synsepalum is up to 3.2 cm long, narrow elliptic, acute, up to 1.8 cm wide, white, with up to 9 green longitudinal veins.

The lateral petals are slightly recurved, oblanceolate, more or less rounded at the apex, 5.5 to 6 cm long and 1.2 to 1.5 cm wide, the basal half green mit 7 oder 8 grünen Längsadern, deren Färbung in das Lila der apikalen Hälfte übergeht. Diese apikale Hälfte ist mattlila überzogen und um 90° nach hinten gedreht. Die Petalen sind, vor allem was die apikale Hälfte betrifft, von dunkelbraun-roten, mehr oder weniger spärlich behaarten Wärzchen besetzt.

Die Lippe ist bis 5 cm lang und 2,2 bis 2,6 cm breit, deutlich sackförmig, unter der Öffnung leicht zusammengezogen. Der den Schuh formende Hauptlappen ist deutlich geohrt. Die Lippe ist grünlich mit glänzendem lederbraunem Überzug und dunklerer grünlich brauner Aderung.

Die Staminodiumplatte ist bis 1 cm breit und 1,1 bis 1,3 cm hoch, fast kreisrund, oben und unten deutlich eingekerbt, unten mit einem kleinen Zähnchen in der Aussparung. Die Platte ist gelblich braun mit deutlicher grüner Zeichnung, überzogen von sehr kurzen Härchen.

Paphiopedilum urbanianum ist nur von der philippinischen Insel Mindoro bekannt. Die Pflanzen wachsen am Waldboden, im tiefen Humus und Blattmoder oder in humusgefüllten Felsspalten an nordöstlichen Hängen in Höhenlagen von 500 bis 800 m ü. NN.

Paphiopedilum urbanianum ist relativ einfach in der Kultur. Die Pflanzen gedeihen im warmen und im temperierten Gewächshaus. Am besten verwendet man ein gut drainiertes durchlässiges Substrat mit einer reichlichen Moosbeimischung. Die Pflanzen müssen reichlich Wasser erhalten, und die Luftfeuchtigkeit sollte andauernd hoch gehalten werden. Ein halbschattiger Standort wird bevorzugt. Im Dezember sollte *Paphiopedilum urbanianum* sehr kühl stehen (10 bis 15 °C unter der nächtlichen Sommertemperatur) und weniger Wasser bekommen. Diese Ruheperiode wird mit üppigem Blühen honoriert.

Die Pflanzen blühen normalerweise zwischen Dezember und April.

with 7 or 8 dark green longitudinal veins, the apical half purple, covered with maroon, sparsely pubescent warts.

The lip is deeply saccate, up to 5 cm long and 2.2 to 2.6 cm across, slightly constricted under the aperture, distinctly auriculate, glossy green, suffused bright leather-brown and with dark greenish-brown venation.

The staminodal shield is up to 1 cm broad and 1.1 to 1.3 cm high, sub-orbicular, distinctly notched above and below, the lower margin with a small tooth in the middle of the sinus, yellowish-brown with a distinct green pattern, shortly pubescent.

Paphiopedilum urbanianum is only known from the Philippine island of Mindoro. The plants grow in humus and leafy debris on the ground or in humus-filled rock cracks and crevices on north-eastern slopes at altitudes between 500 and 800 meters above sea-level.

Paphiopedilum urbanianum is very easy to grow. The plants do best in the intermediate or warm greenhouse, potted in a well-drained open substrate with plenty of moss. Water should be given abundantly and the humidity should be kept as high as possible at all times. *Paphiopedilum urbanianum* should be kept semi-shaded and needs a distinct rest in December with less water and nighttime temperatures 10 to 15 °C lower than the nighttime temperatures in summer.

Paphiopedilum urbanianum flowers between December and April.

Untergattung COCHLOPETALUM (Hallier) Karasawa & Saito

Diese Untergattung wurde durch KARASAWA & SAITO[1] auf der Basis der HALLIERschen Sektion **Cochlopetalum** beschrieben. Obwohl PFITZER[2] diesen Komplex aufgrund des Vorhandenseins von seitlichen Lippenaurikeln als Sektion zu seiner Untergattung **OTOPEDILUM** stellt, und obwohl BRIEGER[3] diese Pflanzengruppe aufgrund der mehrblütigen Infloreszenzen als Sektion der Untergattung **POLYANTHA** ansieht, folge ich KARASAWA & SAITO und betrachte den Komplex als eine gänzlich autonome natürliche Einheit.

Die Pflanzen dieser Gruppe unterscheiden sich von allen anderen Gruppen der Gattung *Paphiopedilum* deutlich durch folgende Merkmale:

1. Die Blüten entfalten sich nacheinander, nie gleichzeitig.
2. Die Blätter sind einfarbig graugrün oder leicht marmoriert.
3. Die seitlichen Petalen sind linealisch, stark gedreht, horizontal abstehend, unregelmäßig gefleckt, mit regelmäßig bewimperten Rändern.
4. Chromosomenzahlen sind 2n = 30, 32, 34, 36 (37).

Obwohl die Arten dieser Gruppe sehr variabel sind (wie die meisten *Paphiopedilum*-Arten), so sind ihre morphologischen Eigenheiten doch sehr einheitlich und in ihrer Gesamtheit unverwechselbar.

Innerhalb dieser Untergattung sind eine Reihe von Taxa beschrieben worden. Einige davon haben die große Vielfalt der nomenklatorischen und taxonomischen Statusmöglichkeiten durchlaufen von Form über Varietät bis zur autonomen Art.

Dies ist gewiß eine schwierige Gruppe für den Taxonomen, aber es gilt nicht nur die Taxa wissenschaftlich abzugrenzen, sondern auch den Laien die Möglichkeit zu geben, sich in dem Wirrwarr zurechtzufinden.

Es gibt zwei Überarbeitungen dieser Problematik. KARASAWA & SAITO *(loc. cit.)*, haben in ihrer Arbeit, die vor allem auf cytogenetischen Untersuchungen basiert, vier Arten anerkannt. Mark W. WOOD kommt in seiner hervorragenden Arbeit[4] zu dem Schluß, nur *Paphiopedilum victoria-regina* als

[1] **Bull. of the Hiroshima Botanical Garden** Nr. 5: 1–69 (1982).

[2] In ENGLER, A.: **Botan. Jahrb. 19:** 42 (1894).

[3] In SCHLECHTER: **Die Orchideen,** 3. Auflage: 174 (1971).

[4] **Orchid Review 84** (995): 133–143 (1976).

Subgenus COCHLOPETALUM (Hallier) Karasawa & Saito

This subgenus, as described by KARASAWA & SAITO[1] was based on the section **Cochlopetalum** of HALLIER. Although PFITZER[2] includes this complex on the sectional level in this subgenus **OTOPEDILUM** on the basis of the lateral ears of the lip, and BRIEGER[3] considers it to be a section of his subgenus **POLYANTHA** because of the tiflowered inflorescences, I follow KARASAWA & SAITO in treating this group as a phylogenetic entity.

The plants of this group differ from all other complexes of the genus *Paphiopedilum* clearly in the following characteristics:

1. The flower open one after the other, never all at once.
2. The leaves are uniformly green or only slightly tesselate.
3. The lateral sepals are linear, strongly twisted, horizontally spreading (not at all drooping), irregularly and roughly blotched, and regularly ciliate at the margins.
4. Mitotic chromosome numbers of 2n = 30, 32, 34, 36 (37).

Although the species of this group are (just like most other Paphiopedilums) very variable, the morphological entity of the group does remain very stable and unmistakable.

A number of different taxa have been described within this group. Some of them have been published (mostly as autonomous species) on the basis of minor deviations from the type and have been set at every level of the taxonomic ladder from form to variety and autonomous species in the various treatments of the subgenus.

I do not deny this to be a difficult group to classify, but it does not suffice to delimit the taxa scientifically; one should furthermore make sure that the amateurs can find their way through the chaos.

There are two reviews of the complex. KARASAWA & SAITO *(loc. cit.)* have recognized four species, mainly on the basis of chromosome numbers. Mark M. WOOD[4] only recognizes *Paphiopedilum victoria-regina* with 5 subspecies, accommodating all other taxa described within the group.

[1] **Bull. of the Hiroshima Botanical Garden** No: 5: 1–69 (1982).

[2] in ENGLER: **Botanische Jahrb. 19:** 42 (1894).

[3] in SCHLECHTER: **Die Orchideen,** 3. Edit.: 174 (1971).

[4] **Orchid Review 84** (995): 133–143 (1976).

Art anzuerkennen, diese aber in fünf Unterarten aufteilen zu müssen.

Obwohl der Argumentation von M. W. WOOD ohne weiteres zu folgen wäre, bin ich der Meinung, daß die einzelnen Taxa doch zu verschieden sind und somit die Anerkennung von mehreren Arten gewährleistet ist. Ich folge damit im großen und ganzen der Argumentation von KARASAWA & SAITO. Ich sehe aber keinen Grund, *Paphiopedilum liemianum* als eigene Art nur der Chromosomenzahl und der undeutlichen Fahnenstreifen wegen anzuerkennen.

Schlüssel zu den Arten der Unterfamilie COCHLOPETALUM

1. Seitliche Petalen grünlich purpurn, nicht gefleckt, mit stark gewellten Rändern, Schuh ungefleckt, etwas zusammengedrückt . . . . *Paphiopedilum victoria-regina*

1a. Seitliche Petalen unregelmäßig mit purpurnen und weißen bis cremefarbenen Flecken, Schuh mehr oder weniger mit kleinen, tief purpurnen Flecken überzogen 2

2. Blattoberfläche blaugrün, Laubblätter dünn, lederig; Fahne mehr oder weniger einheitlich braun überzogen . . . . *Paphiopedilum glaucophyllum*

2a. Blattoberfläche sattgrün, Blätter dick, Fahne mit deutlich abgesetzten dunkelbraunen bis fast schwarzen Längsstreifen . . . . *Paphiopedilum chamberlainianum*

Although, I see no great difficulty in following the views of Mark WOOD, I do consider the single taxa to be different enough as to warrant the recognition of several species. In that respect, I mainly follow KARASAWA & SAITO. I cannot however, recognize *Paphiopedilum liemianum* to be an independent species only on the basis of the chromosome number and the indistinct striping of the dorsal sepal.

Key to the species belonging to subgenus COCHLOPETALUM

1. Lateral petals greenish-purple, unspotted, margins strongly undulate; pouch unspotted, somewhat compressed . . . . *Paphiopedilum victoria-regina*

1a. Lateral petals irregularly suffused with white, cream-coloured or purple spots; pouch more or less spotted with fine, dark purplish dots 2

2. Leaves thin, leathery, upper surface bluish-green; dorsal sepal more or less uniformly suffused brown . . . . *Paphiopedilum glaucophyllum*

2a. Leaves thick, upper surface deep green; dorsal sepal with distinct dark brown to blackish longitudinal stripes . . . . *Paphiopedilum chamberlainianum*

Paphiopedilum chamberlainianum (Sander) Stein

Stein's Orchideenbuch: 461 (1892)

Basionym: *Cypripedium chamberlainianum* Sander
Gardeners' Chronicle, 3. Ser., **11:** 194 (1892)

Synonyma: *Cordula chamberlainianum* (Sander) Rolfe
Orchid Review 20 (1): 2 (1912)

Paphiopedilum victoria-regina (Sander) M. W. Wood subspecies *chamberlainianum* (Sander) Wood
Orchid Review 84 (995): 133–143 (1976)

Paphiopedilum chamberlainianum (Sander) Stein

Paphiopedilum chamberlainianum wurde 1891 in Zentral-Sumatra durch MICHOLITZ gefunden, wahrscheinlich in derselben Gegend oder sogar zusammen mit *Paphiopedilum victoria-regina.*

SANDER inserierte die Pflanzen im **Gardeners' Chronicle** am 13. Februar 1892 und benannte sie als *Cypripedium chamberlainianum* zu Ehren des britischen Parlamentsabgeordneten und Orchideenliebhabers Joseph CHAMBERLAIN. Eine Woche später, am 20. Februar, erschien eine Beschreibung der Art durch James O'BRIEN[1] mit dem Titel „*Cypripedium chamberlainianum* O'Brien, *n. sp.*", und eine ganzseitige Abbildung wurde in derselben Ausgabe auf Seite 241 abgedruckt.

Da aber vor dem 1. Januar 1953 veröffentlichte Handelsangebote als gültige taxonomische Beschreibungen anzusehen sind,[2] ist die Mitteilung von SANDER als Erstbeschreibung einzustufen.

[1] **Gardeners' Chronicle,** 3. Ser., **11:** 234 (1892).
[2] Vgl. Internationaler Code der Botanischen Nomenklatur, Artikel 29 (1983).

Paphiopedilum chamberlainianum was discovered by MICHOLITZ in 1891 in the central part of Sumatra, probably together with *Paphiopedilum victoria-regina.*

SANDER offered the plants for sale in an advertisement in the **Gardeners' Chronicle** for February 13th and named them *Cypripedium chamberlainianum* in honour of the member of Parliament and Orchid enthusiast Joseph CHAMBERLAIN. One week later, on February 20th, James O'BRIEN published in the same Gazette[1], a description of the species as *Cypripedium chamberlainianum* O'Brien, *n. sp.* and a full page illustration of the species was printed on page 241 of the same issue.

However, before January 1st 1953, publications in tradesmen's adverts are considered valid publications.[2] Therefore, the announcement of SANDER is to be considered the basionym.

[1] **Gardeners' Chronicle,** 3. Ser., **11:** 234 (1892).
[2] See ICBN Article 29 (1983).

Die Veröffentlichung durch J. D. HOOKER im **Curtis's Botanical Magazine**[1] zeigt eine untypische Pflanze, die durch FOWLIE[2] als mögliche Naturhybride zwischen *Paphiopedilum tonsum* und *Paphiopedilum chamberlainianum* gedeutet wird.

Paphiopedilum chamberlainianum ist eine als Humusepiphyt am Waldboden oder auf moosbewachsenen Felsen wachsende krautige Pflanze.

Die Laubblätter werden bis 45 cm lang und bis 10 cm breit. Sie sind länglich, mattgrün mit hellgrüner Marmorierung, vorne leicht eingeschnitten, abgerundet zweilappig, mit bewimperten Rändern. Ältere Blätter zeigen wellige Ränder. Die Blattunterseite ist grün, mehr oder weniger mit kleinen dunkelpurpurnen Flecken überzogen.

Die Infloreszenz ist aufrecht, bis 60 cm lang, dunkelbraun, vielblütig, wobei die einzelnen Blüten nacheinander zur Entfaltung kommen. Es können bis 30 Blüten generiert werden.

Die Brakteen sind bis 3,7 cm lang, eiförmig, spitz, behaart. Sie umhüllen die jungen Knospen fast völlig.

Die Blüten sind bis 7,5 cm hoch und bis 8,5 cm breit.

[1] **Curtis's Botanical Magazine 124,** t. 7578 (1898).
[2] **Orchid Digest 34:** 108 (1970).

The article by J. D. HOOKER in **Curtis's Botanical Magazine**[1] displays an atypical specimen considered to be a putative natural hybrid between *Paphiopedilum tonsum* and *Paphiopedilum chamberlainianum* by FOWLIE.[2]

Paphiopedilum chamberlainianum is an herbaceous perennial growing in leaf moulder on the ground in forests or on moss-covered rocks.

The leaves are up to about 45 cm long and up to 10 cm wide, oblong elliptic, buff green with brighter green tesselations, the margins ciliolate, the apex slightly notched and rounded bilobate. The under surface is green, more or less suffused with small purple spots.

The inflorescence is erect, up to 60 cm long, dark brown, multiflorous, the flowers generally opening one at the time. Up to 30 flowers can be generated.

The bracts are up to 3.7 cm long, ovate, acute, hairy. They cover more or less the whole of the juvenile bud.

The flowers are up to about 7.5 cm high and more or less 8.5 cm wide.

[1] **Curtis's Botanical Magazine 124** t. 7578 (1898).
[2] **Orchid Digest 34:** 108 (1970).

Paphiopedilum chamberlainianum (Sander) Stein

A

B

C

D

A *Paphiopedilum chamberlainianum* var. *primulinum* (Wood & Taylor) Braem
B *Paphiopedilum chamberlainianum* var. *primulinum* (Wood & Taylor) Braem
C *Paphiopedilum chamberlainianum* var. *liemianum* (Fowlie) Braem
D *Paphiopedilum glaucophyllum* var. *moquettianum* J. J. Smith

Die Fahne ist bis ungefähr 3 cm lang und ca. 2,5 cm breit, eiförmig, grün bis cremeweiß, mit tiefbraunen Längsstreifen, an der Basis purpurn gefleckt.

Das Synsepalum ist bis 2,7 cm lang, eiförmig, grünlichweiß.

Die Petalen sind abstehend, bis 4 cm lang und bis 0,5 cm breit, linealisch, spiralig gedreht, an den Rändern flaumig behaart und bewimpert, cremeweiß mit purpurner Zeichnung.

Die Lippe ist deutlich sackförmig, bis 4 cm lang, unten breit und gerundet, rosa bis rötlich violett, intensiv purpurn getupft, um die Öffnung weiß berandet.

Die Staminodiumplatte ist rhombisch, braun, der obere Teil heller.

Paphiopedilum chamberlainianum hat ihre natürliche Verbreitung in Zentral-Sumatra. Die Pflanzen besiedeln Kalkfelsen, wobei sie in den Humusansammlungen wachsen.

Paphiopedilum chamberlainianum ist dankbar in der Kultur. Die Pflanzen gedeihen im temperierten oder warmen Gewächshaus bei halbschattigen Lichtverhältnissen. Die Art stellt keine großen Ansprüche an das Substrat, gedeiht aber am besten in einem lockeren, durchlässigen Pflanzstoff mit reichlicher Moosbeimischung. Die Luftfeuchtigkeit sollte hoch gehalten werden.

Paphiopedilum chamberlainianum blüht das ganze Jahr hindurch, schwerpunktmäßig aber von März bis Juni.

The upper sepal is ovate, about 3 cm long and ca. 2.5 cm wide, green to creamy white, with deep-brown longitudinal stripes, the basal portion purple.

The synsepalum is up to about 2.7 cm long, ovate, greenish-white.

The lateral petals are spreading horizontally, up to about 4 cm long and merely about 0.5 cm wide, linear, twisted like a spiral, the margins downy and ciliate, creamy white with purple pattern.

The lip is distinctly saccate, up to about 4 cm long, at the apex broadly rounded, rose to reddish-violet, intensively spotted purple, around the aperture with a white rim.

The staminodal shield is rhomboid, brown, the upper part brighter in colour.

Paphiopedilum chamberlainianum is known only from central Sumatra. The plants grow on a limestone underground, usually in leaf-mould and debris.

Paphiopedilum chamberlainianum is easy to grow and flowers readily. The plants do best in the intermediate or warm house under semi-shaded conditions. They should be potted in a well-drained open substrate with plenty of moss. The relative humidity should be high at all times.

Paphiopedilum chamberlainianum can flower over the entire year, the mean flowering period being March through July.

Varietäten:

Paphiopedilum chamberlainianum var. *liemianum* (Fowlie) Braem, *comb. et stat. nov.*

Basionym: *Paphiopedilum chamberlainianum* subsp. *liemiana* Fowlie **Orchid Digest 33** (2): 55 (1971)

Synonyma: *Paphiopedilum liemianum* (Fowlie) Karasawa & Saito **Bull. of the Hiroshima Botanical Garden** Nr. 5: 57 (1982)

Paphiopedilum victoria-regina subsp. *liemianum* (Fowlie) M. W. Wood **Orchid Review 84** (995): 133–145 (1976)

Die Pflanzen dieses Taxons unterscheiden sich von typischen Pflanzen von *Paphiopedilum chamberlainianum* durch kürzere, bis 20 cm lange, tiefer eingeschnittene Blätter und eine farblich unterschiedliche Fahne und quergestreifte Petalen.

Varieties:

Paphiopedilum chamberlainianum var. *liemianum* (Fowlie) Braem, *comb. et stat. nov.*

Basionym: *Paphiopedilum chamberlainianum* subsp. *liemiana* Fowlie **Orchid Digest 33** (2): 55 (1971)

Synonyma: *Paphiopedilum liemianum* (Fowlie) Karasawa & Saito **Bull. of the Hiroshima Botanical Garden** No. 5: 57 (1982)

Paphiopedilum victoria-regina subsp. *liemianum* (Fowlie) M. W. Wood **Orchid Review 84** (995): 133–145 (1976)

The plants of this taxon differ from the typical specimens of *Paphiopedilum chamberlaninianum* by shorter, up to 20 cm long, at the apex deeply notched leaves, a different colouration of the dorsal sepal and cross-striped lateral petals.

Diese Unterschiede deuten darauf hin, daß es sich hier lediglich um eine geographische Variante aus dem nördlichen Teil des Verbreitungsgebietes von *Paphiopedilum chamberlainianum* handelt. Auf keinen Fall begründen sie die Behandlung des Taxons als Subspecies, geschweige denn als autonome Art.

These differences indicate that this is merely a geographical variant from the northern part of the distribution area of *Paphiopedilum chamberlainianum*. They most certainly do not suffice to consider this taxon to be a subspecies or an autonomous species.

Paphiopedilum chamberlainianum var. *primulinum* (M. W. Wood & P. Taylor) Braem,
comb. et stat. nov.

Basionym: *Paphiopedilum primulinum*
M. W. Wood & P. Taylor
Orchid Review 81 (955): 220 (1973)

Synonyma: *Paphiopedilum chamberlainianum* subsp. *liemianum* forma *primulinum* Fowlie
Orchid Digest 37 (2): 104 (1973)

Paphiopedilum liemianum var. *primulinum* (M. W. Wood & P. Taylor) Karasawa & Saito
Bull. of the Hiroshima Botanical Garden Nr. 5: 57 (1982)

Paphiopedilum victoria-regina subsp. *primulinum* (M. W. Wood & P. Taylor) M. W. Wood
Orchid Review 84 (995): 133–143 (1976)

Eine andere Variante von *Paphiopedilum chamberlainianum*, die sich durch schmalere Blätter, eine etwas kantigere Staminodiumplatte und hauptsächlich gelbliche Blüten mit grünlichen Sepalen unterscheidet.

Paphiopedilum chamberlainianum var. *primulinum* (M. W. Wood & P. Taylor) Braem,
comb. et stat. nov.

Basionym: *Paphiopedilum primulinum*
M. W. Wood & P. Taylor
Orchid Review 81 (955): 220 (1973)

Synonyma: *Paphiopedilum chamberlainianum* subsp. *liemianum* forma *primulinum* Fowlie
Orchid Digest 37 (2): 104 (1973)

Paphiopedilum liemianum var. *primulinum* (M. W. Wood & P. Taylor) Karasawa & Saito
Bull. of the Hiroshima Botanical Garden No. 5: 57 (1982)

Paphiopedilum victoria-regina subsp. *primulinum* (M. W. Wood & P. Taylor) M. W. Wood
Orchid Review 84 (995): 133–143 (1976)

Another geographical variant of *Paphiopedilum chamberlainianum* differing from the type by narrower leaves, a more square staminodal shield and a flower which is primrose to daffodil yellow with greenish sepals.

Paphiopedilum glaucophyllum J. J. Smith

Bulletin de l'Institut Botanique de Buitenzorg 7: 1 (1900)

Synonyme: *Cordula glaucophylla* (J. J. Smith) Rolfe
Orchid Review 20 (1): 2 (1912)
Cypripedium glaucophyllum (J. J. Smith) Masters
Gardeners' Chronicle, 3. Ser., **34:** 405 (1903)
Paphiopedilum victoria-regina (Sander) Wood subsp. *glaucophyllum* (J. J. Smith) Wood
Orchid Review 84 (995): 133–143 (1976)

Paphiopedilum glaucophyllum J. J. Smith

Paphiopedilum glaucophyllum wurde wahrscheinlich durch einen holländischen Pflanzensammler namens VERHEY südöstlich vom Mt. Semeru an der Nordküste Ostjavas entdeckt. Die Pflanze wurde durch den Botaniker J. J. SMITH im Organ des Botanischen Gartens Buitenzorg als neue Art veröffentlicht.

Paphiopedilum glaucophyllum hat bis zu 7 ziemlich weiche, breit riemenförmige, abstehende, am Rande etwas wellige, an der Basis bewimperte Laubblätter. Sie sind ungleich zweilappig, stumpf endend, mit einem kleinen Spitzchen zwischen den Lappen, oben gefurcht, unten scharf gekielt, oben glänzend dunkel bläulichgrün, mit kaum dunkleren Adern durchsetzt, unten etwas hell blaugrün, bis ungefähr 26 cm lang und bis 5,5 cm breit.

Die aufrecht stehende Infloreszenz ist bis 30 cm hoch, stielrund, tief purpurn und besetzt mit flaumigen weißen Härchen. Die Infloreszenz ist mehrblütig, wobei im allgemeinen die Blüten nacheinander zur Entfaltung kommen.

Die Brakteen sind ungefähr 1,5 cm lang, eiförmig, flaumig behaart, grün mit purpurnen Adern, an der

Paphiopedilum glaucophyllum probably was discovered by a Dutch plant collector named VERHEY, to the southeast of Mount Semeru near the north-eastern coast of Java. The species was described by J. J. SMITH, at that time the director of the Botanical Gardens Buitenzorg.

Paphiopedilum glaucophyllum is a short-stemmed herbaceous epiphyte.

The leaves are broadly strap shaped, spreading; the margins somewhat undulate and ciliate in the basal portion; the apex asymmetrically bilobed, obtuse, with a small tooth between the two lobes, the upper surface glossy deep bluish green with slightly darker veins, the under surface somewhat paler, distinctly keeled, up to about 26 cm long and up to 5.5 cm wide.

The inflorescence is erect, terete, up to about 30 cm long, deep purple covered with white downy hairs, multi-flowered, the flowers opening one at the time.

The bracts are about 1.5 cm long, ovate, downy, green with purple veins, somewhat serrate and with a small acute tooth at the broad, blunt apex.

Paphiopedilum glaucophyllum J. J. Smith

breiten stumpfen Spitze etwas ausgerandet, mit einem Spitzchen.

Der Fruchknoten ist bis 4,6 cm lang, grün, dicht abstehend dunkelviolett behaart.

Die Blüten sind bis 8,5 cm breit und bis 6,5 cm hoch.

Die Fahne ist rund eiförmig, etwas wellig, an der Spitze etwas ausgerandet, vorne mit einer untiefen Längsfurche, hinten mit einer stumpfen Längsrippe, beide Hälften etwas konvex, hellgrün, am Rande mit Blaugrün überhaucht, mit bis zu 12 gebogenen, schmalen, dunkel braungrauen Längsstreifen, die vor allem im unteren Teil ziemlich flächendeckend sind, vorne glänzend, am Rande und hinten dicht abstehend dunkelviolett behaart, bis ungefähr 3 cm lang und bis 3,2 cm breit.

Das Synsepalum ist eiförmig oval, stumpf, konkav, grün, vorn glänzend, mit vorstehenden Längsadern, am Rande und hinten abstehend dunkelviolett behaart, bis ungefähr 2,6 cm lang und bis ca. 1,8 cm breit.

Die abstehenden Petalen sind fast lineal, stark wellig und gedreht, an beiden Seiten glänzend, am Rande mit ungefähr 3 mm langen, abstehenden, gebogenen und verzweigten farblosen Haaren, mit

The ovary is about 4.6 cm long, green, densely covered with dark-violet hairs.

The flowers are up to 8.5 cm wide and about 6.5 cm high.

The dorsal sepal is broadly ovate, subcircular, somewhat undulate, the apex somewhat serrate, the inner surface with a shallow median longitudinal groove; on the outside bluntly keeled, both halves somewhat convex, bright green, in the marginal areas suffused with bluish-green; with about 12 bent, narrow, dark-grey to dark-brown longitudinal stripes covering especially the basal area, the inner surface glossy, the margins and the outer surface densely covered with deep violet hairs, up to about 3 cm long and up to about 3.2 cm wide.

The synsepalum is ovate, obtuse, concave, the inner surface glossy with prominent longitudinal veins, the margins and the outer surface with spreading, deep-violet hairs, up to about 2.6 cm long and more or less 1.8 cm broad.

The lateral petals are spreading horizontally, nearly always linear, intensively undulate and twisted, both surfaces glossy, the margins with about 3 mm long, spreading, bent and branched colourless hairs; with about 9 longitudinal veins; whitish,

9 Längsadern. Die Grundfarbe ist weißlich, nach der Basis hin grünlich, nach der Spitze hin blaß braunrot gefärbt, am Rande und entlang den Adern mit zahlreichen kleinen braunroten Flecken, bis ungefähr 4,5 cm lang und bis 1 cm breit.

Die Lippe ist bis ungefähr 4,3 cm lang, dreilappig, an der Basis blaß grünlich weiß, die großen abgerundeten und nach innen geschlagenen Seitenlappen von kleinen hellvioletten Punkten übersät. Der sackförmige Hauptlappen ist abgerundet, ca. 2 cm breit, mit spitzen, aufrecht stehenden Öhrchen, gleichmäßig lila, am Rande blaß gelblich. Die Innenseite ist, mit Ausnahme des Randes, ziemlich kurz dunkelviolett behaart.

Die Staminodiumplatte ist groß, eiförmig, stumpf, bis ca. 1,2 cm hoch und bis 0,9 cm breit, an der Basis grün und kurz abstehend dunkelviolett behaart, sonst stark glänzend schwarzviolett.

Paphiopedilum glaucophyllum wächst an hellen Standorten, an steilen bis fast senkrechten kalksteinhaltigen Felswänden, wobei die Pflanzen humusgefüllte Risse und Falten besiedeln, bis in Höhenlagen von 300 m ü. NN.

Das Klima im Heimatareal ist geprägt durch einen südöstlichen Monsun von Juni bis Oktober. In dieser Periode regnet es oft, aber niemals über längere Zeiträume hinweg. Von November bis April bläst ein westlicher Monsun mit schweren Regenstürmen, die teilweise tagelang andauern. Die jährliche Niederschlagsmenge beträgt ungefähr 4500 mm. Die maximalen und minimalen Temperaturen sind 30 bzw. 20 °C im Sommer und 26 bzw. 17 °C im Winter.

Die Luftfeuchtigkeit ist das ganze Jahr hindurch sehr hoch.

Paphiopedilum glaucophyllum ist sehr dankbar in der Kultur. Die Pflanzen gedeihen am besten im warmen Gewächshaus und brauchen viel Wasser und viel Licht während der sommerlichen Wachstumsperiode. Diese Art stellt keine großen Ansprüche an das Substrat und blüht willig nach einer kleinen Ruheperiode.

Die Blütezeit von *Paphiopedilum glaucophyllum* ist im allgemeinen von März bis Juni, einzelne Pflanzen blühen teilweise sehr lange.

towards the base somewhat greenish, towards the apex pale rose, in the marginal regions and along the veins with multiple, small reddish-brown spots; up to about 4.5 cm long and more or less 1 cm wide.

The lip is trilobate, about 4.3 cm long, at the base pale greenish-white, the large, infolded sidelobes rounded and covered with small bright purple spots; the middle lobe deeply saccate, about 2 cm wide, with acute, erect lateral ears, fairly uniform lilac, the rim of the aperture pale yellowish; the inner surface with the exception of the rim covered with fairly short dark-violet hairs.

The staminodal shield is large, ovate, blunt, up to about 1.2 cm high and about 0.9 cm wide, in the basal portion green with short spreading, deep purple hairs, elsewhere distinctly gloss blackish-purple.

Paphiopedilum glaucophyllum grows in bright spots, on steep to nearly perpendicular limestone cliffs in crevices and rock cracks filled with leaf-mould and debris; in altitudes of up to about 300 m above sea-level.

The climate in the natural distribution area is characterized by a southeastern monsoon with lots of short rain showers between June and October. From November through April the westerly monsoon brings heavy rainstorms, often lasting for days at the time. The yearly rate of precipitation is about 4500 mm. The temperatures are between 20 and 30 °C in the summer and between 17 and 26 °C during the winter.

The humidity is high over the entire year.

Paphiopedilum glaucophyllum is an easy to grow species. The plants do best in the warm greenhouse and need plenty of light and abundant water during the growth period. This species has no special requirements in respect to the substrate and flowers willingly after a short resting period upon the completion of the new growth.

The main flowering season of *Paphiopedilum glaucophyllum* is between March and July, single plants flower very long due to the large amount of blooms that well grown plants can produce.

Varietäten:

Paphiopedilum glaucophyllum var. ***moquettianum*** J. J. Smith
Teysmannia 17: 31 (1906)

Synonyma: *Paphiopedilum moquettianum* (J. J. Smith) Fowlie
Orchid Digest 44 (4): 133 (1980)

Paphiopedilum victoria-regina subsp. *glaucophyllum* var. *moquettianum* (J. J. Smith) Wood
Orchid Review 84 (995): 133–143, 1976

J. J. Smith beschrieb diese Varietät aufgrund von Pflanzen, welche aus dem Südwesten Javas stammten. Die Unterschiede zu den „normalen" Formen von *Paphiopedilum glaucophyllum* bestehen hauptsächlich in einer etwas größeren „aufgeblasenen" Lippe und einer mehr oder weniger konkaven Staminodiumplatte.

Diese Unterschiede reichen meiner Meinung nach nicht aus, um diese Varietät als eigenständige Art zu akzeptieren, um so mehr, da KARASAWA & SAITO[1] keinen karyologischen Unterschied zu typischen Pflanzen von *Paphiopedilum glaucophyllum* feststellen konnten.

[1] **Bull. of the Hiroshima Botanical Garden** Nr. 5: 1–69 (1982)

Varieties:

Paphiopedilum glaucophyllum var. ***moquettianum*** J. J. Smith
Teysmannia 17: 31 (1906)

Synonyma: *Paphiopedilum moquettianum* (J. J. Smith) Fowlie
Orchid Digest 44 (4): 133 (1980)

Paphiopedilum victoria-regina ssp. *glaucophyllum* var. *moquettianum* (J. J. Smith) Wood
Orchid Review 84 (995): 133–143, (1976)

The description of this variety by J. J. SMITH was based on plants originating from the southwestern part of Java. The differences to the typical forms of *Paphiopedilum glaucophyllum* consist in a somewhat "blown-up" lip and a staminodal shield which is more or less concave.

These differences do not, in my opinion, suffice to accept this taxon as an autonomous species, especially as KARASAWA & SAITO[1] did not find any difference in the karyological type compared to the typical *Paphiopedilum glaucophyllum.*

[1] **Bull. of the Hiroshima Botanical Garden** No. 5: 1–69 (1982).

Paphiopedilum victoria-regina (Sander) M. W. Wood

Orchid Review 84 (995): 133–143 (1976)

Basionym: *Cypripedium victoria-regina* Sander
Gardeners' Chronicle, 3. Ser., **11:** 194 (1892)

Synonyma: *Cordula victoria-mariae* (Sander ex M. T. Masters) Rolfe
Orchid Review 20 (1): 2 (1912)
Cypripedium chamberlainianum forma *victoria-mariae* (Sander ex M. T. Masters) Rolfe
Orchid Review 1: 188 (1893)
Cypripedium victoria-mariae Sander ex M. T. Masters
Gardeners' Chronicle, 3. Ser., **13:** 580 (1893)
Paphiopedilum victoria-mariae (Sander ex M. T. Masters) Rolfe
Orchid Review 3: 364 (1896)

Paphiopedilum victoria-regina (Sander) M. W. Wood

Paphiopedilum victoria-regina wurde zusammen mit *Paphiopedilum chamberlainianum* durch Wilhelm MICHOLITZ, Sammler im Auftrage der Firma SANDER, auf Sumatra entdeckt.

MICHOLITZ sandte einen Karton mit getrocknetem Material vorab nach England. Aufgrund dieses getrockneten Materials und in Erwartung der Pflanzenlieferung bot SANDER die Pflanzen in einem Inserat im **Gardeners' Chronicle,** zum Kauf an, und zwar als *Cypripedium victoria-regina.* Die Pflanzen überlebten aber den Transport nicht, und es dauerte ein Jahr, bis eine neue Sammlung organisiert werden konnte. Endlich, am 9. Mai 1893, konnte eine Pflanze der neuen Art anläßlich einer Versammlung der Königlichen Gartenbaugesellschaft vorgestellt werden, wobei SANDER, inspiriert durch die Ankündigung der Verlobung der Prinzessin Victoria Mary mit dem Herzog von York, die Pflanze als *Cypripedium victoria-mariae* umbenannte. Dieser Name wurde dann allgemein gebräuchlich.

Tatsache aber ist, daß die erste Beschreibung als *Cypripedium victoria-regina* erfolgte, und, egal wieviele romantische Gefühle durch die erwähnte Anekdote geweckt wurden oder werden, die Regeln des Internationalen Codes der botanischen Nomenklatur sind eindeutig und unbarmherzig.

Paphiopedilum victoria-regina was discovered together with *Paphiopedilum chamberlainianum* by Wilhelm MICHOLITZ when collecting for SANDER in Sumatra.

MICHOLITZ sent a box with dried specimens back to England and based on this material and awaiting a large shipment of living plants, SANDER offered the species for sale in the **Gardeners' Chronicle** as *Cypripedium victoria-regina.* However, the collection made by MICHOLITZ was lost in transport and it took over a year to get a new shipment on the way to England.

Finally, on the 9th of May 1893, the species could be shown at the meeting of the Royal Horticultural Society and SANDER, inspired by the announcement of the betrothal of Princess Victoria Mary to the Duke of York, renamed the species as *Cypripedium victoria-mariae,* the name generally, but falsely, used up to this time.

The fact remains, that the first publication of this species was rendered under the name of *Cypripedium victoria-regina* and, no matter how many romantic feelings may be initiated by the anecdote cited above, the rules of botanical nomenclature are unmistakable and merciless.

The offer in the **Gardeners' Chronicle** for 1892, is a valid publication of the species and therefore, the

Das Angebot in einem Inserat im Jahre 1892 stellt eine gültige Veröffentlichung dar. Somit ist *Paphiopedilum victoria-regina* als gültige Benennung anzusehen. Auch die Tatsache, daß die Beschreibung durch HOOKER im Botanical Magazin als *Cypripedium victoria-mariae* erfolgte,[1] ändert nichts daran, daß dieser letzte Name als Synonym zu betrachten ist.

Paphiopedilum victoria-regina ist die größte Art der Gruppe. Die 7 oder 8 zweireihig angeordneten, ledrigen Laubblätter sind bis zu 40 cm lang und 6 cm breit. Sie sind länglich linealisch, gerundet, ungleich zweilappig am Ende. Die Blätter sind hell- bis dunkelgrün, mehr oder weniger glänzend, mit einem Netzwerk von dunkleren Adern, die an der Spitze am deutlichsten hervortreten. Die Unterseite ist heller grün gefärbt mit einem leichten purpurnen Überzug an der Basis. Die Blätter sind deutlich gekielt, und der Kiel ist purpurn gefärbt.

Die aufrecht stehende, bis zu 60 cm lange, behaarte Infloreszenz bringt nacheinander bis 8 oder mehr, bis 9 cm breite und bis 8,5 cm hohe Blüten hervor.

Die Brakteen sind gefaltet, länglich, am Ende gespalten.

Der Fruchtknoten ist undeutlich dreieckig, stark behaart, bis 5 cm lang und bis 0,7 mm breit.

Die Fahne ist breit oval, ungefähr 2,6 cm hoch und etwas breiter, mit nach hinten geschlagenen und gewellten Rändern, grün bis olivgrün in der Grundfärbung, mit einem breiten gelblich weißen oder cremefarbenen Rand und bis zu 11 dunkelgrünen Längsstreifen.

Das Synsepalum ist unwesentlich kürzer als die Fahne und ungefähr 2 cm breit, mit gewellten Rändern, grün bis olivgrün mit dunkleren Adern.

Sowohl die Fahne als auch das Synsepalum sind auf der äußeren Seite stark behaart.

Die Petalen stehen fast waagerecht ab. Sie sind bis 4 cm lang und ungefähr 1 cm breit. Sie sind deutlich gedreht über die ganze Länge. Die bewimperten Ränder sind stark gewellt. Die Petalen sind lilalavendel bis rötlich lila an der Basis und tief purpurn an den Enden.

Die Lippe ist sackförmig, beidseitig leicht zusammengedrückt, bis ungefähr 4 cm lang und ca. 1,7 cm breit, allmählich enger werdend zu einer gerundeten Spitze. Die Öffnung ist bis 2 cm breit. Die Farbe der Lippe variiert von einem rötlichen Braun mit einem grünen Rand um die Öffnung bis zu einem zart rötlichem Lila mit einem gelben oder cremefarbenen Rand. Die Lippe ist überzogen von

[1] **Curtis's Botanical Magazine 124,** t. 7573 (1898).

valid name is *Paphiopedilum victoria-regina* and although the detailed description by HOOKER in the **Botanical Magazine**[1] was rendered under the name of *Cypripedium victoria-mariae*, it does not change the fact that this latter designation is to be reduced to synonym status.

Paphiopedilum victoria-regina is the largest species within the group. There are 7 to 8 leaves arranged disticheously along the short stem. They are leathery, up to about 40 cm long and to about 6 cm wide; elongated-linear, rounded, asymmetrically bilobed at the apex, bright to deep green, more or less glossy, with a network of darker veins, more distinct towards the lip. The under surface is brighter green suffused purple at the base, distinctly keeled purple.

The inflorescence is erect, up to about 60 cm high, pilose, bearing up to 8 or more flowers which open one after the other. The flowers are about 9 cm wide and up to about 8.5 cm high.

The bracts are laterally compressed, oblong, acute.

The ovary is indistinctly triangular, densely pilose, about 5 cm long and more or less 0.7 cm wide.

The dorsal sepal is broadly ovate, about 2.6 cm high and somewhat wider, green to olive-green with a broad yellowish-white to cream-coloured margin and up to 11 deep-green longitudinal stripes. The margins are undulate and deflexed.

The synsepalum is slightly shorter than the dorsal sepal and up to about 2 cm wide, green to olive-green with darker veins, the margins undulate.

The dorsal sepal as well as the synsepalum is densely pilose on the outside.

The lateral petals are spreading nearly horizontally. They are about 4 cm long and more or less 1 cm wide, distinctly twisted over the whole length, the margins very undulate and ciliate. The lateral petals are lilac-lavendel to reddish-lilac at the base and deep purple in the apical regions.

The lip is distinctly saccate, somewhat compressed on both sides, to about 4 cm long and about 1.7 cm wide, gradually narrowing to a rounded tip, the aperture ca. 2 cm in diameter. The basic colour of the labellum is very variable from reddish-brown with a green rim around the aperture to a light reddish-lilac with a yellowish to cream-coloured rim. The lip is covered by a network of more or less distinct, darker coloured veins.

The staminodal shield is ovate to sub-rhomboid, ending in a rounded tip; on the base with a ivory-

[1] **Curtis's Botanical Magazine 124,** t. 7573 (1898).

mehr oder weniger deutlich sichtbaren dunkleren Adern.

Die Staminodiumplatte ist eiförmig bis mehr oder weniger eckig, nach unten in einer gerundeten Spitze endend, auf der Basis mit einer elfenbeinfarbenen Erhebung, beidseitig flankiert von behaarten, rötlich gefärbten rundlichen Arealen. Diese Färbung geht über in den dunklen olivgrünen Ton einer zentralen Erhebung.

Paphiopedilum victoria-regina kann von *Paphiopedilum chamberlainianum* und *Paphiopedilum glaucophyllum* unterschieden werden anhand der zusammengedrückten Lippe und der Abwesenheit von länglichen Flecken auf den Petalen sowie durch die meist dunklere Färbung der Laubblätter.

Diese Art wächst auf kalkhaltigen, humusbedeckten Felsen an den westlichen und östlichen Hängen des Mt. Talak Mau auf Sumatra bis in Höhenlagen von 2000 m ü. NN.

Das Klima in dieser Gebirgsregion ist gekennzeichnet durch eine hohe Niederschlagsrate und ständig hohe Luftfeuchtigkeit.

Paphiopedilum victoria-regina gedeiht am besten im temperierten Gewächshaus, in gut drainiertem, durchlässigem Substrat mit reichlich Moos beigemischt. Man sollte viel gießen und immer eine hohe Luftfeuchtigkeit sicherstellen. Die Pflanzen mögen kein saures Substrat, daher ist zeitig umzutopfen. *Paphiopedilum victoria-regina* verträgt kein extrem helles Licht und muß daher halbschattig kultiviert werden.

Paphiopedilum victoria-regina blüht im allgemeinen von März bis Juni.

coloured convexity flanked by hairy, reddish coloured round areas. This colouration fades into a dark olive-green, raised middle part.

Paphiopedilum victoria-regina is easily recognized by the somewhat compressed lip and by the lateral petals failing large longitudinal spots as well as by the more darkly coloured leaves.

Paphiopedilum victoria-regina grows on limestone rock, in humus and leafy debris on the western and eastern slopes of Mount Talak Mau in Sumatra at altitudes of up to 2000 meters above sea-level.

The climate in this mountain area is characterized by heavy rainfall and constant high humidity.

Paphiopedilum victoria-regina does best in the intermediate greenhouse, potted in a well-drained open substrate with plenty of moss. Waterings should be abundant and the relative humidity should be kept high at all times. The plants of this species are sensitive to stale compost and should therefore be repotted regularly. *Paphiopedilum victoria-regina* does not like extreme light conditions and should be kept semi-shaded.

Paphiopedilum victoria-regina generally flowers from March through June.

Liste der Synonyme – List of synonyms

Die hier als korrekt anerkannten Namen stehen in der rechten Spalte.
The correct names as accepted in this book are in the right column.

Cordula amabilis (Hallier) Merrill	=	*Paphiopedilum bullenianum* (Rchb. f.) Pfitzer
Cordula appletoniana (Gower) Rolfe	=	*Paphiopedilum appletonianum* (Gower) Rolfe
Cordula argus (Rchb. f.) Rolfe	=	*Paphiopedilum argus* (Rchb. f.) Stein
Cordula barbata (Lindley) Rolfe	=	*Paphiopedilum barbatum* (Lindley) Pfitzer
Cordula bellatula (Rchb. f.) Rolfe	=	*Paphiopedilum bellatulum* (Rchb. f.) Stein
Cordula boxallii (Rchb. f.) Rolfe	=	*Paphiopedilum villosum* var. *boxallii* (Rchb. f.) Veitch
Cordula bulleniana (Rchb. f.) Rolfe	=	*Paphiopedilum bullenianum* (Rchb. f.) Pfitzer
Cordula callosa (Rchb. f.) Rolfe	=	*Paphiopedilum callosum* (Rchb. f.) Stein
Cordula chamberlainiana (Sander) Rolfe	=	*Paphiopedilum chamberlainianum* (Sander) Stein
Cordula charlesworthii (Rolfe) Rolfe	=	*Paphiopedilum charlesworthii* Rolfe
Cordula ciliolaris (Rchb. f.) Rolfe	=	*Paphiopedilum ciliolare* (Rchb. f.) Stein
Cordula concolor (Bateman) Rolfe	=	*Paphiopedilum concolor* (Bateman) Pfitzer
Cordula curtisii (Rchb. f.) Rolfe	=	*Paphiopedilum superbiens* var. *curtisii* (Rchb. f.) Stein
Cordula dayana (Stone ex Lindley) Rolfe	=	*Paphiopedilum dayanum* (Stone ex Lindley) Stein
Cordula druryi (Beddome) Rolfe	=	*Paphiopedilum druryi* (Beddome) Stein
Cordula esquirolei (Schltr.) Hu	=	*Paphiopedilum hirsutissimum* var. *esquirolei* (Schltr.) Karasawa & Saito
Cordula exul (Ridley) Rolfe	=	*Paphiopedilum insigne* var. *exul* Ridley
Cordula fairrieana (Lindley) Rolfe	=	*Paphiopedilum fairrieanum* (Lindley) Stein
Cordula glanduligera Rolfe, *lapsus*	=	*Paphiopedilum glanduliferum* (Blume) Stein
Cordula glaucophylla (J. J. Smith) Rolfe	=	*Paphiopedilum glaucophyllum* (J. J. Smith) Pfitzer
Cordula godefroyae (Godefroy) Rolfe	=	*Paphiopedilum* × *godefroyae* (Godefroy) Stein
Cordula gratrixiana (Masters) Rolfe	=	*Paphiopedilum villosum* var. *gratrixianum* (Masters) Braem
Cordula haynaldiana (Rchb. f.) Rolfe	=	*Paphiopedilum haynaldianum* (Rchb. f.) Stein
Cordula hirsutissima (Lindley ex Hooker) Rolfe	=	*Paphiopedilum hirsutissimum* (Lindley ex Hooker) Stein
Cordula hookerae (Rchb. f. ex Hooker f.) Rolfe	=	*Paphiopedilum hookerae* (Rchb. f. ex Hooker f.) Stein
Cordula insignis (Wallich ex Lindley) Rafinesque	=	*Paphiopedilum insigne* (Wallich ex Lindley) Stein
Cordula javanica (Reinwardt ex Blume) Rolfe	=	*Paphiopedilum javanicum* (Reinwardt ex Blume) Pfitzer
Cordula lawrenceana (Rchb. f.) Rolfe	=	*Paphiopedilum lawrenceanum* (Rchb. f.) Pfitzer
Cordula lowiana (Lindley) Rolfe	=	*Paphiopedilum lowii* (Lindley) Stein
Cordula mastersiana (Rchb. f.) Rolfe	=	*Paphiopedilum mastersianum* (Rchb. f.) Stein
Cordula nigrita (Rchb.f.) Merrill	=	*Paphiopedilum barbatum* (Lindley) Pfitzer
Cordula nivea (Rchb. f.) Rolfe	=	*Paphiopedilum niveum* (Rchb. f.) Stein
Cordula parishii (Rchb. f.) Rolfe	=	*Paphiopedilum parishii* (Rchb. f.) Stein
Cordula petri (Rchb. f.) Rolfe	=	*Paphiopedilum dayanum* (Stone ex Lindley) Stein
Cordula philippinensis (Rchb. f.) Rolfe	=	*Paphiopedilum philippinense* (Rchb. f.) Stein
Cordula praestans (Rchb. f.) Rolfe	=	*Paphiopedilum glanduliferum* var. *praestans* (Rchb. f.) Braem
Cordula purpurata (Lindley) Rolfe	=	*Paphiopedilum purpuratum* (Lindley) Stein
Cordula rothschildiana (Rchb. f.) Rolfe	=	*Paphiopedilum rothschildianum* (Rchb. f.) Stein
Cordula sanderiana (Rchb. f.) Rolfe	=	*Paphiopedilum sanderianum* (Rchb. f.) Stein

Cordula spiceriana (Rchb. f.) Rolfe	=	*Paphiopedilum spicerianum* (Rchb. f.) Stein
Cordula stonei (Hooker) Rolfe	=	*Paphiopedilum stonei* (Hooker) Stein
Cordula superbiens (Rchb. f.) Rolfe	=	*Paphiopedilum superbiens* (Rchb. f.) Stein
Cordula tonsa (Rchb. f.) Rolfe	=	*Paphiopedilum tonsum* (Rchb. f.) Stein
Cordula venusta (Wallich ex Sims) Rolfe	=	*Paphiopedilum venustum* (Wallich ex Sims) Pfitzer
Cordula victoria-mariae (Sander ex T. M. Masters) Rolfe	=	*Paphiopedilum victoria-regina* (Sander) M. W. Wood
Cordula villosa (Lindley) Rolfe	=	*Paphiopedilum villosum* (Lindley) Stein
Cordula violascens (Schltr.) Rolfe	=	*Paphiopedilum violascens* Schlechter
Cordula virens (Rchb. f.) Rolfe	=	*Paphiopedilum javanicum* (Reinwardt ex Blume) Pfitzer
Cypripedium appletonianum Gower	=	*Paphiopedilum appletonianum* (Gower) Rolfe
Cypripedium argus Rchb. f.	=	*Paphiopedilum argus* (Rchb. f.) Stein
Cypripedium barbatum Lindley	=	*Paphiopedilum barbatum* (Lindley) Pfitzer
Cypripedium barbatum var. *hookerae* (Rchb. f. ex Hooker f.) Regel	=	*Paphiopedilum hookerae* (Rchb. f. ex Hooker f.) Stein
Cypripedium barbatum var. *superbum* Hort. ex Morren	=	*Paphiopedilum superbiens* (Rchb. f.) Stein
Cypripedium barbatum var. *veitchii* Hort. ex Van Houtte	=	*Paphiopedilum superbiens* (Rchb. f.) Stein
Cypripedium bellatulum Rchb. f.	=	*Paphiopedilum bellatulum* (Rchb. f.) Stein
Cypripedium biflorum Williams	=	*Paphiopedilum barbatum* (Lindley) Pfitzer
Cypripedium boxallii Rchb. f.	=	*Paphiopedilum villosum* var. *boxallii* (Rchb. f.) Veitch
Cypripedium bullenianum Rchb. f.	=	*Paphiopedilum bullenianum* (Rchb. f.) Pfitzer
Cypripedium bullenianum var. *appletonianum* (Gower) Rolfe	=	*Paphiopedilum appletonianum* (Gower) Rolfe
Cypripedium burbidgei Rchb. f.	=	*Paphiopedilum dayanum* (Stone ex Lindley) Stein
Cypripedium callosum Rchb. f.	=	*Paphiopedilum callosum* (Rchb. f.) Stein
Cypripedium cannaertianum Linden	=	*Paphiopedilum philippinense* var. *cannaertianum* (Linden) Pfitzer
Cypripedium caudatum Lindley	=	*Phragmipedium caudatum* (Lindley) Rolfe
Cypripedium chamberlainianum Sander	=	*Paphiopedilum chamberlainianum* (Sander) Stein
Cypripedium chamberlainianum forma *victoria-mariae* (Sander ex M. T. Masters) Rolfe	=	*Paphiopedilum victoria-regina* (Sander) M. W. Wood
Cypripedium charlesworthii Rolfe	=	*Paphiopedilum charlesworthii* (Rolfe) Pfitzer
Cypripedium ciliolare Rchb. f.	=	*Paphiopedilum ciliolare* (Rchb. f.) Stein
Cypripedium concolor Bateman	=	*Paphiopedilum concolor* (Bateman) Pfitzer
Cypripedium concolor var. *godefroyae* Collett & Hemsley	=	*Paphiopedilum* × *godefroyae* (Godefroy) Stein
Cypripedium concolor var. *niveum* Rchb. f.	=	*Paphiopedilum niveum* (Rchb. f.) Stein
Cypripedium crossi Morren	=	*Paphiopedilum barbatum* (Lindley) Pfitzer
Cypripedium cruciforme Zoll. & Mir.	=	*Paphiopedilum lowii* (Lindley) Stein
Cypripedium curtisii Rchb. f.	=	*Paphiopedilum superbiens* (Rchb. f.) Stein
Cypripedium dayanum (Lindley) Lindley	=	*Paphiopedilum dayanum* (Stone ex Lindley) Stein
Cypripedium dayi Stone	=	*Paphiopedilum dayanum* (Stone ex Lindley) Stein
Cypripedium delenatii Hort.	=	*Paphiopedilum delenatii* Guillaumin
Cypripedium dilectum Rchb. f.	=	*Paphiopedilum villosum* var. *boxallii* (Rchb. f.) Veitch
Cypripedium druryi Beddome	=	*Paphiopedilum druryi* (Beddome) Stein
Cypripedium elliottianum O'Brien	=	*Paphiopedilum elliottianum* (O'Brien) Stein
Cypripedium ernestianum Castle	=	*Paphiopedilum dayanum* (Stone ex Lindley) Stein

Cypripedium exul (Ridley) O'Brien	=	*Paphiopedilum insigne* var. *exul* Ridley
Cypripedium fairrieanum Lindley	=	*Paphiopedilum fairrieanum* (Lindley) Stein
Cypripedium foerstermanni Hort.	=	*Paphiopedilum sanderianum* (Rchb. f.) Stein
Cypripedium gardineri, nomen nudum	=	*Paphiopedilum glanduliferum* var. *gardineri* Braem
Cypripedium glanduliferum Blume	=	*Paphiopedilum glanduliferum* (Blume) Stein
Cypripedium glaucophyllum (J. J. Smith) Masters	=	*Paphiopedilum glaucophyllum* J. J. Smith
Cypripedium godefroyae Godefroy	=	*Paphiopedilum* × *godefroyae* (Godefroy) Stein
Cypripedium godefroyae var. *leucochilum* Rolfe	=	*Paphiopedilum* × *godefroyae* var. *leucochilum* (Masters) Hallier
Cypripedium gratrixianum Masters	=	*Paphiopedilum villosum* var. *gratrixianum* (Masters) Braem
Cypripedium gratrixianum Warner & Williams	=	Hybride *(Paphiopedilum bellatulum* × *Paphiopedilum Enfieldense)*
Cypripedium haynaldianum Rchb. f.	=	*Paphiopedilum haynaldianum* (Rchb. f.) Stein
Cypripedium hirsutissimum Lindley ex Hooker	=	*Paphiopedilum hirsutissimum* (Lindley ex Hooker) Stein
Cypripedium hookerae Rchb. f. ex Hooker f.	=	*Paphiopedilum hookerae* (Rchb. f. ex Hooker f.) Stein
Cypripedium hookerae var. *amabile* (Hallier) Krzl.	=	*Paphiopedilum bullenianum* (Rchb. f.) Pfitzer
Cypripedium hookerae var. *bullenianum* (Rchb. f.) Veitch	=	*Paphiopedilum bullenianum* (Rchb. f.) Pfitzer
Cypripedium hookerae var. *volonteanum* (Sander) Rolfe	=	*Paphiopedilum hookerae* var. *volonteanum* (Sander) Braem
Cypripedium insigne Wallich ex Lindley	=	*Paphiopedilum insigne* (Wallich ex Lindley) Pfitzer
Cypripedium insigne var. *exul* Ridley	=	*Paphiopedilum insigne* var. *exul* Ridley
Cypripedium javanicum Reinw. non Blume	=	*Paphiopedilum barbatum* (Lindley) Pfitzer
Cypripedium javanicum Reinwardt ex Blume	=	*Paphiopedilum javanicum* (Reinwardt ex Blume) Pfitzer
Cypripedium laevigatum Bateman	=	*Paphiopedilum philippinense* (Rchb. f.) Stein
Cypripedium lawrenceanum Rchb. f.	=	*Paphiopedilum lawrenceanum* (Rchb. f.) Pfitzer
Cypripedium lowii Lindley	=	*Paphiopedilum lowii* (Lindley) Stein
Cypripedium mastersianum Rchb. f.	=	*Paphiopedilum mastersianum* (Rchb. f.) Stein
Cypripedium miteauanum Linden & Rodigas	=	*Paphiopedilum ciliolare* (Rchb. f.) Stein
Cypripedium neo-guineense Linden	=	*Paphiopedilum rothschildianum* (Rchb. f.) Stein
Cypripedium nicholsonianum	=	*Paphiopedilum rothschildianum* (Rchb. f.) Stein
Cypripedium nigritum Rchb. f.	=	*Paphiopedilum barbatum* (Lindley) Pfitzer
Cypripedium niveum Rchb. f.	=	*Paphiopedilum niveum* (Rchb. f.) Stein
Cypripedium orbum Rchb. f.	=	*Paphiopedilum barbatum* (Lindley) Pfitzer
Cypripedium papuanum Ridley ex Rendle	=	*Paphiopedilum papuanum* (Ridley ex Rendle) L. O. Williams
Cypripedium pardinum Rchb. f.	=	*Paphiopedilum venustum* (Wallich ex Sims) Pfitzer
Cypripedium parishii Rchb. f.	=	*Paphiopedilum parishii* (Rchb. f.) Stein
Cypripedium peteri R. Dean	=	*Paphiopedilum dayanum* (Stone ex Lindley) Stein
Cypripedium peteri De Vos	=	*Paphiopedilum dayanum* (Stone ex Lindley) Stein
Cypripedium petri Rchb. f.	=	*Paphiopedilum dayanum* (Stone ex Lindley) Stein
Cypripedium × *petri* (Rchb. f.) Rolfe	=	*Paphiopedilum dayanum* (Stone ex Lindley) Stein
Cypripedium × *petri* var. *burbidgei* (Rchb. f.) Rolfe	=	*Paphiopedilum dayanum* (Stone ex Lindley) Stein
Cypripedium philippinense Rchb. f.	=	*Paphiopedilum philippinense* (Rchb. f.) Stein
Cypripedium pitcherianum W. A. Manda	=	*Paphiopedilum argus* (Rchb. f.) Stein
Cypripedium poyntzianum O'Brien	=	*Paphiopedilum appletonianum* (Gower) Rolfe
Cypripedium praestans Rchb. f.	=	*Paphiopedilum glanduliferum* var. *praestans* (Rchb. f.) Braem

Cypripedium purpuratum Lindley	=	*Paphiopedilum purpuratum* (Lindley) Stein
Cypripedium robinsonii Ridley	=	*Paphiopedilum robinsonii* (Ridley) Ridley
Cypripedium roebbelinii Rchb. f.	=	*Paphiopedilum philippinense* var. *roebbelinii* (Rchb. f.) Braem
Cypripedium rothschildianum Rchb. f.	=	*Paphiopedilum rothschildianum* (Rchb. f.) Stein
Cypripedium sanderianum Rchb. f.	=	*Paphiopedilum sanderianum* (Rchb. f.) Stein
Cypripedium schmidtianum Krzl.	=	*Paphiopedilum callosum* (Rchb. f.) Stein
Cypripedium sinicum Hance ex Rchb. f.	=	*Paphiopedilum purpuratum* (Lindley) Stein
Cypripedium sp. F. K. Ward	=	*Paphiopedilum* × *wardii* (Summerhayes) Braem
Cypripedium spectabile var. *dayanum* Lindley	=	*Paphiopedilum dayanum* (Stone ex Lindley) Stein
Cypripedium spicerianum Rchb. f.	=	*Paphiopedilum spicerianum* (Rchb. f.) Stein
Cypripedium stonei Hooker	=	*Paphiopedilum stonei* (Hooker) Stein
Cypripedium superbiens Rchb. f.	=	*Paphiopedilum superbiens* (Rchb. f.) Stein
Cypripedium superbiens var. *dayanum* (Lindley) Rchb. f.	=	*Paphiopedilum dayanum* (Stone ex Lindley) Stein
Cypripedium tonsum Rchb. f.	=	*Paphiopedilum tonsum* (Rchb. f.) Stein
Cypripedium veitchianum Hort. ex Lemaire	=	*Paphiopedilum superbiens* (Rchb. f.) Stein
Cypripedium venustum Wallich ex Sims	=	*Paphiopedilum venustum* (Wallich ex Sims) Pfitzer
Cypripedium victoria-mariae Sander ex M. T. Masters	=	*Paphiopedilum victoria-regina* (Sander) M. W. Wood
Cypripedium victoria-regina Sander	=	*Paphiopedilum victoria-regina* (Sander) M. W. Wood
Cypripedium villosum Lindley	=	*Paphiopedilum villosum* (Lindley) Stein
Cypripedium virens Rchb. f.	=	*Paphiopedilum javanicum* (Reinwardt ex Blume) Pfitzer
Cypripedium volonteanum Sander	=	*Paphiopedilum hookerae* var. *volonteanum* (Sander) Braem
Cypripedium waltersianum Krzl. ex Desbois	=	*Paphiopedilum appletonianum* (Gower) Rolfe
Cypripedium wardii Anon. non Rolfe	=	*Paphiopedilum* × *wardii* Summerhayes
Cypripedium wolterianum Krzl.	=	*Paphiopedilum appletonianum* (Gower) Rolfe
Paphiopedilum adductum Asher	=	*Paphiopedilum elliottianum* (O'Brien) Stein
Paphiopedilum affine De Wildeman	=	*Paphiopedilum villosum* var. *affine* (De Wildeman) Braem
Paphiopedilum amabile Hallier	=	*Paphiopedilum bullenianum* (Rchb. f.) Pfitzer
Paphiopedilum× *ang thong* Fowlie	=	*Paphiopedilum* × *godefroyae* var. *ang-thong* Braem
Paphiopedilum barbatum var. *argus* Hort.	=	*Paphiopedilum argus* (Rchb. f.) Stein
Paphiopedilum barbigerum Tang & Wang	=	*Paphiopedilum insigne* var. *barbigerum* (Tang & Wang) Braem
Paphiopedilum birkii Fowlie, *nomen nudum*	=	*Paphiopedilum callosum* (Rchb. f.) Stein
Paphiopedilum bodegomensis, nomen nudum	=	*Paphiopedilum glanduliferum* (Blume) Stein
Paphiopedilum bodegomii, nomen nudum	=	*Paphiopedilum glanduliferum* (Blume) Stein
Paphiopedilum bougainvilleanum Fowlie	=	*Paphiopedilum violascens* Schlechter
Paphiopedilum boxallii (Rchb. f.) Pfitzer	=	*Paphiopedilum villosum* var. *boxallii* (Rchb. f.)
Paphiopedilum burbidgei (Rchb. f.) Pfitzer	=	*Paphiopedilum dayanum* (Stone ex Lindley) Stein
Paphiopedilum celebesensis Fowlie & Birk	=	*Paphiopedilum bullenianum* (Rchb. f.) Pfitzer
Paphiopedilum ceramensis Birk *nomen nudum*	=	*Paphiopedilum appletonianum* (Gower) Rolfe
Paphiopedilum chamberlainianum ssp. *liemiana* Fowlie	=	*Paphiopedilum chamberlainianum* var. *liemianum* (Fowlie) Braem
Paphiopedilum chamberlainianum ssp. *liemianum* forma *primulinum* Fowlie	=	*Paphiopedilum chamberlainianum* var. *primulinum* (Wood & Taylor) Braem
Paphiopedilum chiwuanum Tang & Wang	=	*Paphiopedilum hirsutissimum* (Lindley ex Hooker) Stein
Paphiopedilum ciliolare var. *miteauanum* (Linden) Pfitzer	=	*Paphiopedilum ciliolare* (Rchb. f.) Stein

Paphiopedilum cothurnum (Velloso) Pfitzer	=	CATASETUM sp.
Paphiopedilum curtisii (Rchb. f.) Stein	=	*Paphiopedilum superbiens* (Rchb. f.) Stein
Paphiopedilum devogeli, nomen nudum	=	*Paphiopedilum supardii* Braem & Löb
Paphiopedilum dianthum Tang & Wang	=	*Paphiopedilum parishii* var. *dianthum* (Tang & Wang) Karasawa & Saito
Paphiopedilum dilectum (Rchb. f.) Pfitzer	=	*Paphiopedilum villosum* var. *boxallii* (Rchb. f.) Veitch
Paphiopedilum epidendricum (Velloso) Pfitzer	=	*EULOPHIA* sp.
Paphiopedilum esquirolei Schltr.	=	*Paphiopedilum hirsutissimum* var. *esquirolei* (Schltr.) Karasawa & Saito
Paphiopedilum exul (Ridley) Kerchove	=	*Paphiopedilum insigne* var. *exul* (Ridley)
Paphiopedilum fowliei Birk	=	*Paphiopedilum lawrenceanum* (Rchb. f.) Pfitzer
Paphiopedilum gardineri, nomen nudum	=	*Paphiopedilum glanduliferum* var. *gardineri* Braem
Paphiopedilum godefroyae (Godefroy) Stein	=	*Paphiopedilum* × *godefroyae* (Godefroy) Stein
Paphiopedilum godefroyae var. *leucochilum* (Masters) Hallier	=	*Paphiopedilum* × *godefroyae* var. *leucochilum* (Masters) Hallier
Paphiopedilum gratrixianum (Masters) Guillaumin	=	*Paphiopedilum villosum* var. *gratrixianum* (Masters) Braem
Paphiopedilum hainanensis Fowlie	=	*Paphiopedilum appletonianum* (Gower) Rolfe
Paphiopedilum hartwegii (Rchb. f.) Pfitzer	=	*Phragmipedium hartwegii* (Rchb. f.) L. O. Williams
Paphiopedilum hennisianum (Schoser ex Wood) Fowlie	=	*Paphiopedilum lawrenceanum* (Rchb. f.) Pfitzer
Paphiopedilum hennisianum var. *fowliei* (Birk) Cribb	=	*Pahiopedilum lawrenceanum* (Rchb. f.) Pfitzer
Paphiopedilum hinksianum (Rchb. f.) Pfitzer	=	*Phragmipedium longifolium* (Rchb. f. & Warsz.) Rolfe
Paphiopedilum johorensis Fowlie & Yap	=	*Paphiopedilum appletonianum* (Gower) Rolfe
Paphiopedilum laevigatum (Bateman) Pfitzer	=	*Paphiopedilum philippinense* (Rchb. f.) Stein
Paphiopedilum leucochilum (Rolfe) Fowlie	=	*Paphiopedilum* × *godefroyae* var. *leucochilum* (Masters) Hallier
Paphiopedilum liemianum (Fowlie) Karasawa & Saito	=	*Paphiopedilum chamberlainianum* var. *liemianum* (Fowlie) Braem
Paphiopedilum liemianum var. *primulinum* (Wood & Taylor) Karasawa & Saito	=	*Paphiopedilum chamberlainianum* var. *primulinum* (Wood & Taylor) Braem
Paphiopedilum linii Schoser	=	*Paphiopedilum bullenianum* (Rchb. f.) Pfitzer
Paphiopedilum longifolium (Rchb. f. & Warsz.) Pfitzer	=	*Phragmipedium longifolium* (Rchb. f. & Warsz.) Rolfe
Paphiopedilum moquetteanum (J. J. Smith) Fowlie	=	*Paphiopedilum glaucophyllum* var. *moquettenaum* J. J. Smith
Paphiopedilum nigritum (Rchb. f.) Pfitzer	=	*Paphiopedilum barbatum* (Lindley) Pfitzer
Paphiopedilum palawense, nomen nudum	=	*Paphiopedilum philippinense* (Rchb. f.) Stein
Paphiopedilum papuanum Hort. non (Ridley ex Rendle) L. O. Williams	=	*Paphiopedilum violascens* Schlechter
Paphiopedilum pardinum (Rchb. f.) Pfitzer	=	*Paphiopedilum venustum* (Wallich ex Sims) Pfitzer
Paphiopedilum paulistanum (Barb.-Rodr.) Pfitzer	=	*Phragmipedium vittatum* (Velloso) Rolfe
Paphiopedilum petri (Rchb. f.) Pfitzer	=	*Paphiopedilum dayanum* (Stone ex Lindley) Stein
Paphiopedilum praestans (Rchb. f.) Pfitzer	=	*Paphiopedilum glanduliferum* var. *praestans* (Rchb. f.) Braem
Paphiopedilum primulinum Wood & Taylor	=	*Paphiopedilum chamberlainianum* var. *primulinum* (Wood & Taylor) Braem
Paphiopedilum purpurascens Fowlie	=	*Paphiopedilum javanicum* (Reinwardt ex Blume) Stein
Paphiopedilum reflexum Hort. ex Stein	=	*Paphiopedilum callosum* (Rchb. f.) Stein

Paphiopedilum regnieri Hort. ex Stein	=	*Paphiopedilum callosum* (Rchb. f.) Stein
Paphiopedilum roebbelinii (Rchb. f.) Pfitzer	=	*Paphiopedilum philippinense* var. *roebbelinii* (Rchb. f.) Cribb
Paphiopedilum roezlii (Regel) Pfitzer	=	*Phragmipedium roezlii* (Rchb. f.) Garay
Paphiopedilum schlimii (Rchb. f.) Stein	=	*Phragmipedium schlimii* (Rchb. f.) Rolfe
Paphiopedilum schmidtianum Hort.	=	*Paphiopedilum callosum* (Rchb. f.) Stein
Paphiopedilum sinicum (Hance ex Rchb. f.) Stein	=	*Paphiopedilum purpuratum* (Lindley) Stein
Paphiopedilum sublaeve (Rchb. f.) Fowlie	=	*Paphiopedilum callosum* (Rchb. f.) Stein
Paphiopedilum superbiens subsp. *ciliolare* M. W. Wood	=	*Paphiopedilum ciliolare* (Rchb. f.) Stein
Paphiopedilum thailandense Fowlie, *nomen nudum*	=	*Paphiopedilum callosum* (Rchb. f.) Stein
Paphiopedilum tortipetalum Fowlie	=	*Paphiopedilum appletonianum* (Gower) Rolfe
Paphiopedilum victoria-mariae (Sander ex M. T. Masters) Rolfe	=	*Paphiopedilum victoria-regina* (Sander) M. W. Wood
Paphiopedilum victoria-regina ssp. *chamberlainianum* (Sander) Wood	=	*Paphiopedilum chamberlainianum* (Sander) Stein
Paphiopedilum victoria-regina ssp. *glaucophyllum* (Smith) Wood	=	*Paphiopedilum glaucophyllum* J. J. Smith
Paphiopedilum victoria-regina ssp. *glaucophyllum* var. *moquetteanum* (J. J. Smith) Wood	=	*Paphiopedilum glaucophyllum* var. *moquetteanum* J. J. Smith
Paphiopedilum victoria-regina ssp. *liemianum* (Fowlie) Wood	=	*Paphiopedilum chamberlainianum* var. *liemianum* (Fowlie) Braem
Paphiopedilum victoria-regina ssp. *primulinum* (Wood & Taylor) Wood	=	*Paphiopedilum chamberlainianum* var. *primulinum* (Wood & Taylor) Braem
Paphiopedilum violascens var. *gautierense* J. J. Smith	=	*Paphiopedilum violascens* Schlechter
Paphiopedilum virens (Rchb. f.) Pfitzer	=	*Paphiopedilum javanicum* (Reinwardt ex Blume) Stein
Paphiopedilum vittatum (Velloso) Stein	=	*Phragmipedium vittatum* (Velloso) Rolfe
Paphiopedilum volonteanum (Sander) Stein	=	*Paphiopedilum hookerae* var. *volonteanum* (Sander) Braem
Paphiopedilum wallisii (Rchb. f.)	=	*Phragmipedium wallisii* (Rchb. f.) Garay
Paphiopedilum wardii Summerhayes	=	*Paphiopedilum* × *wardii* Summerhayes
Paphiopedilum warscewiczii (Rchb. f.) Pfitzer	=	*Phragmipedium caudatum* (Lindley) Rolfe
Paphiopedilum wentworthianum Schoser & Fowlie	=	*Paphiopedilum violascens* Schltr.
Paphiopedilum wilhelminiae L. O. Williams	=	*Paphiopedilum glanduliferum* var. *wilhelminiae* (L. O. Williams) Cribb
Paphiopedilum wolterianum (Krzl.) Pfitzer	=	*Paphiopedilum appletonianum* (Gower) Rolfe
Paphiopedilum zieckianum Schoser	=	*Paphiopedilum papuanum* (Ridley ex Rendle) L. O. Williams
Selenipedium elliottianum Gower	=	*Paphiopedilum elliottianum* (O'Brien) Stein
Selenipedium laevigatum (Bateman) May	=	*Paphiopedilum philippinense* (Rchb. f.) Stein
Selenipedium parishii E. André	=	*Paphiopedilum parishii* (Rchb. f.) Stein
Stimegas venustum (Wallich ex Sims) Rafinesque	=	*Paphiopedilum venustum* (Wallich ex Sims) Pfitzer

Verzeichnis spezifischer und infraspezifischer Taxa
Index to specific and infraspecific taxa

Die anerkannten Arten sind **kräftig** gedruckt.
The accepted species are in **strong**.

Taxon

Paphiopedilum acmodontum
Paphiopedilum appletonianum
Paphiopedilum argus
Paphiopedilum armeniacum
Paphiopedilum barbatum
Paphiopedilum bellatulum
Paphiopedilum bullenianum
Paphiopedilum callosum
Paphiopedilum chamberlainianum
Paphiopedilum charlesworthii
Paphiopedilum ciliolare
Paphiopedilum concolor
Paphiopedilum dayanum
Paphiopedilum delenatii
Paphiopedilum druryi
Paphiopedilum elliottianum
Paphiopedilum emersonii
Paphiopedilum fairrieanum
Paphiopedilum glanduliferum
Paphiopedilum glaucophyllum
Paphiopedilum × godefroyae
Paphiopedilum haynaldianum
Paphiopedilum hirsutissimum
Paphiopedilum hookerae
Paphiopedilum insigne
Paphiopedilum javanicum
Paphiopedilum kolopakingii
Paphiopedilum lawrenceanum
Paphiopedilum lowii
Paphiopedilum malipoense
Paphiopedilum mastersianum
Paphiopedilum micranthum
Paphiopedilum niveum
Paphiopedilum papuanum
Paphiopedilum parishii
Paphiopedilum philippinense
Paphiopedilum purpuratum
Paphiopedilum randsii
Paphiopedilum robinsonii
Paphiopedilum rothschildianum
Paphiopedilum sanderianum
Paphiopedilum sangii
Paphiopedilum spicerianum
Paphiopedilum stonei
Paphiopedilum sukhakulii
Paphiopedilum supardii
Paphiopedilum superbiens
Paphiopedilum tonsum
Paphiopedilum urbanianum
Paphiopedilum venustum
Paphiopedilum victoria-regina
Paphiopedilum villosum
Paphiopedilum violascens
Paphiopedilum × wardii

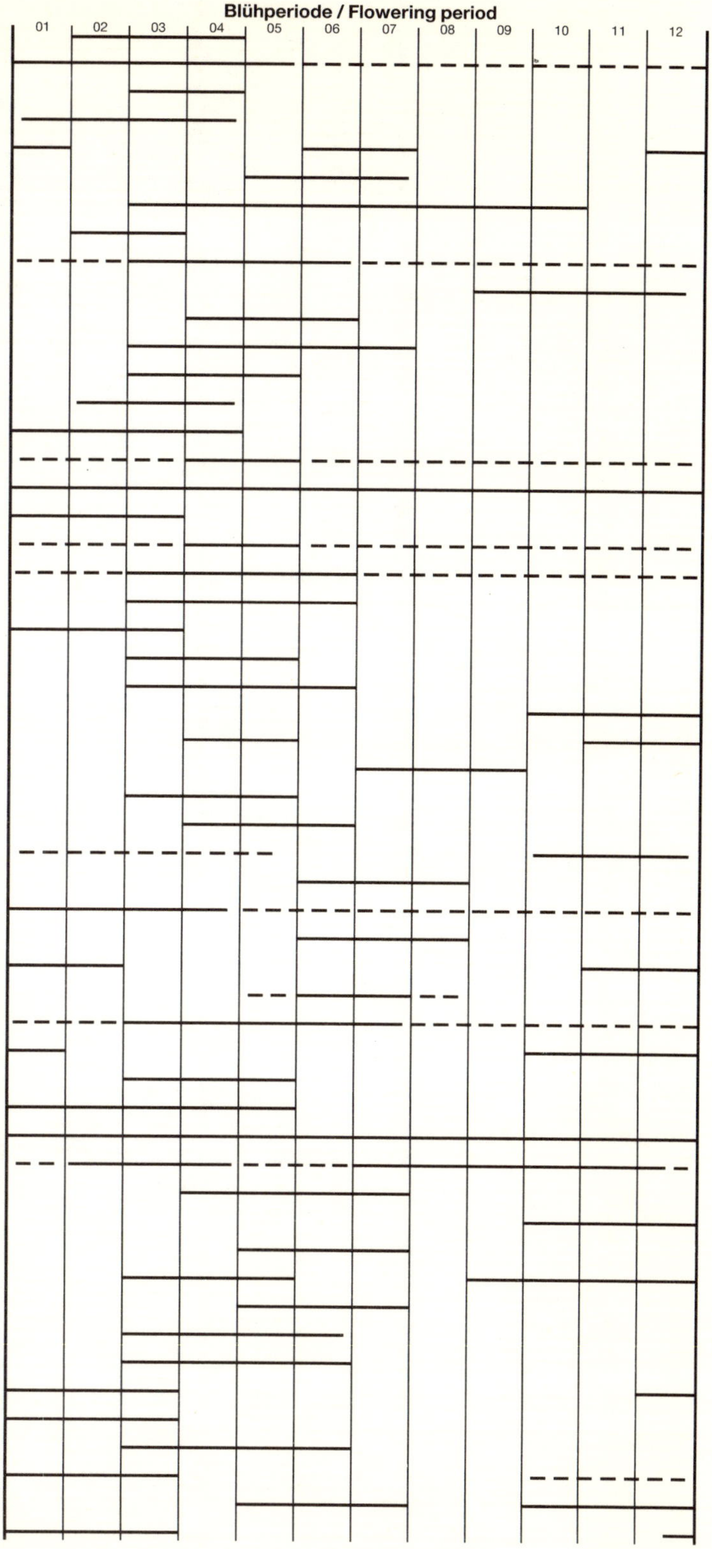